Business Skills for Engineers and Technologists

£2.20

Business Skills for Engineers and Technologists

Harry Cather
Richard Morris
Joe Wilkinson

Newnes
OXFORD AUCKLAND BOSTON JOHANNESBURG MELBOURNE NEW DELHI

Butterworth-Heinemann
Linacre House, Jordan Hill, Oxford OX2 8DP
225 Wildwood Avenue, Woburn, MA 01801-2041
A division of Reed Educational and Professional Publishing Ltd

A member of the Reed Elsevier plc group

First published 2001

British Library Cataloguing in Publication Data
A catalogue record for this book is available from the British Library

ISBN 0 7506 5210 1

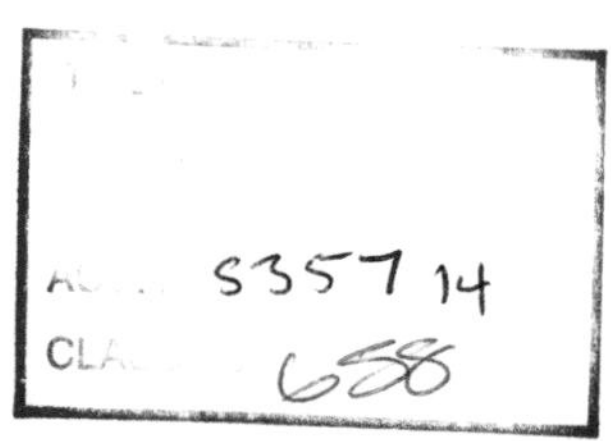

Composition by Genesis Typesetting, Laser Quay, Rochester, Kent
Printed and bound in Great Britain

Contents

Series Preface

'There is a time for all things: for shouting, for gentle speaking, for silence; for the washing of pots and the writing of books. Let now the pots go black, and set to work. It is hard to make a beginning, but it must be done' – Oliver Heaviside, *Electromagnetic Theory*, Vol 3 (1912), Ch 9, 'Waves from moving sources – Adagio. Andante. Allegro Moderato'.

Oliver Heaviside was one of the greatest engineers of all time, ranking alongside Faraday and Maxwell in his field. As can be seen from the above excerpt from a seminal work, he appreciated the need to communicate to a wider audience. He also offered the advice 'So be rigorous; that will cover a multitude of sins. And do not frown.' The series of books that this prefaces takes up Heaviside's challenge but in a world which is quite different to that being experienced just a century ago.

With the vast range of books already available covering many of the topics developed in this series, what is this series offering which is unique? I hope that the next few paragraphs help to answer that; certainly no one involved in this project would give up their time to bring these books to fruition if they had not thought that the series is both unique and valuable.

This motivation for this series of books was born out of the desire of the UK's Engineering Council to increase the number of incorporated engineers graduating from Higher Education establishments, and the Insitution of Incorporated Engineers' (IIE) aim to provide enhanced services to those delivering Incorporated Engineering courses and those studying on Incorporated Engineering Courses. However, what has emerged from the project should prove of great value to a very wide range of courses within the UK and internationally – from Foundation Degrees or Higher Nationals through to first year modules for traditional 'Chartered' degree courses. The reason why these books will appeal to such a wide audience is that they present the core subject areas for engineering studies in a lively, student-centred way, with key theory delivered in real world contexts, and a pedagogical structure that supports independent learning and classroom use.

Despite the apparent waxing of 'new' technologies and the waning of 'old' technologies, engineering is still fundamental to wealth creation. Sitting alongside these are the new business focused, information and communications dominated, technology organisations. Both facets have an equal importance in the health of a nation and the prospects of individuals. In preparing this series of books, we have tried to strike a balance between traditional engineering and developing technology.

The philosophy is to provide a series of complementary texts which can be tailored to the actual courses being run – allowing the flexibility for course designers to take into account 'local' issues, such as areas of particular staff expertise and interest, while being able to demonstrate the depth and breadth of course material referenced to a framework. The series is designed to cover material in the core texts which approximately corresponds to the first year of study with module texts focusing on individual topics to second and final year level. While the general structure of each of the texts is common, the styles are quite different, reflecting best practice in their areas. For example *Mechanical Engineering Systems* adopts a 'tell – show – do' approach, allowing students to work independently as well as in class, whereas *Business Skills for Engineers and Technologists* adopts a 'framework' approach, setting the context and boundaries and providing opportunities for discussion.

Another set of factors which we have taken into account in designing this series is the reduction in contact hours between staff and students, the evolving responsibilities of both parties and the way in which advances in technology are changing the way study can be, and is, undertaken. As a result, the lecturers' support material which accompanies these texts, is paramount to delivering maximum benefit to the student.

It is with these thoughts of Voltaire that I leave the reader to embark on the rigours of study:

'Work banishes those three great evils: boredom, vice and poverty.'

Alistair Duffy
Series Editor
De Montfort University, Leicester, UK

Further information on the IIE Textbook Series is available from
bhmarketing@repp.co.uk
www.bh.com/iie

Please send book proposals to:
rachel.hudson@repp.co.uk

Other titles currently available in the IIE Textbook Series

Mechanical Engineeing Systems	0 7506 5213 6
Design Engineering	0 7506 5211 X

1 Organizations and organizing

Summary

Organizations exist in an intricate, continually changing world, with many pressures on it that influence its behaviour. The organization structures itself to survive and prosper by planning and forming links with other organizations. At the same time, it must present an acceptable persona by working within society's constraints.

Objectives

By the end of this chapter, the reader should:

- appreciate the complexity of the business world and the forces acting on it and be aware of the different forms of ownership and the processes involved in starting a company (Section 1.1);
- understand why organization structures arise and the links made internally and externally in the supply chain (Section 1.2);
- understand the application of organizational analysis using methodology such as the EFQM model and the Balanced Scorecard (Section 1.3);
- understand the need for ethics in organizations expressed through codes of conduct especially where applied to environmental management (Section 1.4).

1.1 The business world

This section examines the business world by starting with a look at the influencing on a business through the PEST analysis. It then describes the different types of organizations, including the different legal forms that can be taken. It goes through the business plan essentials for a start-up situation and then concludes by looking at the external money transactions that are made.

The business world is a complex situation. Each organization exists in a world of opportunities and constraints. The best way to describe this environment is to carry out a PEST analysis, to consider the ways that outside forces impinge on an organization. PEST stands for Political, Economic, Social and Technological factors. There are a few similar acronyms such as STEP, STEEP, LE PEST C, which cover similar factors.

PEST analysis

This analysis looks at a variety of present factors which presently affect the business world and gauges the probability of changes arising. The areas examined include:

Political factors

- Legislative structures of the EC, national and local government and how they are changing.
- Monopoly restrictions such as the Office of Fair Trading and the Monopolies Commission in the UK and similar structures in the EC commission and other countries and how they are interpreting events in a worldwide context as well as domestically.
- Political and government stability – not only extreme cases such as armed insurrection, but even in the UK a change of government may herald changes in legislation such as deregulation and privatization of state controlled industries, which affect businesses – as happened in Western Europe in the later decades of the twentieth century.
- Political orientations of governments' attitude towards business, trade unions and the environment often drive tax and other economic policies.
- Pressure groups such as Friends of the Earth and the various consumer groups can influence not only buyers' and investors' attitudes, but also government policy. This affects matters such as the environment legislation and delays in the construction and other industries by public enquiries. Other examples include the fuel price escalator in the UK during the 1990s.
- Taxation and grant policies can change over a period. For example, the various UK and EC grant schemes and changes in policy on capital allowances for factory construction in areas of high unemployment.
- Employment legislation in areas such as equal opportunities, working time directive, etc.
- Foreign trade regulations both at home and in a country targeted for exports, for example many countries, have restrictions on the importation of certain goods to protect indigenous industries. This happens not just in the developing countries, but also in the developed countries, as can be seen from the regular meetings of GATT (General Agreement on Tariff and Trade) where discussions have been ongoing since 1947 without completely eradicating national tariffs and quotas.

Economic factors

- Business cycles: These are a natural rise and fall in demand for products which have been observed throughout history. They are caused by a complex mixture of factors and can be quite severe on occasions, i.e. the depression of the 1930s. Unfortunately these cycles vary considerably and are not easy to forecast. Governments have attempted to do so in the hope of controlling them, with limited success.

- Money supply: Reflects government action on government tax and spend policy and actions such as credit availability controls. The saving behaviour of individuals can affect the circulation of available money as can be seen in Japan where there is a high tendency towards saving when future trends look pessimistic.
- Inflation rates, i.e. the rate at which prices of products, services and wages change. This can affect people's attitude to credit and savings and the ability of people on fixed incomes to have surplus money for spending. It can affect the flow of imports and exports as the exchange rates vary in relation to other countries.
- Investment levels: Ties in with business confidence about the future, perhaps based on an assumption regarding the stage of the business cycle. This cycle especially affects the construction and machine tool industries.
- GNP (Gross National Product), i.e. the value of the productive efforts of the nation as a whole.
- Pattern of ownership – especially trends in same. Examples include home computers, mobile phones and the growth in car ownership.
- Energy costs: Can affect the cost of making products. Also affects the consumer's attitude towards various products and their functions. A good example is the drive for lower fuel consumption in cars following the oil price rises of the 1970s.
- Unemployment. Is a double-edged factor. Low unemployment means a healthy demand from consumers, but it also leads to a shortage in certain key skill areas which tends to drive up wage costs for these jobs.

Socio-cultural factors

- Demographics, i.e. the make-up of the population – the age groups, where they live, etc.
- Lifestyles: As well as fashion trends, this also looks at habits such as eating out, holiday preferences, central heating, etc.
- Education levels: This will determine the ability of staff to take on new ideas, equipment and processes. It will also affect job seekers' expectations towards company policy and pay.
- Consumerism: This reflects the growing power of special interest groups towards policies. Consumer panels often advise the regulator on privatized industries regarding service delivered and allowed charges.

Technological factors

- Levels and focus of R&D expenditure within the industry and especially by your competitors.
- Speed of technology change in processes and products.
- Product life cycles.

In addition to the PEST analysis, there are always the competitors' actions which will directly affect a business in its relationships with its customers. Chapter 6 deals further with the market and an organization's competitive position therein.

Organizations

Organizations come in many shapes and forms, from the one-man business to the large multinationals employing thousands of people in many countries. Although most are profit making there are also some which do not have making a profit as one of their aims, such as the National Health Service, one of the largest employers in the UK.

For a group of people to become an organization requires more than just casual contact – it requires a formal relationship of the participants in working towards a set goal. So what are the goals for organizations?

There are probably as many goals as there are organizations. The first one must be continual survival, and money is the important component in this. Even all the not-for-profit organizations such as charities, trade unions, the civil service, the National Health Service and state education, which are very large organizations, must have money to complete their function and ensure their continued existence.

In this textbook, we are mainly considering the organizations which supply services or products for a profit. If the profit does not materialize, then they will have to cease trading – either voluntarily or by being put in the hands of the official receivers, i.e. by being made bankrupt.

It is therefore important to recognize the legal status of those companies you deal with – especially if you are supplying them with goods on payment terms. It is illegal for a business to continue to trade when they know they cannot meet existing debts, but that fact does not guarantee your payments will be made. You need to determine your own risk.

Types of ownership

Sole trader/proprietor

The basic one-person business – although it may employ others. This is the majority of businesses where one person raises the investment capital and takes all the profit. The owner also takes all the risk and remains personally liable for the business debts.

Most small businesses do not grow and many fail because:

- Many ideas are not commercially viable.
- Many owners have little commercial understanding and training.
- Capital raised is insufficient to cover the initial time until customers build up.

This size of business does, however, have its advantages:

- Personal involvement of the owner.
- Quick decision making and action.
- Normally communications are easy.
- Details of earnings required by tax authorities are limited, in effect they are treated as self-employed and do not require to submit audited accounts.

The latter does not mean that sole traders should not keep detailed accounts for their own purposes.

Partnership

Basically similar to the sole proprietor, but involving more than one person – up to twenty people can be partners, although more are allowed in certain professions. All partners have the same ultimate liability as the sole trader, even a sleeping partner, i.e. someone who contributes to funding but takes no part in the day-to-day management of the partnership. Working, and sleeping, partners do not necessarily take the same share of the profits, but cannot take a secret share.

This is a common arrangement amongst professions such as accountants and consultant engineers. It has the advantage of more skills and money available initially.

Again the details of earnings submitted for tax purposes are limited although it is highly likely that detailed audited accounts will be required to demonstrate a proper share out of the profits.

The registered company

This is an artificial legal individual under the law, i.e. it is legally separate from its owners. There are three types:

- Unlimited company: Whilst still a separate entity, the owners retain full liability for the business debts. This has advantages in certain circumstances that full accounts need not be disclosed.
- Limited company: The owners have a liability for debts only to the extent of their agreed shareholding. If they have not paid for all their shares when the company ceases trading, they will have to pay up – but only for the amount they have outstanding. A limited company can be private, or public:
  - Private limited company: The shares for this type of company cannot be offered on the stock market, but it is still a limited company. It has the advantage that owners can take decisions without considering how their share price fluctuates on the stock market, but has the disadvantage that selling of shares is more controlled, i.e. they cannot be offered on the stock market only by personal contact with the buyer.
  - Public limited company: The shares for this type of company are bought and sold on the stock market and as such the share price has an influence on the operation of the company. Legally these companies are controlled by their owners, i.e. the shareholders, but in practice the day-to-day operation is very much in the hands of the company directors.

The accounts of registered companies do need to be audited independently annually and submitted for company tax purposes (see Chapter 5).

The limitation of liability does make dealing with a limited company more risky, but most companies do have a long life, therefore as long as you keep track of the company's performance these risks can be kept small.

Without this limitation in liability, it would be extremely difficult to persuade investors to risk participation in companies. However, it is also said that the limitation deprives creditors of a full comeback if the company is mismanaged – after all the shareholders are legally supposed to be the ultimate controllers.

Setting up a business

Sole traders and partnerships are relatively easy to set up as they require no legal stages to come into existence. They do, however, require registration for tax, VAT (Value Added Tax) and National Insurance contributions. The main problem with their formation tends to be the raising of capital.

Although it is not a requirement, most partnerships do complete a legal Deed of Partnership, i.e. a personal contract between the partners. This should spell out their relationship on matters such as capital input, share of the profits, management of the partnership, signatories for contracts and cheques, and how partners are changed, e.g. addition of a new partner or a present partner leaving the partnership.

Where a partnership wishes to trade under a name other than that of the partners, they need to have the name registered with the Registrar of Business Names.

When forming a company, the procedure is more arduous and registration under the Companies Act 1980 is required. The following need to be presented.

Memorandum of association

This is a document explaining the external workings of the company for prospective subscribers. It must include:

- The company name – including Limited or Ltd if a limited company.
- The registered office.
- The objects clause: This sets out the purpose that the company is being formed to do. Legally this limits what the company can be involved with and may make some contracts void, although this is often mitigated if a contract has been entered into with good faith.
- Limited liability – if appropriate.
- The nominal share capital at registration, i.e. the amount available and the unit of issue, e.g. £100 000 in £1 shares. Note a private company has no minimum amount, but a public company has an authorized minimum of £50 000 under the Companies Act, although this need not be all issued, i.e. sold, at the start-up. This section should also denote the type of shares.

The articles of association

This document shows the internal workings of the company, i.e. its rule book. It covers matters such as annual meetings, election of directors and some matters relating to the day-to-day management.

The costs of setting up a company, even a limited company, are quite small and the steps are simple, although it is advisable to employ lawyers skilled in this type of work to avoid unnecessary complications later.

Starting up your own company

Every organization should be working to a moving series of business plans which lay out the direction in which it proposes to move over the following years. These tend to be split into short term (tactical) for a year or so and longer term (strategic) for five to ten years. Even for the long-term plans, we need to prepare matters so that there is a firm base.

A business plan is especially necessary for the start-up situation as it can be the basis on which it may attract financing. Many of the tools you require to complete a business plan are contained within the chapters of this textbook and the main pertinent chapters are shown with the headings of the plan.

For example, when you decide to start up a new business you will need to approach a bank to open a business account, even if you are not borrowing or arranging credit facilities through them. The bank will want to see your business plan before it decides on taking you as a customer.

There are many packs available from banks which guide you through the process, and you are advised to get a few from various banks. In the main they will require a properly presented plan which will contain the following details.

A summary of the plan

This will outline what the business is and the market potential for its product or service. It should state the dates of starting and include a profit forecast and what additional investment is required. If you have been trading, it should include the accounts for that period.

Management (see Chapters 1 and 2)

Management should denote the legal form of the business, i.e. sole trader, partnership or limited company, including the Memorandum and Articles of Association for the latter.

Starting with yourself you should detail your proposed role in the business and your business record and achievements, especially those directly related to the present business. The latter should give a clear idea of past responsibilities, and personal skills and competence.

You should do the same with any partners and other key skilled people involved, so that the bank can see the breadth and depth of the management. This should also indicate what management weaknesses exist and what is planned to address them.

Product or service (see Chapter 6)

This should describe your product or service in clear simple terms. If it is a complex product then you should include the technical detail in a separate appendix. Any unique selling points and price need to be highlighted, including any after-sales requirements.

A brief summary should be made of your nearest competitor's products/services and price for their nearest equivalent offering. This should indicate why you believe your offering can compete.

Market description (see Chapter 6)

Needs to describe:

- Who are the potential buyers – by size, number, business sector, etc.?

- Is the market static or changing?
- Are there identifiable sectors which require different offerings?
- How often does a buyer purchase your product, i.e. a single sale or repeat customers?
- What are the market qualifying and order winning criteria?
- Number of competitors and their position in the market.
- What your selling and advertising plan will be, including who will be doing the sales contacts.
- What are the sales targets, tied to dates to be achieved?

Operations (see Chapter 5)

Here you need to detail the premises where you intend to operate in and trade from and the associated equipment and vehicles you need to purchase, or already possess. The age and expected usage/life may need spelling out with a replacement policy if equipment is second-hand.

This should include any processes that will be operated at particular locations, so that you clearly have a right to do so, and have considered environmental matters such as safety, noise and other nuisances.

It will not be sufficient to say merely that you propose to have your product entirely produced and packed by a sub-contractor. You must include actual quotes if this is proposed and show how you will prevent your supplier marketing a similar product.

Record system (see Chapters 5 and 7)

As it is important that a close watch is kept by the company over matters such as cash flow and outstanding debts, it is imperative that you demonstrate that proper records will be kept. The intended use of software would look good here.

Finance (see Chapter 5)

You will need to state where you are getting the funds to start your business. In addition to your own money, you may be borrowing from friends and family or perhaps a venture capital company if you have a good product or service. Banks like you to be risking your own personal money, especially if you wish to borrow some extra from them.

You will have to include as appendices cash flow and budgeted profit and loss statements for the next two years. The first year normally needs to be on a month-by-month basis, but the second year may only need to be on a quarterly basis. This should indicate what assumptions have been used to arrive at the figures.

Costs must be clearly identified in value and timing. These will include:

- Source of capital and repayment and interest terms.
- Start-up and development costs.
- Premises cost including any lease terms.
- Equipment purchase and running expenses.
- Material costs, both direct and other consumables, including any arrangements for trade credit.
- Wage costs and basis of application. Include your own anticipated drawings.

- Cost of services such as telephone, lighting, heating and water/ sewage.
- Arrangements for VAT, tax and National Insurance payments.
- Vehicle purchase and running.
- Miscellaneous expenses such as stationery and postage.
- Insurance of persons, property, employees and product liability.
- Pension arrangements.

Legal matters (see Chapters 1 and 3)

In addition to the particulars of what type of business it is, i.e. sole trader, partnership or a company, you will need to demonstrate you have considered all other legal aspects such as:

- Health and safety requirements.
- Environmental/Trading Standards requirements.
- Registration for VAT and National Insurance contributions.
- Income tax arrangements.
- Company tax arrangements, including audit proposals.

Owners' objectives

Although this may appear an extra, it is often necessary to include the long-term plans of the owners to see how the business fits in with this.

Who to consult

There are many sources who it would be useful to consult both before and when starting a business. Some offer assistance with training and others will arrange the appointment of mentors to aid you through your start-up. The sources include:

- Business Links (in Scotland Business Shops).
- Training and Enterprise Councils (TECs).
- Local enterprise and development agencies.
- Department of Trade and Industry.
- Banks.
- Accountants.
- Solicitors.
- Other businessmen.

The last is a very useful source of potential pitfalls and advice on many matters. It could prove useful to join a local small business club to share problems with others – and incidentally the added possibility of picking up business.

Money transactions

In Chapter 5 we examine how the organization looks at money within itself, i.e. investment appraisal, costing, book-keeping and budgeting. What we are concerned with here are the financial transactions that the organization has with the outside world.

In a business contract, there must be consideration, i.e. in return for a product or service something of value will be given in return. In past times, a system of barter did exist, but money became the universal medium of exchange many centuries ago. It could be said that if money did not exist, then neither would the industrial world, as the barter system could not accommodate the many transactions which take place today.

The organization will have to make the following payments out of their funds on a continual basis and needs a detailed record of them and any associated paperwork:

Pay to employees, after deductions
Expenses to employees. Note petrol allowances can vary on a mileage basis
Dividends to owners
Payment to suppliers of products, services and facilities, including details of delivery notes and credit for returned purchases
Payments under leases for plant or premises
Payment for plant and equipment purchased
Payment for service cover and spare parts used
Payment to sub-contractors
Payment to professional advisers
Royalty payments
Rent and industrial rates
Company tax
Employer's National Insurance contribution (NIC)
Employer's contribution to pension funds
Insurance
Owner's income tax
Statutory Sick Pay
Statutory Maternity Pay

In addition organizations act as a collecting arm of the government by calculating out, collecting at source and paying to the government the following:

Employees' PAYE income tax contributions
Employees' National Insurance Contributions (NIC)
Value Added Tax

Income tax, NIC and VAT

Employees have income tax and National Insurance contributions deducted from their wages under PAYE, i.e. pay as you earn. Calculation of income tax and NI is complex and varies from year to year. The amounts due are based on a stepped percentage of the pay after reaching a base figure. The percentages and bases are different for both systems and ready reckoning tables are used which are easily incorporated into computer software. However, this means that the software has to be updated each year with changes introduced in the annual government budget.

For many years there have been moves to amalgamate employees' income tax and NIC and this is now going ahead.

Income tax

Almost everyone has to pay this on their total income less any tax allowances. If you are an employee the government works out your tax allowance and gives you a code. Your employer works out your income tax based on this code which gives you an amount before you pay tax. On the remainder you pay a progressive tax, i.e. an initial rate for a set amount, then a variety of increasing rates above that.

Example using 2000–2001 rates:

Earnings	£32 000	
Tax allowance	£6 000	
Taxable pay	£24 000	
10% on 0–£1500	10% × 1500	= £150
22% on £1501–28 400	22% × 22 500	= £4950
40% on over £28 400	Nil	
Total due		= £5200

This figure is not calculated at the end of the year, but has to be done on each pay day.

If you are self-employed, i.e. a sole trader or a partner, you have to agree with the tax office what allowances you can offset before paying tax. If you are self-employed, you can elect to calculate out the tax due yourself.

You may also be due to pay income tax on other earnings such as capital gains.

National insurance

With National Insurance there are different bases for calculating the contributions due from the employee, the employer and the self-employed. If you are an employee in an organization and earn additional money on a self-employed basis, you may have to pay both as an employee and self-employed.

Class 1 contributions are paid by the employee and the employer on employee earnings in stepped stages between two amounts. These contributions are not the same. The contributions further depend on whether the employee is contracted out or into the government SERPS (State Earning Related Pension Scheme). Where there is an occupational pension most employees will be contracted out.

Example 1.1.1

Example of Class 1, contracted-out, contributions using 2000–2001 figures:

	Employee pays		Employer pays	
Employee's earnings:	£32 000			
Up to £3948	0% =	0	0% =	0
£3948 to £4380	8.4% =	£36	0% =	0
£4380 to £27 828	8.4% =	£1970	9.2% =	£2157
£27 828 and over	0% =	0	12.2% =	£509
Total due	£2006		£2666	

Again this calculation has to be done on each pay day.

Self-employed people pay two contributions. In 2000–2001, these were Class 2, a fixed amount of £2, and Class 4, equal to 7% of taxable income between £4385 and £27 820. This appears to be slightly more than an employee, but is based on taxable income whereas the employee's is based on total income. Note that if you are self-employed you will not be eligible for all the same state benefits as an employee.

VAT

The VAT payment (standard rate is 17.5%, but can be 5% or 0%) appears initially to be more simple. You first keep records of all the VAT you pay on items and services purchased – it normally appears as a separate amount. Secondly you keep a note of all the VAT you charge your customer. You pay the government the difference if you charge more than you paid, or collect a refund if you pay more than you charge. This has to be done quarterly unless you are a very small business.

The difficulty is that you may not be registered for VAT. Non-registration is a choice you make if your turnover is below £52 000, note this is turnover – not profit. You may also be zero rated or exempt in certain dealings. In non-registration and exempt cases, no VAT is charged and you cannot claim back any VAT paid.

As this demonstrates, calculation of income tax, National Insurance and VAT can be complex. There are penalties attached for late payment, avoiding payment or supplying false information – hence the need to employ a good accountant.

Payment can also be collected and passed on for items such as:

Employees' pension fund contributions
Trade union subscriptions
SAYE (save as you earn) contributions
Charity deductions
Wage attachment orders for a set amount to be paid to another person, e.g. for child support

In addition, the organization has to keep tabs on all incoming money, including any VAT paid with the account, money owed to them from their customers or suppliers (i.e. by credit notes), and any grants or subsidies due from central or local government.

The control of customers' credit is especially important to ensure that money is received for products despatched or services rendered. There are several credit reference agencies which you can use to check out the credit worthiness of potential customers. Remember that this needs to be periodically rechecked to ensure that the credit rating remains in force.

The wait for payment can be quite long. Your invoice may state a maximum period but it is common for this to be exceeded, sometimes by months. This is a common reason for small companies to have cash flow problems. The government has attempted to reduce these delays and has passed a law limiting the amount of credit taken by late payment. However, if you are a small business with a lot of your income coming from another business, you are in a weak position to force matters. You may lose future orders.

It is possible to factor out the receivable accounts, i.e. get another organization to do the actual collection. You will then receive money due quickly from the agent, but at a cost. The factoring agent will charge a fee – usually up to 2.5% of total turnover.

If you are dealing directly with the general public and offer credit terms, you need to follow the conditions laid down in the Consumer Credit Act 1974. This includes being licensed under credit control regulations.

This means that the organization has to keep careful notes of all collections and payments as they become liable for any mistakes made. Therefore the organization has to know all the regulations and laws involved so that no penalties are incurred, this is where the real knowledge of the accountant proves valuable.

In order to keep tabs on the money flow, the use of accounting software makes it easy to ensure that all transactions are recorded, an easy audit trail is made and end-of-year summaries such as profit and loss accounts and balance sheets are readily compiled. Figure 1.1.1 shows show screen layouts from a PC accounts system for small companies.

Problems 1.1.1

(1) We are mostly members of a family. So are we members of a legal partnership in this case?

(2) Do you consider it fair that a company can go bankrupt through management paying themselves excessive salaries with no comeback against their property?

(3) What new business do you feel may succeed in your local town?

(4) Do you think that registering for VAT as a small business, and charging customers this extra, puts you at a disadvantage compared to another business not being registered and not charging?

(5) As practice for operating accounts in a small business, try keeping a detailed record of personal income and spending over a three-month period.

1.2 Structures and links

In small organizations, people often tackle a wide variety of tasks. As the organization grows, people tend to deal with a limited number of the available tasks. This leads to problems in communications and focusing which organizational structures attempt to deal with. Links and relationships now extend from inside the organization outwards towards its suppliers and customers.

Internal structures

When organizations start, they have very few people – often only one, the owner. This means that this person (people) has to do every job that arises. However, as organizations grow, the volume of work also grows. It is natural then that people stop doing just any of the many tasks that arise and start to concentrate on doing only a selection of the tasks, i.e. a division of labour takes place.

This division of labour is important to the efficiency of the organization as it enables employees to become skilled in doing a limited number of

Sage Instant Accounting 2000

Bank statements, invoices and VAT returns appear exactly the same on screen as they do in real life.

Easy-to-understand prompts will quickly help you to create, file, access and update all the financial information you need.

E-mail and web addresses of your customers, suppliers and your bank are stored within the program, allowing you to either send e-mails or access web sites directly.

Paying staff made easy with Sage Instant Payroll

Easy to follow task bar eliminating errors.

Look up, re-order and locate employee records quickly and easily from your employee listings.

Choose the employees you wish to process at the touch of a button.

Total pay and statutory deductions for tax and National Insurance calculated on screen to show you your employees' net pay.

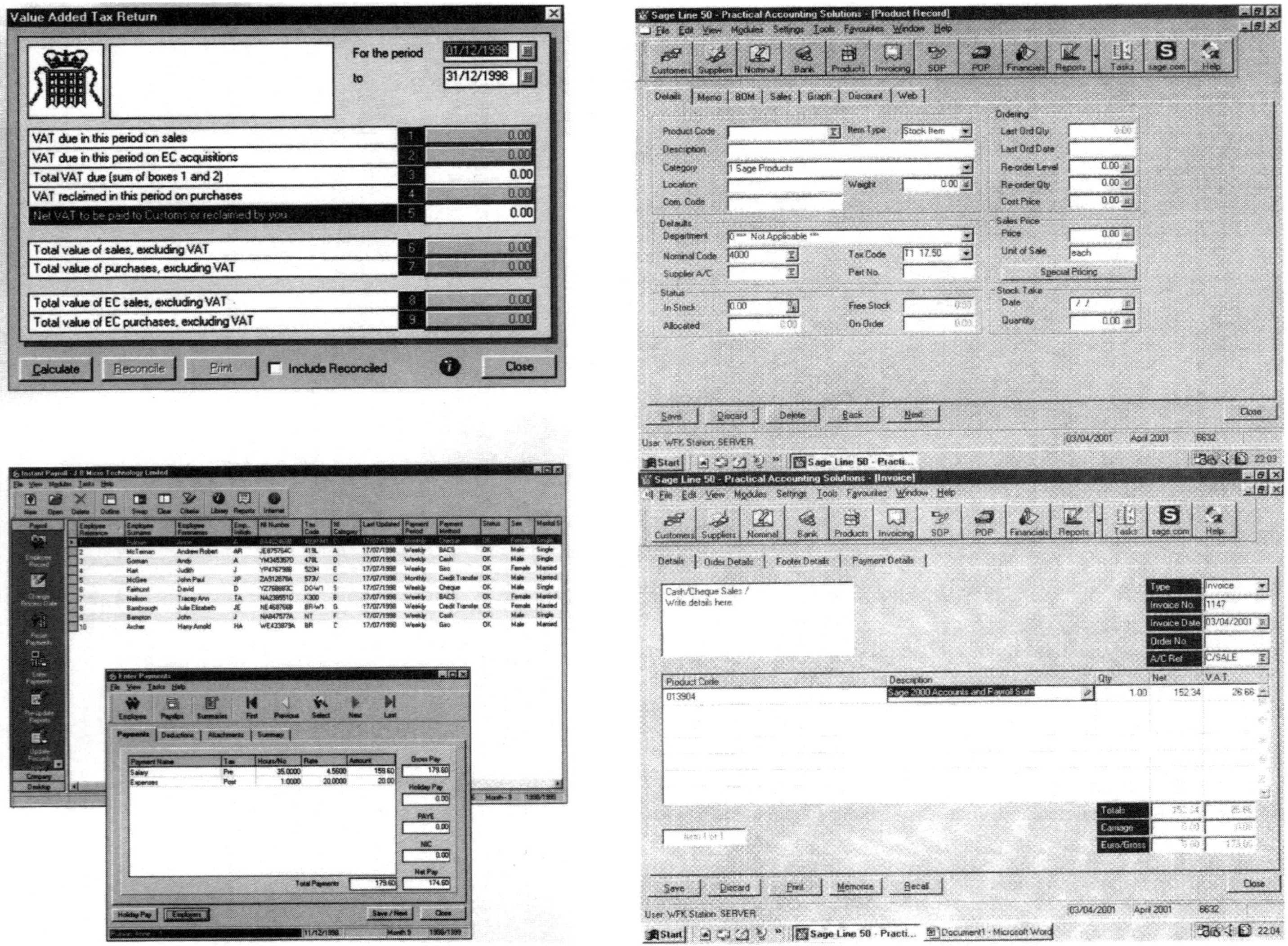

Figure 1.1.1 *Screenshots from PC-based accounting system. (Reproduced courtesy of Sage Group plc.)*

tasks through continual repetition, i.e. they undergo a learning curve. They then have the opportunity and motivation to seek ways to become more knowledgeable in these tasks so that they can perform them to a higher level and hence become more useful to the organization. Sometimes it may be so they can become more secure in their jobs, or even gain more influence within the organization.

As the organization continues to grow, the work for these individuals increases; individuals become sections and then departments within the organization. People are then employed with a deeper level of knowledge and experience in these particular areas. This, unfortunately, also means that individuals have a smaller overall breadth of knowledge about all the jobs being done within the organization.

These departments tend to fall into recognizable functional areas. However, just as people grow up differently, so does each organization. A department in one organization therefore often does not cover exactly the same detailed functions as a similar named department in another organization.

We are going to examine the typical functions, i.e. jobs that need to be done within any organization. Most organizations start with two basic functions:

- Selling the products and/or service.
- Making the products and/or providing the service.

From within these grows a multitude of other functions:

Cash management
Purchasing
Stock keeping
Product design and development
Estimation of cost and prices
Process development
Quality
Recruitment
Disputes
Training
Maintenance
Transportation
Information technology
Legal

The list will continually grow and diversify as the organization increases in size. There is a disadvantage in too much specialization as it creates problems of communication and ensuring that everyone is moving in the same direction. Large organizations therefore require the integration of management to:

- Plan ahead.
- Give directions to the separate groupings.
- Ensure groups are operating correctly, i.e. controlling.

Each organization finds different structures to link these together. This is partially determined by the industry and their market, and partially by the characteristics and competencies of their staff as they are growing. If we look at an organization's environment, as in Figure 1.2.1, we see that there is a multitude of influences and constraints on how they operate.

Local government UK government

New technology Advisors Distributors

Material suppliers **The organization** Customers

Plant and equipment suppliers Competitors

Local community Pressure groups

Trade unions

Figure 1.2.1 *The environment of an organization*

This also has an effect on the growth and composition of their internal structures as they form interfaces with the outside world.

Historically the only organizations that were large enough to have separate functions within them were governments and armed forces. When other organizations became large enough to support internal specialization, they tended therefore to imitate those models available. This gave rise to the concept of line and staff posts (see Figure 1.2.2),

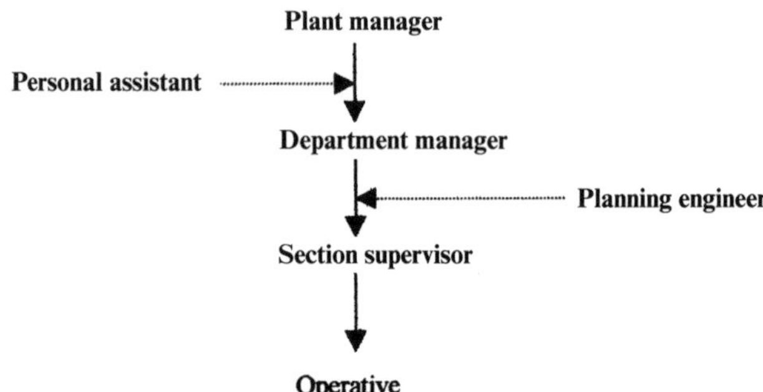

Figure 1.2.2 *Line and staff relationships. (Line relationships are indicated by the full lines and staff relationships by dotted lines)*

where line personnel are directly in the chain of command and staff personnel are in an advisory position.

The first stage in internal structuring arises with the change from anyone doing any job to people having set jobs.

As the organization grows then the number of people performing similar duties grows. If this is only two or three they may be still part of the overall team. Once numbers grow it becomes difficult for the remainder of the overall team to communicate with each person and usually one person becomes the communication focus and a specific section is formed. With continual growth this focus becomes that person's main role and you now have a section leader, or supervisor. This role then has responsibility for the productivity and quality of the section.

If the number in the section continues to grow, then the span of control may become too much for one person. You are then left with a choice:

- Do you form two/three separate sections, splitting the duties between these sections? or
- Do you retain the section as a whole under the present leader/ supervisor, but with others put in a similar, but junior role reporting

to the leader/supervisor? The latter would then be classified as a manager.

Your choice is to increase the number of different junctions or to add levels of management. However, there probably will not be too much difference eventually as you may have to have someone co-ordinating the several separate sections!

In this way as the organization grows, the levels of management and their span of control change. Two problems tend to arise in this situation:

- The sections become specialized and lose focus on the organization's overall objectives.
- Top management become increasingly distanced from the detail of the day-to-day operations.

This has led to most organization's changing their internal structure to try to keep direction and focus. Some organizations appear to be continually changing their internal structures – perhaps on the basis that by pure chance, the ideal must be hit on sometimes. Many centuries back, a Roman general was reported to have complained that the reaction to any problem is to reorganize.

The various structures that can be seen are:

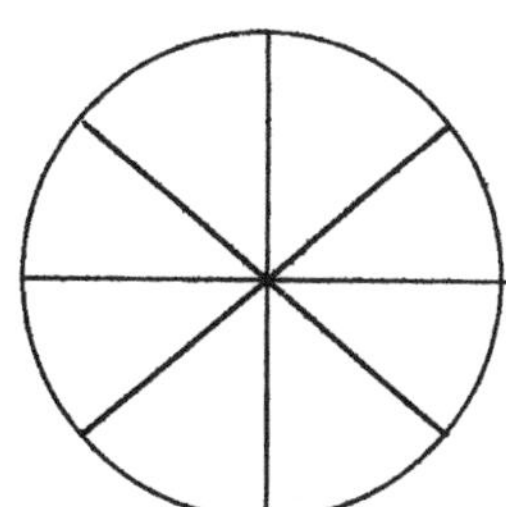

Figure 1.2.3 *Entrepreneurial organizational structure. (The Chief Executive is at the centre of all decision making)*

- Entrepreneurial as in Figure 1.2.3.
- Functional as in Figure 1.2.4: The actual make-up of the functions will be dependent on the importance of different tasks. For example, the purchasing function may be under accounts in one organization, but under operations in another. The former may give better financial management, the latter better ties to production schedules.
- Regional as in Figure 1.2.5: Here the lower levels could be a mixture of any of the other structures. In sales, it is often a lower level copy of the main structure.
- Product as in Figure 1.2.6: This is common where products are diverse and focus is required at product level.
- Business sub-units: Each smaller unit is an organization by itself with its own internal structure. Very common in conglomerates.
- Matrix as in Figure 1.2.7: Here there are several main different focuses at work. Can create problems when each focal point is under conflicting pressure from the different axis.
- Centralization: Here the centre attempts to control certain aspects.
- Decentralization: Here the centre plays a specialized role. This is normally financial, but could include research and development activities.

There is no universal panacea that an organization can just impose on an existing structure. Any change must be well thought through and tested, and needs very careful management during implementation.

The problems in communication and different viewpoints that tend to develop with specialization by functions have led many organizations back to reforming project teams for various tasks with members coming from the different functions. In effect they are reforming the entrepreneurial structure.

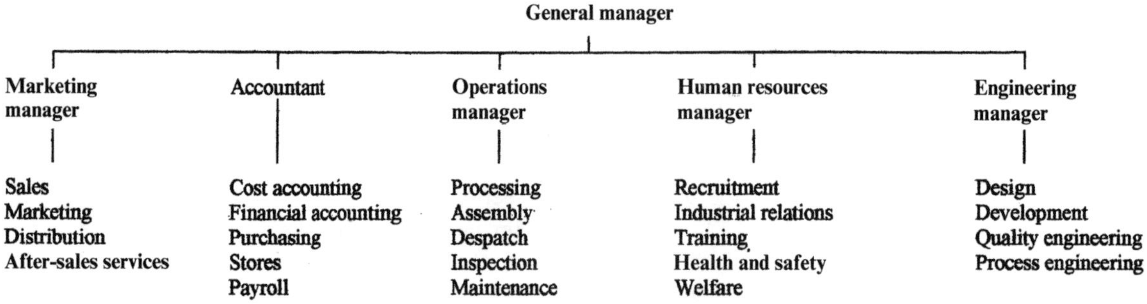

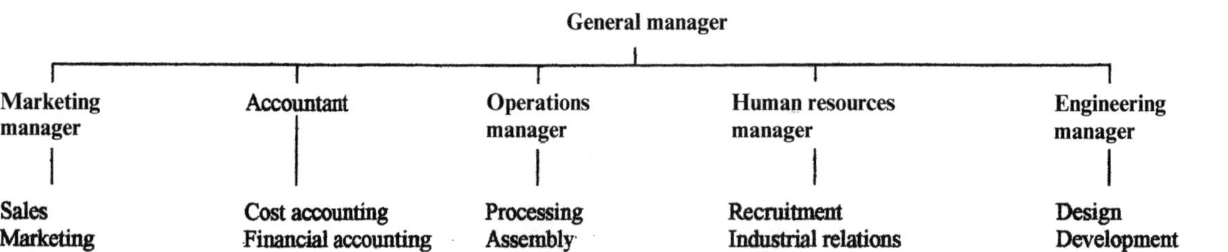

Figure 1.2.4 *Functional organizational structure*

Figure 1.2.5 *Regional organizational structure*

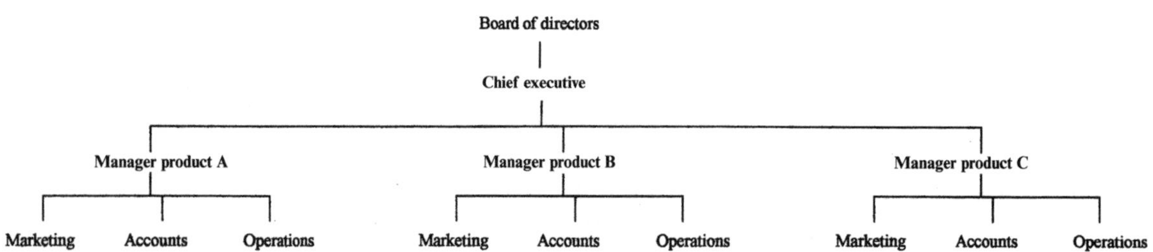

Figure 1.2.6 *Product organizational structure*

		UK operations director		
		Glasgow factory manager	Bristol factory manager	Sheffield factory manager
European sales and marketing director	Product A brand manager	X		
	Product A brand manager	X	X	
	Product A brand manager	X	X	X
	Product A brand manager		X	X

Figure 1.2.7 *Matrix organization. (The factory managers have to satisfy the requirements of the competing brand managers. Where problems arise they may have to go to discussion between the UK operations director and the European sales and marketing director)*

This leads to three advantages:

- Communications and feedback are quick.
- People in close working relationships tend to be more prepared to listen to the differing viewpoints of others and adjust their own stance.
- The team can easier come to an agreed focus on the task.

This has led to substantial cuts in project times and tends to produce results which are both effective and more acceptable to the people involved.

Supplier relationships

In many modern manufacturing industries, the manufacturer of the final product is just an assembler of bought-in components and assemblies. In these companies, the cost of the bought-in items can often be more than 50% of the direct cost of the products.

This means that purchasing cannot be thought of just as a cost. It must be managed in a way that maximizes its potential to contribute towards profit. This means sourcing the best suppliers that fit into the organization's operational demand of cost, time, and quality (see Chapter 6 Section 4, for a discussion on order-winning criteria and competitive analysis).

In addition there are many recent adoptions within organizations of ISO 9000 quality registration and changes in material procurement to a JIT (just-in-time) system that increase the need to have high quality suppliers delivering error-free items direct into the production process.

ISO 9000 Quality System Registration

ISO 9000 refers to an internal Quality Assurance System with procedures covering all aspects of managing design, procurement, materials, processing and inspection. To be registered under the scheme requires the organization to prove it is using high quality suppliers – ideally also registered organizations.

The objective of the JIT methodology is to operate with a minimum of internal stock. This means that the factory makes only what is required for immediate delivery and only brings in materials and components required for immediate use by the process lines. This incoming material needs to be of high quality and adherence to delivery times is very important.

It used to be sufficient to have incoming goods retained in a reception area and then to have inspection carry out either 100% inspection or apply an acceptance sampling scheme to batches of incoming material and reject batches that exceeded a set number of defects. Nowadays, this method cannot cope with the stockless operations with their tight time schedules and high quality demands. The most important process of ensuring one has suitable material available as and when is to carry out a complete assessment of all suppliers' abilities to supply the correct material.

Vendor assessment

This assessment can be an initial stage prior to investigatory visits to suppliers to examine their processing facilities and audit their quality procedures. When examining suppliers, one should take into account all the factors which are of importance to the company, namely:

- QUALITY, i.e. the level of rejected batches.
- PRICE per item, relative to other suppliers.
- DELIVERY, i.e. deliveries made on time.
- SERVICE, i.e. amount of back-up and flexibility demonstrated.

Vendor assessment is a simple rating of different suppliers against each other under headings so that their overall value to the company can be assessed. It is a matter of selecting the particular factors, such as these above, which are considered important and applying a weighting to them in relation to their agreed impact on the company.

Example 1.2.1

An organization may decide on the following weightings out of 100 points available. They can compare each supplier against these factors:

- Maximum points for QUALITY = 40
- Maximum points for PRICE = 25
- Maximum points for DELIVERY = 20
- Maximum points for SERVICE = 15

If a supplier has the following profile:

- 90% of his deliveries are made on time.
- Their price is 105p per item against the cheapest quote of 90p.
- 95% of their delivered batches are accepted on inspection.
- His reactions meet our requirements on 80% of occasions.

This supplier would be rated as:

Delivery rating = 90% of 20 = 18
Price rating = (90p/105p) × 25 = 21

Quality rating	= 95% of 40	= 38
Service rating	= 80% of 15	= 12
Total rating		= 89

We then can compare this supplier's total rating against all other suppliers for the same material or components to decide on which better serves our needs. It can also be used in conjunction with selected suppliers to show where their total service is behind their competitors to drive up the general rating of all suppliers combined. We wish to end up with a few high value suppliers.

Vetted suppliers

Once suppliers have been selected to be the main, or sole, supplier and we are convinced that the quality of their supply consistently meets our requirements, it should be possible to reduce the need to inspect their incoming goods.

It will still be necessary to monitor their quality which can be partially from occasional quality audits and partially from records of their material/components service within the plant and in service with our customers. To ensure that this can take place it is important to ensure that full traceability of all materials used is possible and maintained.

When examining a potential supplier, the areas which should be examined are:

Management

- Organization structure and responsibilities.
- HRM policies on recruitment, training, promotion, health and safety. Can be judged on number of disputes, stability/turnover, absenteeism and accidents.
- Customer satisfaction statistics.
- Business integration – especially towards their customers.
- Supplier management.

Delivery

- Delivery system integration with processing.
- On-time manufacturing systems – percentage of orders late, average days late statistics.
- Packaging, shipping and receiving operations.

Technology

- Process development, e.g. age of plant, skill level of employees.
- Product design process.
- Manufacturing and test equipment.

Quality

- Quality policy and procedures.
- Process control, e.g. measuring automation and use of statistical quality control.
- Problem prevention, detection and correction procedures.
- Supplier quality process and audit statistics.

Cost

- Labour and materials allocation systems.
- Rates used for accounting and other costs:
  – Performance statistics on processes.
- Processes for estimating costs on proposal and pricing policy.
- Financial condition of the organization.

Technical support

- Organization.
- Logistics support development.
- Field maintenance support.

This will give an indication of the suppliers' ability now. It must be continually addressed so that improvements are made to increase quality and lower costs.

The supply chain

Supply chains

As in Figure 1.2.8, there is a complete supply chain from the basic, or primary industries, through to the ultimate consumer through which we can see material progressing and changing shape. A few companies are vertically integrated and own parts of their supply chain, but this is rare.

Every organization supplying services or products exists within a mixture of different supply chains as in Figure 1.2.9. Some of these chains will be of great importance to the organization; others will be less so. The organization should identify which chains are important, and then fully engage with them to gain the maximum advantage for themselves.

Today the supply chain for many products is worldwide and organizations' destinies are becoming more interwoven and interdependent in an international setting. The earthquake in Kobe, Japan, in 1999, for example, caused shortages in components for organizations throughout the world.

In 1961, a model demonstrating the effect of a change in demand rippling back through the supply chain was demonstrated by J. M. Forrester in *Industrial Dynamics*. This showed that the further upstream, i.e. away from the end user, the higher the fluctuations in demand experienced.

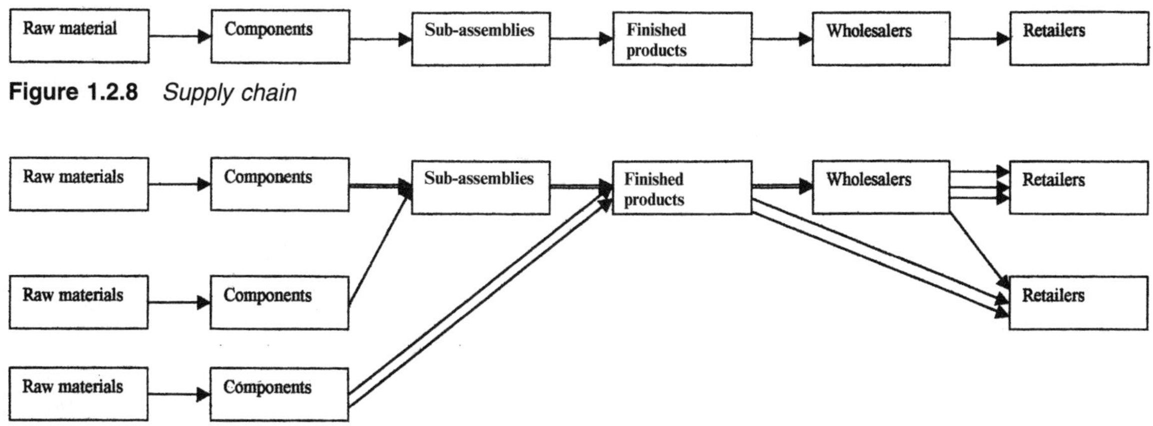

Figure 1.2.8 *Supply chain*

Figure 1.2.9 *Mixture of supply chains. (Where multiple arrows are shown there can be several different supplier-customers involved)*

The Forrester Effect

To show the effect of a change in demand of 10% from the end customer on the subsequent participants in the supply chain, a simple example of the flow of orders and products is shown in Table 1.2.1. The rules in the table are:

Each point in the chain aims to have at each period-beginning, a stock equal to the orders received during the previous period.

- Orders passed on to the next stage in the supply chain are based on a sufficient quantity to bring the closing stock up to the desired opening stock quantity for the next period.
- Orders are placed at the end of a period and are instantaneously delivered so that the items are available in the opening stock for sale in the next period.
- Stock shortages are allowed and taken off the quantity delivered.

The first period shows a stable market, therefore opening stock for each stage is set at 100 units. As 100 units are sold, the closing stock drops to zero and an order is placed for 100 units at each stage.

In the second period, the end sales reduce by 10 units to 90 units. The retailer therefore resets his stock target to 90 units (same as orders received). Order quantity then becomes 90 – 10 (closing stock), i.e. 80 units.

The manufacturer receives an order for 80 units and therefore has a closing stock of 20 units left from the opening stock of 100 units. As the new opening stock is 80 the order placed for components is 60 units, i.e. 80 – 20 (closing stock).

The component supplier receives the order for 60 units which leaves them with a closing stock of 40. They reset their opening stock to 60 (as sales) and therefore make only 20 units.

In the next period, as we can see from Table 1.2.1, both the manufacturer and the component supplier have a stock-out, shown as a minus figure in the closing stock. This means they would have to make special arrangements (i.e. incur extra costs) to quickly make these.

As can be seen in Figure 1.2.10, a change is doubled at each stage initially. Thereafter the further up the supply chain the more severe the fluctuations become before settling down as the effect of the change is dampened after four periods.

As all markets have fluctuations, continual small changes in end demand will tend to send constant ripples through the whole supply chain, forcing the lower level suppliers to continually vary their manufacturing orders much more radically than the actual changes in the end demand.

In order to level out these disturbances large amounts of safety stocks are installed to buffer the effect and ensure no stock-outs arise. This means a large amount of safety stock at all points in the chain:

- Retailers would hold stock in case the manufacturers failed to deliver sufficient products.
- The manufacturer would hold stock of the end product in case the retailer suddenly requested more than normal and at the same time would hold stocks of raw materials and components in case their supplier failed to meet their orders.
- The component supplier, of course, would also hold stocks of components and raw materials in case the manufacturer suddenly increased their order.

The advent of EDI (see Chapter 7) has meant that organizations can share data not only within themselves but with partners along the chain, i.e. customers and suppliers. It is possible to have point of sales information collated in a computer and then relayed to all points in the supply chain as in Figure 1.2.11. This means that only replacement items need to be produced so that any changes in the sales pattern are immediately catered for.

Eliyahu Goldratt discusses the handling of this integration in his book *It's not luck*. In his latest book *Necessary But Not Sufficient* he in fact goes one step further in advocating that all supply chain members only get paid when the final sale is made.

It is natural that supplier and customer follow on to their developed relationships in trading together over factors such as quantities and delivery time, to other aspects such as developing products and services together.

Manufacturers of bought-in components will have a wealth of experience in making their own products. For their customer to ask them

Table 1.2.1 Order pattern demonstrating the Forrester effect

	Retailer			*Manufacturer*			*Component supplier*		
End demand	*Opening stock*	*Closing stock*	*Orders for products*	*Opening stock*	*Closing stock*	*Orders for components*	*Opening stock*	*Closing stock*	*Make component*
100	100	0	100	100	0	100	100	0	100
90	90	10	80	100	20	60	100	40	20
90	90	0	90	80	–10	100	60	–40	140
90	90	0	90	90	0	90	100	10	80
90	90	0	90	90	0	90	90	0	90
a	*b*	*c* = *b* – *a*	*d* = *a* – *c*	*e*	*f* = *e* – *d*	*g* = *d* – *f*	*h*	*j* = *h* – *g*	*k* = *g* – *ju*

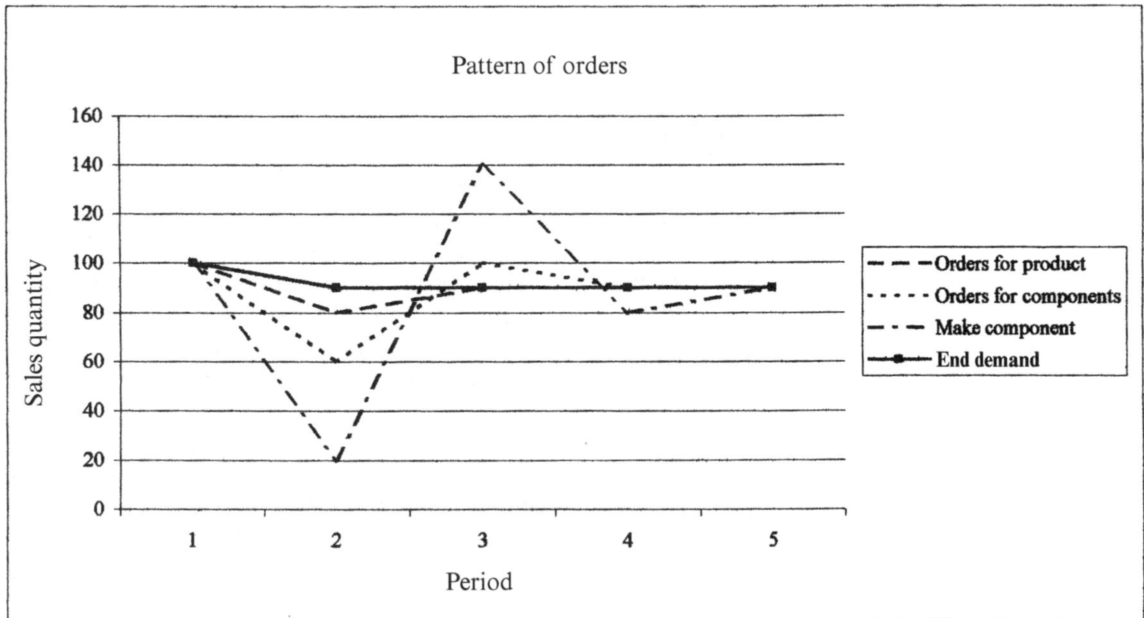

Figure 1.2.10 *Graph showing Forrester effect*

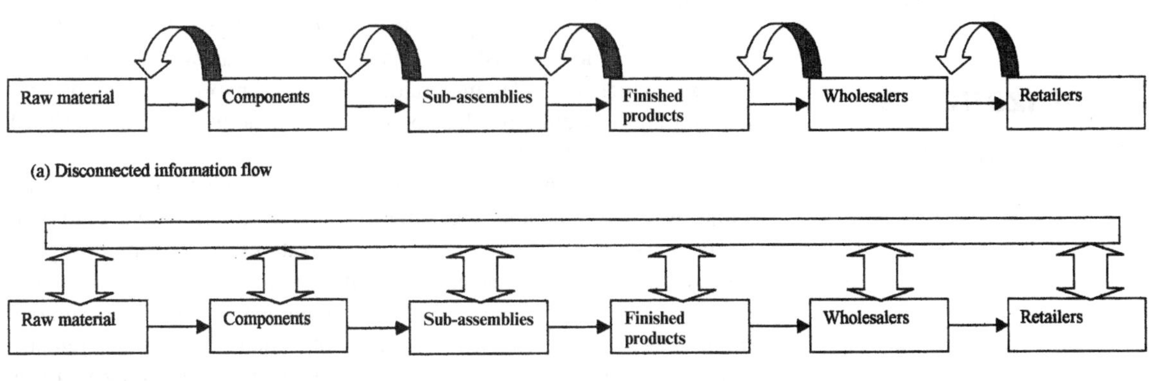

Figure 1.2.11 *Information flow in the supply chain*

to make an item to an imposed design may not produce the best component in terms of function, quality or cost.

In the same way as cross discipline teams are formed within organizations, it is easy to apply the same concept between organizations to develop a rapid response and design functionally effective products and processes at a minimum cost whilst protecting profits in both organizations.

This does require openness and trust between organizations. As their long-term future will be tied up together, a lack of fairness and honesty in the setting up of a relationship will leave both open to future difficulties.

An example of trust between organizations is the on-site stock of a supplier's materials that is only charged for when it is withdrawn for use by the customer. This was initially set up in spare parts for maintenance, but has now spread to low value items which are used in manufacture, such as bolts and screws.

Problems 1.2.1

(1) Think about a hockey team. Is it important that each player fulfils a set role? Should there be occasions when they can step outside that role?
(2) Think about the last group project you were involved with – did everyone carry out the same task? Why not?
(3) If you were asked to investigate a supplier, how would you persuade them to allow you to closely examine their operations?
(4) How easy would it be to make the consumer demand available to everyone in the chain, when there are many suppliers serving many intermediates in the chain?
(5) What problems do you see in Goldratt's proposition of all supply chain members getting paid only when the final sale is made?

1.3 Improving the organization's performance

All organizations need to be continually improving themselves, especially when they are exposed to keen competition in their markets. To achieve this is no easy task and this section looks at two methodologies that exist which attempt to integrate all aspects of an organization towards it becoming fully effective. It also examines the modification to ISO 9004 introducing Quality Management Principles.

As the twentieth century ended, many organizations were being subjected to tremendous pressures due to the international nature of many markets. During the 1990s organizations had been through a process of ensuring that high quality products and services were delivered to the customer through techniques such as continuous improvement, company-wide quality and total qality management. These have now become so ingrained in organizational culture that it is difficult to demonstrate their uniqueness, except when they are missing.

What was required during the twentieth century were management approaches which integrated all functions and aspects of the organization towards its key result areas. We are going to examine three of these:

- Business Excellence Awards
- The Balanced Scorecard
- The ISO 9004:2000 Quality Management Principles.

Business Excellence Awards

The history of Excellence Awards

These awards started as part of the national quality initiative in Japan in the 1950s with the Deming Prize, and spread to the USA with the Malcolm Balridge National Quality Award in 1988. Europe followed suit in 1992 with the European Quality Award.

In the UK, the British Quality Foundation (BQF) introduced the British Quality Award (BQA) using the Business Excellence model which has since been adapted as the framework for most countries within Europe and is also used for regional awards within the UK.

Although the original criteria of these awards was biased towards design and manufacturing quality aspects, the models have expanded to include all aspects of a business and hence the BQA award has been renamed the EFQM (European Foundation for Quality Management) Excellence Model. It was revised to its present format in 1999.

The Business Excellence Model

The model comprises nine criteria covering all aspects of the organization. It can be applied to the organization as a whole, or to parts of it. The criteria are grouped into two sectors as in Figure 1.3.1. The first sector is concerned with inputs and contains five *enablers*. The second sector is concerned with criteria outputs and contains the remaining four *results* criteria.

The organization is assessed under each criterion and given a score for that criterion. The individual criterion scores are then summated, using weighting indicating the criterion's contribution, to give an

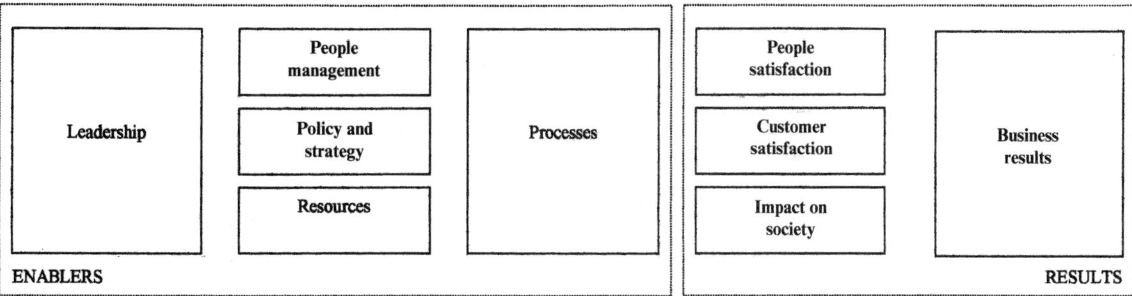

Figure 1.3.1 *The Business Excellence Model. Reproduced courtesy of the European Foundation for Quality and Management*

overall score for the organization. The individual criterion score and the overall score can then be used to determine the organization's present standing using a process of comparison, i.e. benchmarking, against known organizations. These exemplar organizations can come from various business sectors, not just the sector in which the organization is active.

Benchmarking

Benchmarking is the systematic search for the best practices which, if implemented, lead to superior performance.

By conducting a survey, you can see how you are doing relative to other organizations as in Table 1.3.1:

Table 1.3.1 Results from a benchmarking survey

Criteria	*Best*	*You*
Lead time	3 days	28 days
Orders met from stock	85%	55%
New products last year	7	2
Defects	0.05%	3%
Material yield	98%	87%
Customer returns	0.5%	7.5%

More importantly, the assessment process highlights key areas that can then be improved to increase the organization's all-round effectiveness. If assessments are carried out over time they also indicate any trend that is happening to judge the result of action taken.

We will examine the enabler criteria first and then see what the results criteria are and how the enablers contribute towards the performance in the result criteria.

The enabler criteria

These are the criteria in which the organization's performance should be reflected in the results criteria.

Leadership. A key area as the leaders drive all the other process by their behaviour, communicating purpose through relevant systems and motivating staff.

Purpose: How the leader develops visions and values and forms a role model:

- Defines mission and desired results.
- Sets and reinforces an ethical culture.
- Gathers feedback on impact of style and amends behaviour.
- Seeks and supports improvements.
- Encourages collaboration.
- Encourages learning.
- Operates a no-blame culture.

Management systems:

- Matches structure to support strategy.
- Reviews and updates administration processes.
- Aligns systems to strategy and desired results.

External contacts:

- Involved in partnership and community projects.
- Works with customers and authorities to determine their needs.
- Supports staff involvement in community projects.

Behaviour:

- Are they accessible?
- Do they communicate effectively?
- Do they recognize individual and team achievement?

Policy and strategy. Looks at how the organization develops implements and reviews its policy and strategy.

Establishing needs: Here we are looking at what data is used to establish needs and expectations of all possible stakeholders:

- Customers, suppliers and other partners.
- Shareholders.
- Staff.
- Representatives of local and national government.
- Authorities and regulators.
- Other indicators, e.g. a PEST analysis.

Develop and review process:

- Match to short- and long-term objectives of organization and partners.
- Balance between the various stakeholders.
- Assumptions made, risks identified and contingency plans.
- Alignment with critical success factors.
- Review process.

Deployment:

- Do resources match requirements?
- Do processes deliver objectives?
- The review process for future needs.

Communications:

- Effectively communicated?
- Is employee awareness evaluated?
- Tie into lower level tactics.

People management. This looks at how the staff are developed and motivated to effectively perform towards the business goals.

Planning:

- Is there a manpower plan prepared on the basis of a business plan?
- Are there clear career development plans?
- How will assessing of morale be done?
- Are equal opportunity, promotion policy, etc. in place?

Development:

- Is training need analysis used to develop training and development programmes?
- How are individuals' targets set and reviewed?
- Is there a system of staff appraisal?

Participation:

- What decisions are people empowered to do themselves?
- How are staff encouraged to participate in continuous improvement?

Communication:

- How is two-way communication encouraged?
- How is knowledge shared?

Rewards:

- What tie into organizations objectives is there?
- Additional facilities, e.g. child-care, sports facilities.
- Opportunity for changing work patterns.

Partnerships and resources. Here we look at how the organization adds to its results through effectively using partners and its available resources.

External partners:

- Are they actively sought and developed?
- Does relationship add mutual value?
- Are joint developments encouraged?

Finances:

- Support to policies.
- Review and development procedure.
- Key financial ratios.
- How are investments evaluated?
- How is empowerment monitored and controlled?
- How are risks identified and managed?

Buildings and equipment:

- How is security managed?
- How are resources utilized and maintained?
- How is waste generation and power usage monitored?

Technology:

- How is technology used to improve efficiency?
- How are new technologies evaluated?
- How are staff skills updated in new technology?

Information:

- How is information collected and used?
- How are information systems linked to user's needs?
- How is information validity and security maintained?
- How accessible is the information?

Processes. If the people are to deliver the main objectives, the key processes must support them in their roles in satisfying customers and other stakeholders.

Management:
- Are there measures of capability and output?
- Are appropriate standards applied, e.g. ISO 9000 and ISO 14 000?
- The review and updating procedures.

Innovation:
- Is there identification of potential improvements?
- Number and degree of changes.
- Change management and communication of the changes.
- Training in new processes.

Design:
- Is it driven by research of customer needs?
- What is the evaluation procedure?
- What co-operation is there with partners?

Process:
- Match to customer and other stakeholders' needs.
- Are some processes outsourced?
- The communication with customers.
- What efficiency measures are employed?

Customer relationships:
- Is feedback from customers collected and acted upon?
- What monitoring of customer perceptions is carried out?

The results criteria

There are two areas of measures on the first three criteria – perception and performance indicators. On the fourth criterion, the performance results, the measures are outcomes and leading indicators.

Customer satisfaction. What the organization is doing in relation to the complete chain of its external customers. We must ensure that the measures used are of real importance to the customer.

Perception measures: Can be collected from surveys, interviews, trade analysts' views, vendor assessments, and complaint records:

- Overall perception: Accessibility, flexibility, responsiveness and co-operation.
- Product/service: Functionality, specification range available, quality, value for money, reliability, design innovation, environmentally friendly.
- Pre-after-sales: Staff friendliness and competence, technical/sales advice and support, match to customer requirements, response time, problem solving, complaint procedure, product training, sales and customer literature and manuals.
- Future relationship: Intention to continue to trade/looking for other suppliers, willingness to recommend you to others.

Performance indicators: Many of these can be collated from internal data:

- Product/service performance: Defect/reject rate, complaints, returns, warranty claims, delivery delays.
- Design: Number of innovations, product life cycle, time to market, conversation/rejection of quotations into orders.
- Sales/after-sales: Number of complaints, speed of response, demands for information/training.
- Customer loyalty: Length of relationship, number/value of repeat orders, lost business/retention rate, value of order per customer.
- Customer recommendation: Prizes, award nominations, trade press coverage.

People satisfaction. Here we are looking at how the organization's actions are perceived by their own staff:
Perception measures: Taken from surveys, focus groups, interviews and staff appraisals:

- Motivating factors: Career development and opportunities, communication, involvement, empowerment, equal opportunities, leadership issues, recognition for achievement, target setting and appraisal, learning and training opportunities, commitment to organization, sharing organization goals.
- Satisfaction measures; Condition, pay and benefits, job security, change management, health and safety, peer relationships, administration.

Performance indicators: Again most of these can be collated from internal data:

- Employee achievement: Competency gap, productivity, number of training initiatives, training uptake and completion, post-training evaluation.
- Evidence of participation: Suggestion scheme responses, involvement in improvement teams, recognition of individual/teams, response to employee surveys.
- Satisfaction indicators: Absenteeism, sickness levels, accidents, turnover, recruitment response, number of grievances, use of welfare and social facilities.
- Administration: Accuracy in data held, response time to queries, health and safety audits, environment audits, community projects.

Society satisfaction. What the organization is doing in relation to the local, national and international communities – especially in environmental matters and with authorities which affect or regulate its business:
Perception measures: Taken from surveys, newspaper reports, public meetings, dealing with local/national authorities:

- Corporate citizenship: Disclosure of relevant information, equal opportunity practice, impact on local economy, ethical behaviour, relationship with authorities.

- Supporting local community: Schools and colleges, hospitals, community projects, voluntary work, sport and leisure facilities, medical and welfare provision.
- Nuisances: Noise, smell, sound, traffic, pollution, health risks and general safety.
- Environment: Product recyclability, waste produced and collection efficiency, usage of utilities, material usage, use of resources, power requirements.

Performance indicators:

- Handling of changes in employment levels.
- Usage of formal environmental systems.
- Relationships with planning authorities.
- Staff involvement with voluntary and other local activities.
- Number of visits to local schools and colleges.
- Donations to and assistance with local causes.

Key performance results. These are the normal business measures which organizations use to denote their performance:
Performance outcomes:

- Financial: Profit and loss, margins, sales, share price, dividends, borrowing, return on capital/assets, cash flow.
- Non-financial: Market share, volume, size of orders, time to market, quotation conversions, product launch date achievement.

Performance indicators:

- Process: Performance/productivity, defect rate, innovations and improvements, cycle times, time to market, late orders, delivery times, project completion rates.
- Suppliers: Prices, quality, performance, late deliveries, joint initiatives
- Financial: Cash flow, working capital requirements, returns on investments made, bad debts, inventory turns, maintenance costs, debtor ratios, creditor ratios, asset utilization, credit ratings.
- Technology: Patents, royalties, new product launches.

The assessment process

The EFQM model can be used in different ways to assess an organization's performance and give pointers for improvement. The initial assessment is normally carried in a self-assessment mode. Ultimately the organization can carry out the detailed award style assessment itself, or it may wish to be formally assessed by external, independent assessors appointed by the BQF.

Initially assessment can be by a series of questionnaires and workshops to determine how an organization's management determines its performance on the criteria. Figure 1.3.2 demonstrates the BQC Business Improvement Matrix used as a broad-based questionnaire to gauge performance on each criterion.

	Leadership	Policy and strategy	People management	Resources	Processes	Customer satisfaction	People satisfaction	Impact on society	Business results
10	All managers are proactive in sustaining continuous improvement	Mission and business policy statements cover the whole business	All actions are directed towards realizing the full potential of all employees	The organization's resources are deployed effectively to meet policy and strategy objectives	Key value added processes are understood. Formally managed and continuously improved	There is a positive trend in customer satisfaction. Targets are being met. There are some benchmarking targets across the industry	Regular comparison with external companies show employee satisfaction is comparable with other companies and has improving trends	Views of local society are proactively canvassed. Results are fed back into the company and have improving trends	There are consistent trends of improvement in 50% of key result areas. Some results are clearly linked to approach
9	Managers are able to demonstrate their external involvement in the promotion of total quality as a business philosophy based on their own experience	A process is in place to analyse competitor business strategy and modify plans as a result in order to develop and sustain a competitive advantage	Employees are empowered to run their business processes	A process is in place to identify additional resources which can be used to strengthen competitive advantage	The existence of a formal quality management system can be demonstrated	75% of customer satisfaction targets are being met	Results indicate that employees and their families feel integrated into the work environment	Benchmarking has started for 25% of impact on society targets	All targets are being met and showing continuous improvement in 25% of trends
8	Managers have a consistent approach towards continuous improvement across the unit	The policy and strategy process are benchmarked	The human resource plan for the unit supports the company's policy and strategy for continuous improvement	A system is in place to review and modify the allocation of resources based on changing business needs	Process performance is demonstrably linked to customer requirements	50% of customer satisfaction targets are being met	Results indicate that people feel valued for their contribution at work	50% of impact on society targets are being met	75% of targets have been achieved. Able to demonstrate relevance of key result area to business
7	etc.	etc.	etc.	etc.	etc.	etc.	etc.	etc.	etc.

Figure 1.3.2 *BQC Business Improvement Matrix. Reproduced courtesy of the European Foundation for Quality and Management*

Elements	Attributes	SCORE				
		0%	**25%**	**50%**	**75%**	**100%**
Approach	**Sound:** ● Approach has a clear rational ● Well defined and developed processes ● Approach focuses on stakeholder needs	No evidence or anecdotal	Some evidence	Evidence	Clear evidence	Comprehensive evidence
	Integrated: ● Approach supports policy and strategy ● Approach linked to other approaches	No evidence or anecdotal	Some evidence	Evidence	Clear evidence	Comprehensive evidence
	Total score	0 10	20 30	40 50 60	70 80	90 100
Deployment	**Implemented:** ● Approach is implemented	No evidence or anecdotal	Implemented in 25% of areas	Implemented in 50% of areas	Implemented in 75% of areas	Implemented in all areas
	Systematic: ● Approach deployed in a structured way	No evidence or anecdotal	Some evidence	Evidence	Clear evidence	Comprehensive evidence
	Total score	0 10	20 30	40 50 60	70 80	90 100
Assessment and Review	**Measurement:** ● Regular measurement of effectiveness of approach and deployment carried out	No evidence or anecdotal	Some evidence	Evidence	Clear evidence	Comprehensive evidence
	Learning: ● Learning activities are used to identify and share best practice and improvement	No evidence or anecdotal	Some evidence	Evidence	Clear evidence	Comprehensive evidence
	Improvement: ● Output from measurement and learning is analysed and used to identify, prioritize, plan and implement improvements	No evidence or anecdotal	Some evidence	Evidence	Clear evidence	Comprehensive evidence
	Total score	0 10	20 30	40 50 60	70 80	90 100

Figure 1.3.3 *Scoring enablers against elements and attributes under RADAR. Reproduced courtesy of the European Foundation for Quality and Management*

When we come to the detailed award style assessment, we can examine each criterion as a whole, or break it down into its component parts, for assessment. We then use the criterion scores arrived at, and apply their weightings to build up an overall score for the organization. The score can be expressed as a percentage or out of a maximum of 1000 points.

When assessing the enabler criteria, we award each sub-criteria a RADAR card score as shown in Figure 1.3.3, based on the degree of hard evidence available for:

- Approach taken: Is it sound and integrated?
- Deployment within the organization: Is the approach fully implemented in a systematic way?
- Assessment and review: Is the effectiveness regularly measured, and improvements sought?

Example 1.3.1

People management criterion:

Approach:	Soundness	60%
	Integrated	50%
	Average	55%
Deployment:	Implemented	30%
	Systematic:	60%
	Average	45%
Assessment:	Measurement	50%
	Learning	40%
	Improvement	30%
	Average	40%

People management overall average of (55% + 45% + 40%)/3 = 47%.

The weighting in the overall score for the people management criterion is 9% or 90 points, therefore the people criterion contribution to the overall score is (9% × 47%) = 4.2%, or 42 points.

When assessing the results criteria, the RADAR card score (see Figure 1.3.4) is determined by:

- Trends: Positive and sustained.
- Targets: Appropriate and achieved.
- Comparisons: i.e. bench-markings against averages and best in class organizations.
- Causes: Tie into approach.
- Scope: Addresses relevant key areas.

Elements	Attributes	SCORE				
		0%	**25%**	**50%**	**75%**	**100%**
Results	**Trends:** ● Trends are positive and/or there is sustained good performance	No results or anecdotal information	Positive trends/ satisfactory performance in some results	Positive trends/ satisfactory performance in many results over three years	Positive trends/ satisfactory performance in most results over three years	Positive trends/ satisfactory performance in all areas over five years
	Targets: ● Targets are achieved ● Targets are appropriate	No results or anecdotal information	Favourable and appropriate in some areas	Favourable and appropriate in many areas	Favourable and appropriate in most areas	Excellent and appropriate in all areas
	Comparisons: ● Comparisons with external organizations take place and results compare well with industry averages or acknowledged 'best in class'	No results or anecdotal information	Comparisons in some areas	Favourable in some areas	Favourable in many areas	Excellent in most areas and 'world class' in many areas
	Causes: ● Results are caused by approach	No results or anecdotal information	Some results	Many results	Most results	All results and these will be maintained
	Total score	0 10	20 30	40 50 60	70 80	90 100
	Scope: ● Results address appropriate areas	No results or anecdotal information	Some areas addressed	Many areas addressed	Most areas addressed	All areas addressed
	Total score	0 10	20 30	40 50 60	70 80	90 100

Figure 1.3.4 *Scoring results against elements and attributes under RADAR. Reproduced courtesy of the European Foundation for Quality and Management*

Example 1.3.2

People satisfaction

Results:	Trends	70%
	Targets	80%
	Comparison	30%
	Causes	60%
	Average	60%
Scope		80%

People satisfaction overall average of

(60% + 80%)/2 = 70%.

The weighting in the overall score for the people satisfaction criterion is 9% or 90 points, therefore the people results criterion contribution to the overall score is (9% × 70%) = 6.3%, or 63 points.

Table 1.3.2 Results from self-assessment under the EFQM criteria

Criterion	*Raw score*	*Weighting*	*Weighted score*
Leadership*	50	10%	25
Policy and strategy	60	8%	48
People management*	47	9%	42
Partnerships and resources*	30	9%	27
Processes	55	14%	77
People satisfaction	70	9%	63
Society satisfaction*	40	6%	24
Customer satisfaction	55	20%	110
Key performance results	60	15%	90
TOTAL			506

The results for the individual criterion are then summated to give an overall organization score as shown in Table 1.3.2. In the table, the points are awarded out of a total of 1000 overall.

Although an overall score of 506 out of a possible 1000 indicates that there is room for improvement, it is still a reasonable score as even the recognized 'best' organizations seldom achieve more than 700 points.

The areas highlighted for improvement are those indicated by an asterisk (*) which are below industry averages.

The Balanced Scorecard

This is a method developed by Robert S. Kaplan and David P. Norton of the Harvard Business School in the 1990s. Its key approach is to ensure all decisions are made on how they contribute towards the main aims of the organization.

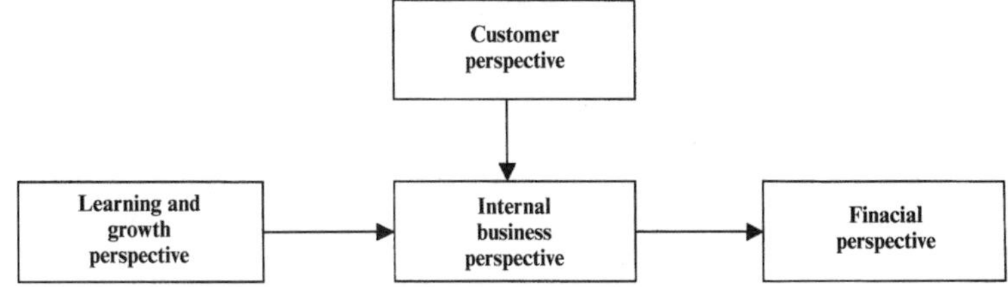

Figure 1.3.5 *The Balanced Scorecard Perspectives*

The Balanced Scorecard (BSC) has four perspectives tied together (see Figure 1.3.5). These can be common across an organization, but more usually differ within different areas.

- Financial: ROI; value added; profitability; revenue growth and mix; cost reduction; time to pay invoices; risk reduction.
- Customer: Market share; new customers; retaining customers; customer profitability; customer satisfaction.
- Process: Innovations; operations; post-sales – from identifying needs to satisfying them.
- Learning and growth: Employee satisfaction; employee retention; employee productivity.

These perspectives enable the organization to identify how each links together to meet strategic aims using a cause-and-effect relationship as shown in Figure 1.3.6. It identifies the performance drivers that will produce the desired outcomes – e.g. on-time deliveries producing customer loyalty, in turn giving profitability.

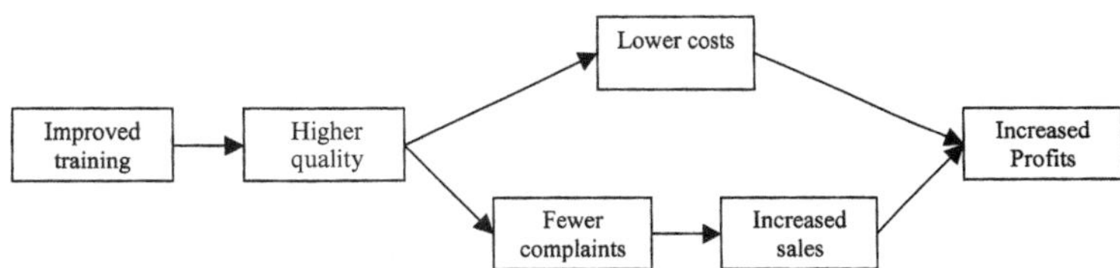

Figure 1.3.6 *Cause-and-effect relationships throughout the organization*

Financial perspective

Businesses, or products, can be in various stages of the market life cycle, each of which calls for differing strategies as shown in Table 1.3.3.

In addition, organizations are interested in reducing risk and variability which may guide strategy by broadening markets served.

Customer perspective

Market segmentation. Customers are different and value different product characteristics. Some want low cost, others high value. Matching each segment calls for different strategies.

Table 1.3.2 Strategies in relation to state of market

	STRATEGIC THEME		
MARKET STRATEGY	Revenue growth and mix	Cost reduction/productivity improvement	Asset utilization
GROWTH	Sales growth by segment. Percentage revenue from new products, services and/ or customers	Revenue/employee	Investment (% of sales). R&D (% of sales)
SUSTAIN	Share of target's customers. Cross-selling. Percentage revenue from new applications. Customer and product line profitability	Cost versus competitors. Cost reduction rates. Indirect expenses	Working capital ratios. ROCE by key asset. Asset utilization
HARVEST	Customer and product line profitability	Unit costs (per unit of output, per transaction)	Payback. Throughput

Core measurement:

- Market share: Proportion of business in a given market.
- Customer acquisition: rate of attracting new customers.
- Customer retention: Rate of retaining ongoing relationships.
- Customer satisfaction: Level on specific performance measures.
- Customer profitability: Net profit after allowing for unique expenses required for that customer group (see Table 1.3.4 for strategies).

Table 1.3.4 Strategies in relation to profitability of segments

CUSTOMERS	*Profitable*	*Unprofitable*
Targeted segment	Retain	Transform
Untargeted	Monitor	Eliminate

Measuring customer value propositions:

- Product attributes: Function; price; quality; delivery; image.
- Relationships: Knowledgeable people; convenient access; responsiveness; EDI, supplier integration.
- Image and reputation: Of product, brand or service.

Internal business process perspective

The internal value chain covers the three principal processes of innovation, processing and post-sales.

Innovation

Important both in short and long product life because it determines the future value creation and even sets the basic costs involved, as these are designed in.

There are two components:

- What range of products will customers need?
- How can we preempt development?

These are achieved by:

- Performing basic research to develop radically new products.
- Performing applied research to exploit existing technology.
- Focusing development to bring new products to the marketplace.

Measures are:

- Percentage of sales from new products.
- Percentage of sales from proprietary products.
- New product introduction versus competitors and plan.
- Manufacturing capability improvements.
- Time to develop next generation.
- Yields.
- Numbers passing through each process stage.
- Successful products.
- Redesigns and modifications – before and after launch.
- Time to recoup investment.

Operations

This is the present value creation. Starts with a customer order and ends with delivery of the product. Stresses efficient, consistent and timely delivery. They tend to be repetitive and scientific management can often be applied to improve them. Measurements are:

- Process time: Processing time/throughput time; waiting times; material handling; batch policy; set-up times.
- Quality: Defect rates; yields; waste; scrap; rework; returns; process under SPC.
- Cost: Process cost; overhead costs – ABC; interruptions; inaccurate information; priorities.

Post-sales

Includes warranty and repair activities and processing of payment. Measures can be:

- Time to react.
- Single call service.
- Disposal of waste, by-products and end-of-life plant.
- Number of claims.
- Predictive behaviour.

Learning and growth perspective

Employee capabilities

Changes in IT mean that employees no longer need to perform limited work and in fact the organization requires that they deliver ideas for improvement. This requires both reskilling and attitude changes.

Core measurement used to be only:

- Employee satisfaction.
- Employee retention.
- Employee productivity.

But now broadens to include:

- Critical skill coverage.
- Suggestions made.
- CPD.
- Matching of personal goals to company objectives.
- Continuous improvement (half-life – time to reduce defects/errors by half).

Reskilling scenarios:

- Strategic: A focused portion needs a high level of new skills.
- Massive: A large portion needs massive skills renovation.
- Competency upgrade: Some portion requires core skill upgrading.

IT capabilities

Employee skills alone may be insufficient to deliver necessary improvements. They will probably need additional IT systems to give timely and accurate information. This will include expert systems, databases and CAL to quickly bring themselves up to speed on new processes and products.

Team measurements:

- Survey of cross support.
- Team-based activities.
- Time of team working.

Missing measurements

There will not be available all the measures initially identified as being leading indicators. This means that these, at present, are not being managed effectively.

Linking BSC measures to strategy

Cause-and-effect relationships are the key (see Figure 1.3.7). The BSC shows up the cause-and-effect relationships between outcome measures and performance drivers. Every measure should be an element of this chain.

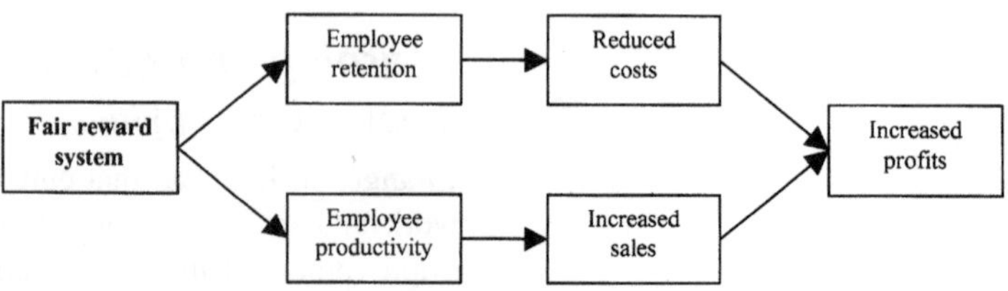

Figure 1.3.7 *Further cause-and-effect relationships*

Outcomes and performance drivers

Outcomes tend to be lagging indicators of performance. Performance drivers are the leading indicators that will deliver the value propositions to customers. BSC requires both and they form the cause-and-effect relationships which can be tested during operations.

Linkage to financials

BSC retains a strong connection to financial outcomes. Each area will have 15 to 25 mixed financial and non-financial measures, grouped into the four perspectives.

BSC can be used by individual SBU to create its own BSC to meet corporate objectives depending on their own circumstances and markets. For example, the various units can be in different growth scenarios or against different aggressiveness in competitors. These can reflect common corporate themes such as safety, environment and innovation. They can also follow mandatory instructions such as cross-selling, sharing technology or using central sources, e.g. group purchasing.

Joint ventures

With the difficulty of serving more than one master, the selection of BSC within the joint venture can create a common purpose that all partners can subscribe to.

Support departments

Corporate staff should contribute added value to their internal customers to give a competitive advantage. Where they do not then their function should either be assumed by the operating department or outsourced to more competitive and responsive suppliers.

Barriers

The barriers to achieving the connections between strategy formulation and implementation are:

- Visions and strategies that are not actionable – no consensus.
- Strategies not linked to departmental, team or individual goals.
- Strategies not linked to long- and short-term resource allocation.
- Feedback that is tactical rather than strategic.

Any strategy, such as BSC, must have top-level management involvement and support, as it is a change to the entire management process, not just measurement.

Many measures only reflect what all companies should be doing – the BSC should throw up unique strategies, targeted customers and critical internal processes and development.

The organizational defects are:

- BSC delegated to mid-management, rather than top level.
- BSC used to broaden benchmarks rather than develop own criteria.
- Seeking a perfect BSC before implementation.

Quality Management Principles in ISO 9004:2000

The ISO 9000:1994 Series on Quality Systems has been augmented by modifying the ISO 9004 standard to include the following Quality Management Principles.

Quality Management Principles

A quality management principle is a comprehensive and fundamental rule or belief, for leading and operating an organization, aimed at continually improving performance over the long term by focusing on customers while addressing the needs of all other stakeholders.

Principle 1 – Customer-focused organization

Organizations depend on their customers and therefore should understand current and future customer needs, meet customer requirements and strive to exceed customer expectations.

Principle 2 – Leadership

Leaders establish unity of purpose and direction of the organization. They should create and maintain the internal environment in which people can become fully involved in achieving the organization's objectives.

Principle 3 – Involvement of people

People at all levels are the essence of an organization and their full involvement enables their abilities to be used for the organization's benefit.

Principle 4 – Process approach

A desired result is achieved more efficiently when related resources and activities are managed as a process.

Principle 5 – System approach to management

Identifying, understanding and managing a system of interrelated processes for a given objective improves the organization's effectiveness and efficiency.

Principle 6 – Continual improvement

Continual improvement should be a permanent objective of the organization.

Principle 7 – Factual approach to decision making

Effective decisions are based on the analysis of data and information.

Principle 8 – Mutually beneficial supplier relationships

An organization and its suppliers are interdependent, and a mutually beneficial relationship enhances the ability of both to create value.

It is possible to see this modification as a move towards the EFQM model with traces of the Balanced Scorecard.

Problems 1.3.1

(1) Do you agree with the 50/50 split into enablers and results in the EFQM model?
(2) What do you think is the most important aspect that affects you as an employee, or student, in your organization?
(3) What factors would you use to measure how your present organization performs against its customers', or students', expectations?
(4) What would you say was your present organization's staff attitude to their organization?
(5) Looking at the financial outcomes, what benchmarks would you use to judge how well your organization is performing?

1.4 Business and society

This section looks at the reasoning behind codes of conduct and the changes in legislation caused by changes in the general public's perception of what constitutes ethical behaviour by organizations. It looks briefly at health and safety before concluding with environmental policy guidelines.

Codes of conduct

Throughout history society has had rules of conduct for its members. Some of these rules have developed into laws, and some have remained desired norms of conduct rather than laws. There are even actions which are considered wrong in one circumstance and right in another, or perhaps legally wrong but morally right by some members of the public.

Most people would rate lying as being against the general rules of conduct, but no country has yet drafted a law against lying *per se*. There are, however, laws which cover redressing wrongs arising out of some particular types of lies:

- Perjury, i.e. false evidence.
- Reputations, i.e. slander and libel.
- Contracts under the Trades Description Act and the Misrepresentation Act.

In business, it is easy to determine which actions are illegal, but not so easy to define those which are not ethical. Laws and ethics overlap to some extent, but they should not be thought of as the same – especially when considered on an international basis.

Quite often laws lag behind public perceptions on ethical issues. In addition special interest groups may have their own reasons for launching action, e.g. human rights activists, environmental activists and animal rights activists. In some cases the law may offer a degree of protection and avenues for recovering damage, but public opinion may be persuaded otherwise. This can affect trading or the value of shares.

Historically a business has been thought of as having only one aim – to make money for the owners – and has tended to ignore ethical matters except where they impinged on this duty. They have been accused of almost using a definition of anything that is not illegal, is legal and therefore can be carried out.

Society does not take that view today, and looks for businesses having an ethical dimension. Where businesses do not self-regulate, they find that governments, influenced by pressure groups, are laying down constraints within which organizations must work.

It is therefore in the organization's own interest that it produces its own code of conduct in many issues. The Code of Conduct (see Figure 1.4.1) of the Institute of Management is a good starting point for any organization.

As you can see the Code covers a wide area, some of which can be linked into legislation and other codes, e.g. the Equal Opportunity Commission's model for an Equal Opportunity Policy (see page 68) and the Investor in People (IIP) initiative (see page 78). However, that still leaves areas which are not set by laws and may appear difficult to fully determine what action to take.

One problem is that codes of conduct, and laws to a degree, are not the same throughout the world. This means that an organization has to be careful when dealing with people from different cultures.

Ethics

Debates on ethics and moral rights have been going on throughout history. In ancient Greece philosophers such as Aristotle and Plato attempted to define the ideal of virtue and citizenship which many subscribe to still. Yet in their society slave owning was the acceptable norm and women had no rights.

Is there a universal code that can be followed to guide organizations when considering ethical dilemmas? Unfortunately there is not, but we shall examine several codes which may give guidance.

Immannuel Kant (1724–1804) put forward a theory stating that there are three main criteria which should be used to determine the ethical approach:

- Reversibility, i.e. what if someone did it to you?
- Universality, i.e. what if everyone acted like this?
- Intrinsic human dignity.

The problem arises when you have to deal with someone with blatant egoism, i.e. self-interest, who believes in being free to do as he/she pleases. In the short term at least, this type of person often wins but organizations that allow this action to foster set up their own conflicts and often the individual's self-interest may not coincide with the interest of the organization.

Guides to Professional Management Practice

1. As regards the Individual Manager

The Professional Manager should:

- Pursue managerial activities with integrity, accountability and competence.
- Disclose any personal interest which might be seen to influence managerial decisions.
- Practise an open style of management so far as is consistent with business needs.
- Take active steps for continuing development of personal competence.
- Adopt a reasoned approach to the identification and resolution of conflicts of values, including ethical values.
- Safeguard confidential information and not seek personal advantage from it.
- Exhaust all available internal remedies for dealing with matters perceived as improper, before resorting to public disclosure.
- Encourage the development and maintenance of quality in all management activities.

2. As regards others within the organization

The Professional Manager should, in addition to the above:

- Ensure that others are aware of their responsibilities, areas of authority and accountability.
- Encourage and assist others to develop their potential.
- Consider the mental and physical health, safety and well being of others.
- Have regard for matters of conscience of others.
- Have regard for the needs, pressures and problems of others and not discriminate on grounds other than those demonstrably necessary for the task.

3. As regards the organization

The Professional Manager should, in addition to the above:

- Uphold the lawful policies and practices of the organization.
- Identify and communicate relevant policies, practices and information.
- Keep under review organization structure, objectives, procedures and controls.
- Seek to balance departmental aims in furtherance of the organization's overall objectives.
- Safeguard the assets and reputation of the organization.

4. As regards others external to but in direct relationship with the organization

The Professional Manager should, in addition to the above:

- Ensure that the interests of others are properly identified and responded to in a balanced manner.
- Establish and develop continuing and satisfactory relationships based on mutual confidence.
- Avoid entering into arrangements which unlawfully or improperly affect competitive practice.
- Avoid entering into any agreement or undertaking any activity which may give rise to a conflict of interest with the organization or prejudice professional management performance.
- Neither offer nor accept gifts, hospitality or services which could, or might appear to, imply an improper obligation.

5. As regards the wider community

The Professional Manager should, in addition to the above:

- Have due regard to the short and long term effects and possible consequences of present and proposed activities, taking action where appropriate.
- Ensure truthfulness in all public communications.
- Seek to conserve resources wherever possible and preserve the environment.
- Respect the customs, practices and reasonable ambitions of other peoples which may differ from the manager's own.

6. As regards the Institute of Management

The Professional Manager should, in addition to the above:

- Promote the mission, aims and objectives of the Institute.
- Uphold the integrity and good name of the Institute and refrain from conduct which detracts from its reputation.
- Promote the Institute's professional image and standing.

Figure 1.4.1 *Institute of Management's Code of Conduct (reproduced with permission from the Institute of Management)*

Kant's criteria are similar to the Four-way Test adopted by the Rotary movement which states that in all business transactions the following questions should be asked:

- Is it the truth?
- Is it fair to all concerned?
- Will it build goodwill, and better friendship?
- Will it be beneficial to all concerned?

All these appear to be a little theoretical and idealistic in values. Can this be put into more businesslike terms?

Ronald Green in *The Ethical Manager* introduces the concept of stakeholders into the equation. He suggests the following sequence:

- Define your self-interest and the interest for other directly involved parties.
- Identify other stakeholders, i.e. others who may be affected, and the minimum rights and duties involved.
- Determine the best outcome for the interested parties only.
- Test that outcome against the other stakeholders' interests.
- Where conflicts arise determine if the strategic outcome outweighs the morale consequences.
- Where conflict is unresolved attempt to find another outcome more equitable to the direct parties and the other stakeholders.

At the end of the day, you may still be left with an outcome you wish to proceed with and that you feel is in your self-interest but against other stakeholders, although you should have amassed an argument in your favour. However, remember that being legally right does not protect you from public disquiet and attacks from those whose interests you have gone against. The 'fat cats' arguments of the 1990s regarding directors' pay and bonuses are an example of the media attention that you may be subjected to.

Constant disregard of public disquiet by organizations has led over the years to changes in the laws of the UK and other countries.

Trends in laws

Due to changes in public acceptance of various practices, governments during the twentieth century have introduced a series of laws and regulations constraining the freedom of organizations to act as they see fit.

Contract law

The historic position was caveat emptor – let the buyer beware.

This led to unethical behaviour on the part of some organizations where the buyers were not in a position of power, nor had the necessary skills or knowledge, to fully take up their right to examine the contract or the product before buying.

This led to changes to protect the weaker parties such as the Sale and Supply of Goods Act 1994, the Unfair Contracts Act 1977, the Consumer Protection Act 1987 and Unsolicited Goods and Services Acts 1971 and 1975. These are dealt with in Chapter 3.

There are other acts which limit the action of organizations to operate cartels, unfair competition and restrictive practices. There are even laws which protect the interests of minority shareholders in companies.

It is expected that the EC will introduce Europe-wide legislation which will further constrain organizations.

Labour law

In Chapter 2 on human resource management we continually address legislation concerning employment and employee rights that were enacted in the last quarter of the twentieth century. This period has seen changes in legislation that have weakened the trade union organizations, but at the same time have increased individual's rights and benefits.

The latest of these is the Public Interest Disclosure Act 1998. This gives protection to employees engaged in whistle blowing, i.e. making public information where:

- A criminal offence has been committed, is being committed or is likely to be committed.
- A person has failed, is failing or is likely to fail to comply with any legal obligation to which he is subject.
- A miscarriage of justice has occurred, is occurring or is likely to occur.
- The health or safety of any individual has been, is being or is likely to be endangered.
- The environment has been, is being or is likely to be damaged, or
- Information tending to show any matter falling within any one of the preceding paragraphs has been, is being or is likely to be deliberately concealed.

This means that an employee can make public any illegal action taken by an organization. The problem that can arise is in the grey areas where the general public may consider that although a certain action is legal it is unethical, e.g. the use of child labour by one of your suppliers.

The report from the Nolan Committee clearly states that an organization requires to formulate a procedure for handling this.

Blowing the whistle – the view from the Committee on Standards in Public Life

All organizations face the risk of things going wrong or of unknowingly harbouring malpractice. Part of the duty of identifying such a situation and taking remedial action may lie with the regulatory or funding body. But the regulator is usually in the role of detective, determining responsibility after the crime has been discovered. Encouraging a culture of openness within an organization will help: prevention is better than cure. Yet it is striking that in the few cases where things have gone badly wrong in local public spending bodies, it has frequently been the tip-off to the press or the local Member of Parliament – sometimes anonymous, sometimes not – which has prompted the regulators into action.

Placing staff in a position where they feel driven to approach the media to ventilate concerns is unsatisfactory both for the staff member and the organization. We observed in our first report that it was far better for systems to be put in place which encouraged staff to raise worries within the organization, yet allowed recourse to the parent department where necessary. In the course of the present study, we received evidence from the independent charity, Public Concern at Work, which specialises in this area. They proposed that an effective internal system for the raising of concerns should include:

- A clear statement that malpractice is taken seriously in the organization and an indication of the sorts of matters regarded as malpractice.
- Respect for the confidentiality of staff raising concerns if they wish, and the opportunity to raise concerns outside the line management structure.
- Penalties for making false and malicious allegations.
- An indication of the proper way in which concerns may be raised outside the organization if necessary.

We agree. This approach builds on some aspects of existing practice. For example the duty of accounting officers in education bodies to notify the funding councils of the misuse of public funds. It goes further by inviting all staff to act responsibly to uphold the reputation of their organization and maintain public confidence. It might help to avoid the cases when the first reaction of management faced with unwelcome information has been to shoot the messenger.

The Nolan Committee
Second Report of the Committee on Standards in Public Life, p. 22 Cm 3270–1 (May 1996). (Crown Copyright is reproduced with the permission of the Controller of Her Majesty's stationery office).

Health and safety

There are a multitude of acts controlling the workplace. These apply to offices and factories (see also Chapter 3). The EC has even introduced working conditions which previously came under labour law in the UK, e.g. the Working Time Directive, under the heading of Health and Safety.

The Health and Safety at Work Act 1974 is the main one concerning safety and is rigorously imposed. This Act lays down a general duty on everyone at work – including the self-employed – that is backed by criminal liability. This duty is owed to employees, sub-contractors, customers and other non-employees on site and the general public who may be affected by anything we do.

The duty is 'To ensure, so far as is reasonably practical, that the process, article or substance is so designed and managed as to be safe and without risk to health when properly used, stored and transported.' Improper use is further defined as 'without regard to any relevant information'.

Specific duties laid down are:

- To carry out any necessary testing and examination.
- To provide adequate information for the process operation or the article's use.
- To ensure that any necessary research is carried out to eliminate, or minimize, any risk to health or safety.

Internally this means that organizations have to take positive steps to protect their employees and the public in general from the risk of injury from a multitude of potential hazards. In 1992, the following specific sets of regulations came into force:

- Management of Health and Safety at Work Regulations: Identify hazards and assess risks to eliminate or control them to an acceptable level.
- Workplace Health, Safety and Welfare Regulations: Lighting, heating, washing and sanitary facilities, traffic, glazing, etc.
- Provision and Use of Work Equipment Regulations: Work equipment must be safe, including use, maintenance and instructions.
- Personal Protective Equipment at Work Regulations: Personal protective clothing selection, use and maintenance.
- Manual Handling Operations Regulations: Avoid heavy handling but, if required, assess the risk.
- Health and Safety (Display Screen Equipment) Regulations: Use of computer workstations, including screen, desk, chair, keyboard, etc. and general environment. Establish eye care programme.

Formal risk assessment checks must be carried out, where risk refers to the likelihood that a harm will arise and the severity of its occurrence:

- Identify all hazards, i.e. the potential to do harm: Noise, dust, electricity, hazardous material, sharp edges, fire/explosion, tripping, handling, vehicle movement, moving machinery, storage, cranes and lifting equipment, etc.
- Identify everyone who could be harmed by these hazards.
- Evaluate risks arising and decide if they can be eliminated (preferred) or reduced with adequate precautions installed.
- Record findings and decisions/action taken.
- Review assessments periodically and revise if necessary.

HAZOP (hazard and operability) studies involve examining all operational plant and processes to determine the possibility of hazards arising and designing procedures to eliminate them, reduce the possibility of them occurring or to reduce the possibility of harm arising to employees or the environment.

There has been a code of practice issued by the Health and Safety Commission covering the Act (1974) and the subsequent regulations. Whereas this has no legal status, if not observed, it would weaken your position in any court proceedings. This code covers:

Risk assessment
Health and safety management arrangements

Health surveillance of employees and the general public
Health and safety assistance
Procedures for serious and imminent danger and for danger areas
Information for employees
Co-operation and co-ordination of shared sites
Persons working in host employer's undertakings
Capability and training
Employees' duties
Temporary workers

Action on all of these has all to be recorded if you have more than five employees. The health and safety aspects are growing and breaches can have severe consequences such as:

- An order not to use a machine, or process.
- Criminal charges which could lead to a fine, or in extreme cases imprisonment.

You are advised to consult the Health and Safety Commission or the Health and Safety Executive whose inspectors may visit you at any time unannounced.

Environmental management

Nowadays environment aspects are becoming more important to an organization as governments continually expand the legislation dealing with this matter. For example, the following are active in the UK:

Clean Air Act 1968
Control of Pollution Act 1974
Health and Safety at Work Act 1974
Control of Industrial Major Accident Hazard (CIMAH) Regulations 1985
Road Traffic Act 1986
Consumer Protection Act 1987
Control of Substances Hazardous to Health (COSHH) 1988
Water Act 1989
Control of Industrial Air Pollution Regulations 1989
Town and Country Planning Act 1990
Water Resources Act 1991
Environmental Information Regulations 1992
Council Regulation (EEC) No. 1838/93
The Environment Act 1995
Control of Major Accident Hazards Regulations (COMAH) 1995
Landfill Tax – in the Finance Act 1996
Producer Responsibility Obligations (Packaging Waste) Regulations 1997
Integrated Pollution Prevention and Control, EC Directive 96/61

At present the EC and national governments are in discussion with industry with the aim of structuring legislation and setting targets for the recycling of products which, if enacted, will have considerable further impact. It is necessary that an organization carries out an immediate environmental impact assessment and if there is an environmental impact envisaged takes action to reduce or minimize it.

In the UK and the EC, there are four key principles that govern the approach to environmental protection:

- Prevention: Priority should be given to anticipating and preventing environmental harm. This includes worst case scenarios and contingency planning.
- Precaution: Here the concept is to minimize possible regrets, i.e. err on the pessimistic side, when considering possible harm. This has fuelled the debate on genetically modified foodstuff.
- Polluter-pays principle: The producer of any environmental damage has to meet the financial cost of that damage. This has led to problems when the present owner of land that has been contaminated is not the contaminator.
- Integration: Care should be taken that improvements in one area do not lead to reductions in another.

Benefits

There is often a business case to be made for an organization that is concerned with the environment but is unable to have an effect thereon. The case considers:

- Cost savings: Through reduced raw materials, reduced energy used, reduced transport and disposal cost, reduced risk of litigation and lower insurance.
- Marketing benefits: Improved public image, safeguards existing market and offers new opportunities in environmental products.
- Benefits for customer: Increased confidence, competitive pricing, better value, less risk.
- Benefits for employees: Improved working conditions, improvement to health and enhanced personal commitment.
- Benefits to environment: Reductions in raw material extraction, emissions, waste, odours, noise, energy use and the impact on the ecosystem.
- Better relations with regulators and other authorities.

Organizations can be registered under various schemes such as BS 7750, ISO 14001 and the EC's Eco-Management and Audit Scheme (EMAS) – see Figure 1.4.2 for a comparison table between these. They all follow similar stages.

Environment review

The initial review requires top-level commitment through the formation of a team whose task will be to:

- Determine any significant impact on the environment.
- Identify any breaches of existing, or proposed, legislation.
- Quantify the emissions, discharges and wastes presently arising.
- Identify improvement possibilities and prioritize them.

A useful initial step is to demonstrate the sources of waste using a process flow diagram (as in Figure 1.4.3) then calculate out a mass balancing calculation (see Figure 1.4.4.), quantifying complaints and summating disposal costs.

Requirements of BS EN ISO 9000 Series	Requirements for BS 7750										
	Management system	Environmental policy	Organization and personnel	Environmental effects	Objectives and targets	Management programme	Manual and documentation	Operational control	Records	Audits	Reviews
4.1 Management responsibility	*	*	*								*
4.2 Quality system	*						*				
4.3 Contract review				*	*	*					
4.4 Design control						*	*	*			
4.5 Document control							*				
4.6 Purchasing				*				*			
4.7 Customer supplied product				*							
4.8 Product identification									*		
4.9 Process control								*			
4.10 Inspection and testing								*			
4.11 Inspection, measuring and test equipment								*			
4.12 Inspection and test status								*			
4.13 Control of non-conforming product								*			
4.14 Corrective action								*			
4.15 Handling, storage, packaging and delivery				*				*			
4.16 Quality records									*		
4.17 Internal quality audits										*	
4.18 Training			*								
4.19 Servicing				*				*			
4.20 Statistical techniques								*			

COMMON FEATURES IN ENVIRONMENT REGISTRATION PROCEDURES

ISSUE	ISO 14001	EMAS	BS 7750
Environmental policy	v	v	v
Environmental effect	aspects/ impacts	v	v
Environmental review	Advised	v	Advised
Identification of significant environmental effects	v	v	v
Legislative and other requirements	v	v	v
Objectives and targets	v	v	v
Draw up environmental improvement programme	v	v	v
Prepare Environmental Statement	x	v	x
Environmental Statement validated	x	v	x
Define responsibilities	v	v	v
Identify training needs	v	v	v
Procedure of operational control	v	v	v
Procedures for internal communication	v	x	x
External communication	Consider processes for significant aspects	Give information to public	
Contractors	Communicate relevant procedures	Consider environmental performance	Communicate relevant procedures
Maintain documented systems	v	v	v
Procedure for controlling all required documents	v	x	v
Records to be kept	v	v	v
Conduct audits	v	At least every three years	v
Certification of company	v	x	v
Verification of site	x	v	x

Figure 1.4.2 *Comparison between quality and environmental standards. (GG43 Environmental Management Systems in Foundries. Crown Copyright is reproduced with the permission of the Controller of Her Majesty's Stationery office)*

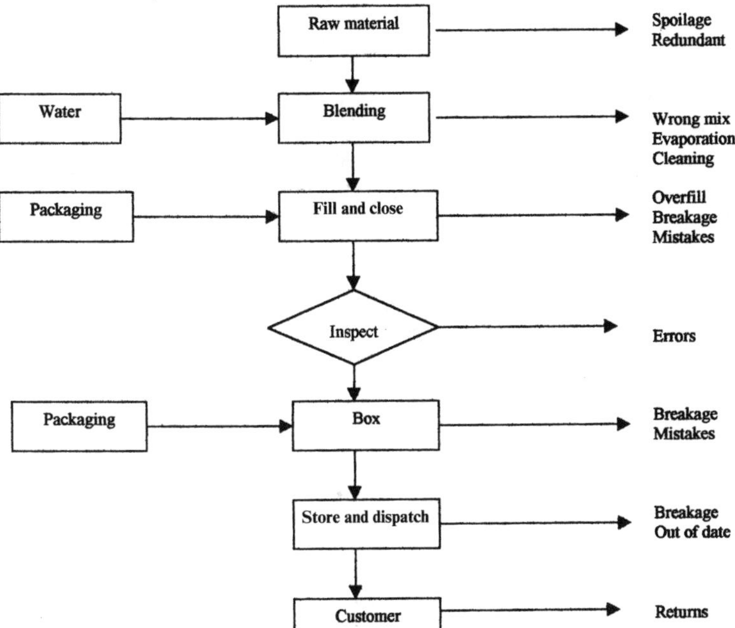

Figure 1.4.3 *Flow process chart. (GG43 Environmental Management Systems in Foundries. Crown Copyright is reproduced with the permission of the Controller of Her Majesty's Stationery office)*

Figure 1.4.4 *Mass balance calculation. (Input – output = losses. All units must be in same value, e.g. weight). (GG43 Environmental Management Systems in Foundries. Crown Copyright is reproduced with the permission of the Controller of Her Majesty's Stationery office)*

Input to process (kg)		Output to process (kg)	
Raw material purchased Decrease in stock	1000 200	Products sold (by standard weight) Increase in product stock Poor quality product scrap Waste to landfill	900 50 70 100
Total	1200	Total	1120
Estimated loss			80 (6.7%)

Detailed data gathering

The first stage is to gather information on all materials, processes and controls, and appropriate legislation, including present authorizations and complaints. Then it is a matter of carrying out a site survey and measuring all waste and effluent flows:

- Site: Main processes; site history; major incidents; warnings or prosecutions; complaints.
- Utilities: Where used and consumption of steam, water, electricity, gas, oil, etc.
- Raw materials: Amount used, where stored and the storage medium.
- Aqueous effluent: Description, source, quantity, treatment, disposal method.
- Emissions to atmosphere: Description, source, quantity, treatment, disposal.
- Solid waste: Description, source, quantity, treatment, storage, disposal.

Assessing significance

Matrices are available to assess the environment effect of the processes under both operating (Figure 1.4.5) and other conditions (Figure 1.4.6).

	Score				Weighting factor			
	3	2	1	0				
Legislation	Existing	Impending		None	*	2	=	a
Environmental impact	Known detriment	Possible detriment	Limited detriment	No detriment	*	3	=	b
Interested parties	Considerable	Moderate	Little	None	*	2	=	c
Quantity	High	Medium	Low	None	*	3	=	d
Normal operating conditions total score = a + b + c + d								

Figure 1.4.5 *Matrix to score environmental effects under normal operating conditions. (GG43 Environmental Management Systems in Foundries. Crown Copyright is reproduced with the permission of the Controller of Her Majesty's Stationery office)*

	Score					
	12	6	3	0		
Abnormal operations		Increased environmental impact	No change	Reduced environmental impact	=	e
Accident/emergency		Increased environmental impact	No change	Reduced environmental impact	=	f
Past activities	Evident and requires action	Possible damage or damage difficult to evaluate		No damage	=	g
Planned activities		Increased environmental impact	No change	Reduced environmental impact	=	h
Other operating conditions total score = e + f + g + h						

Figure 1.4.6 *Matrix to score environmental effects under other operating conditions. (GG43 Environmental Management Systems in Foundries. Crown Copyright is reproduced with the permission of the Controller of Her Majesty's Stationery office)*

Key point

Fault tree analysis (FTA)

FTA is the logical analysis of the chain of lower level events which have to occur to cause another event to happen. In failure analysis it is used to trace back to find the root causes that allow systems or components to fail.

Key point

Failure mode effects and analysis (FMEA)

FMEA is a design procedure in which we investigate the consequences that the potential failures of a component or sub-system may have. The procedure starts from the opposite end to the FTA, which starts with an overall system failure. It can be applied to a product or a process.

The process can score a maximum of 30 points under each of normal operations conditions and possible variations aspects giving an overall maximum of 60. The higher the score, the more impact on the environment. These also can be used to prioritize the processes to be tackled – the higher the score the more significant the effect.

Thereafter processes similar to FTA and FMEA should be carried out to determine root causes and preventative measures that can be taken.

Environment policy

Following onto the environment review, a policy should be drawn up listing the organization's environmental objectives, staff responsibilities and their targets for improvement. An environment management manual listing all the above and associated procedures needs to be drawn up.

The procedures should be subjected to audit at set intervals, a minimum of once each year. Any report of non-compliance should conclude with the action to be taken.

A validated environmental statement is required to be produced and made available to the public under the EMAS scheme.

Certification and/or validation

Under BS 7750 and ISO 14001 an independent certifier certifies the Environmental Management System. Under EMAS an environmental verifier carries out a similar process of accreditation. Being certified to

BS 7750 normally suffices for EMAS, but some slight additional work may be required if the certification is under ISO 14001.

The process starts with a report from a desk-top study and is concluded by a site visit by the certifier/verifier.

Problems 1.4.1

(1) Look at the code for your profession and compare it with that for the Institute of Management. Do you see any major omissions in either?

(2) Your company employs over 200 people. It is in severe difficulties and will go out of business if new orders are not gained. You are involved in sales negotiations when it becomes clear that you will win an order that will save the jobs of your employees, if you pay a substantial bribe to the buyer's main negotiator. What do you do?

(3) Look up details of the Pinto small car made in the USA in the 1970s and comment on Ford's cost-benefit analysis regarding the fault discovered before its launch.

(4) Who do you think should pay to clean up if an organization owns land that was polluted by another organization which has gone out of business?

(5) Do you think it is ever ethical for an organization to bind its employees in a non-disclosure contract?

2 Human resource management

Summary

It is often said that an organization's main resource is its people and it is certainly true that without people the organization will neither exist nor function.

As both a manager and as an employee you will need to deal with selecting and getting the best out of people. This chapter will introduce you to the many aspects of human resource management that you will need to know.

Objectives

By the end of this chapter, the reader should:

- be able to determine future manpower needs and how to meet them by recruitment and terminations (Section 2.1);
- understand how to make the most of people through motivation, job design, training and development (Section 2.2);
- understand the way people are rewarded by payment schemes and other benefits, including job evaluation and productivity payment (Section 2.3);

2.1 The manpower plan – recruitment and termination

The aim of human resource management (HRM) is to meet the manpower needs of the company, now and in the future. To this end it must be effectively linked into the processes of recruitment and termination.

Manpower planning

The various stages of manpower planning are designed to ensure that the organization has manpower resources to meet the business needs in skills, number of employees and cost. It does this by:

- Determining future recruitment needs.
- Making provision for training.
- Anticipating wastage and redundancies.

In order to do this, we start with the future business plan and break it down into the activities which will need to be carried out (Figure 2.1.1).

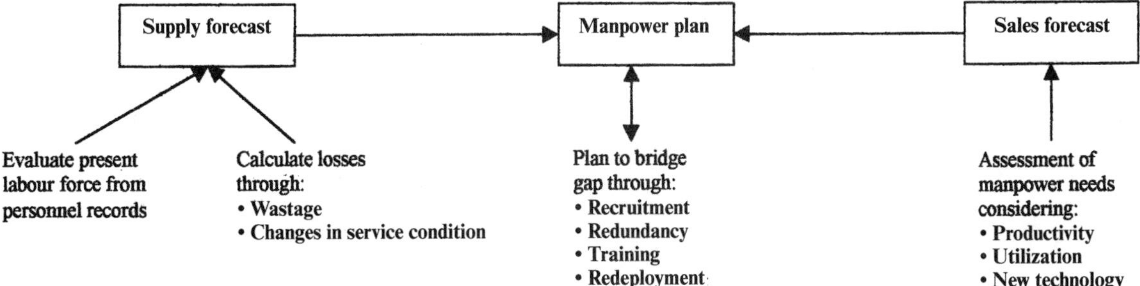

Figure 2.1.1 *Development of manpower plan*

These activities then need to be expressed in the skills required and the number of people possessing them.

We then have to evaluate the existing manpower in skills and numbers, calculate the expected losses and compare the resultant balance to that required by the business plan. We have to bear in mind any major organizational or technological changes expected both within and outside the organization.

Evaluate the existing workforce

Assuming that adequate personal records are available, it should be possible to draw up a profile of the existing workforce:

- Skills available: Each skill used within the organization needs identifying. This will include qualifications and experience of using the skill.
- Present training plan: This gives an idea of how present skill levels are changing.
- Number of staff: Linked to skills will be the number of staff having that skill, and how they are at present structured.
- Levels: This relates to the structure within the organization, i.e. supervisors and managers.
- Age analysis: This is important and is tied to the above four factors.

Age Analysis

Few skills and departments within an organization will have an identical age structure. Analysing the age of the existing workforce will not only give advance warning of any looming retirement, but will also affect present policy and perhaps even training capability.

Examining Figure 2.1.2, which is the age profile of engineers within a particular department, gives rise to two observations:

- There is a distinct low in the age group 25–35. Perhaps due to recent restructuring, or it may denote an inability to retain younger staff.
- A peak in the age profile in the 55–60 age groups. This shows a potential retirement problem in the next five to ten years.

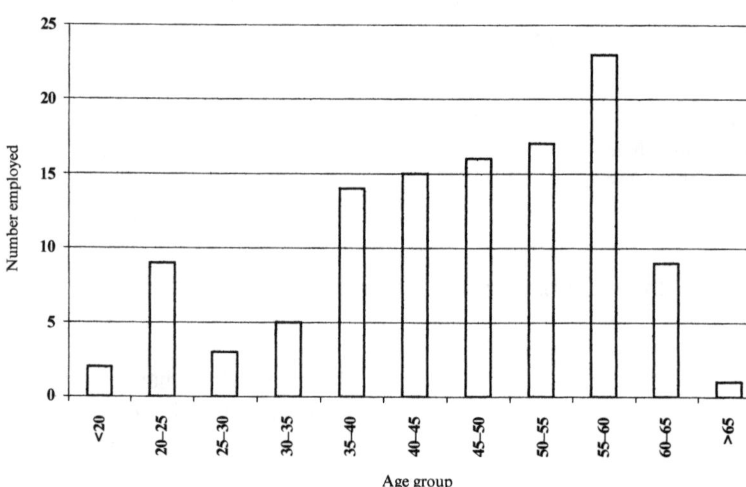

Figure 2.1.2 *Age analysis of engineers within an organization*

Allowing for change

It is possible by analysing past and present records, coupled to any anticipated changes in conditions, to calculate what the probable changes in the present labour force over the planning period will be.

Stability

Any group of people joining an organization will stay varying lengths of time. Some will leave quickly, some will be promoted and the others will spend differing times with the organization, perhaps to retirement age or leaving due to health problems. It is common that a pattern is set up within a particular job role, which will repeat itself in the future unless a change is made.

We calculate the number of survivors over time by comparing the people at the start of a time period with those still there at the end of that period. We can then take the leavers during this period to find the average. This, combined with the age analysis, will enable us to make predictions about the numbers leaving by natural wastage during periods of change.

$$\text{Stability} = \frac{\text{people employed at beginning of period and still there at the end} \times 100}{\text{people employed at beginning of period}}$$

This can be shown graphically as in Figure 2.1.3, and demonstrates that the stability of the assembly operators is less than the other categories.

$$\text{Turnover} = \frac{\text{people leaving} \times 100}{\text{average number employed}}$$

The figures on stability and turnover can also indicate where a change in HRM policy may be required. It is costly in management time in

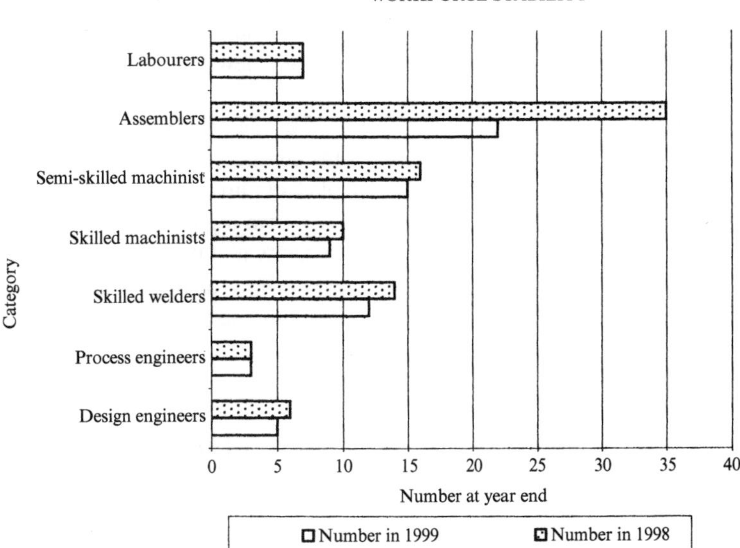

Figure 2.1.3 *Workforce stability chart showing numbers remaining after one year's service*

recruitment and training to be continually dealing with newcomers (see Section 2.2).

Changes in conditions of employment

In addition to leavers, changes in conditions of employment such as hours of work and holiday entitlement may alter the available working hours to meet the demand.

Some of these changes will be by negotiation with employee representatives. Some may be due to changes in the general labour market. Others may be dictated by government decree – national or EC such as the EC Directive on working hours.

Other internal changes

Factors such as the technology used and the organizational mode will effect numbers of staff required, skill requirements and even the levels of supervision and management needed.

The productivity and motivation of the employees can be affected by mode of management and/or linkage to their pay (see Section 2.3). This can also be factored into the equation.

External changes

In addition to government induced changes in conditions of employment, the external availability of labour is affected by:

- Social factors: For example, the attitude and aspirations of school leavers.
- Education: There has been a rapid growth in both further and higher education for school leavers.
- Changes in age dispersion in general population: In western society the life expectancy is increasing.
- Changes in migration within, and outwith, the country.
- General economic conditions, locally, nationally and internationally.
- New legislation.

- Specific changes in the organization's market:
  - new competition
  - new technology.

All of the above requires careful analysis to determine the potential of the existing staff resources and the labour market to meet the business plan requirements.

Where a gap has been identified between the existing labour force's capacity and that required by the business plan, the manpower plan lays down how to bridge that gap. This will include recruitment, training, redeployment and perhaps redundancy.

Example 2.1.1

The following figures were calculated for an assembly shop:

1999
Output: 750 000 units
Total hours worked by operatives: 70 000
Employees: 38
Annual hours per employee: 1842
Standard available hours: 47 weeks at 40 hours per week
= 71 440
Lost hours: 71 440 – 70 000 = 1440 = 2%

This led to the following plan for 2001:

Expected output: 850 000 units
Expected productivity gain: 10%
Required hours: 72 121
Required hours plus 2% for losses: 73 564
New standard annual hours per employee: 38 × 46 = 1748
Number of employees required: 42
Expect leavers and retirees: 8
Number of new employees required: 12

Equal opportunity

At the manpower planning stage, it is useful to categorize employees by sex, ethnic origin and disability. This should include their level of authority, and will enable you to determine if there is under-representation of any of these group which may call for some positive action to redress this.

Finding the right person – recruitment

For an organization to be successful, it is important to have the right person in any job. The recruitment process is the entry point for all personnel, therefore this is the first opportunity to match what the organization requires in its people.

The positive reasons are:

- The job will be carried out effectively in time, cost and quality.
- Training time will be short.
- The employee will remain with the organization.
- The employee can change with the job.

The negative reasons are:

- The job will be carried out ineffectively, increasing cost and time and reducing quality.
- Training will be long and perhaps ineffective.
- The employee will leave within a short period – voluntarily or otherwise:
  - requiring more recruitment time and cost
  - requiring further training time and cost.
- Other staff may be overworked to cope with one person's shortcomings.
- Constant turnover effects employee morale.

It is important therefore that the recruiting process be:

- Effective: Finds and selects the correct person.
- Efficient: Cost effective in staff time and advertising.
- Fair: To all potential candidates, especially in legal terms.

The process involves identifying the requirements, attracting suitable applicants and selecting the most suitable.

The job's requirement

What is the job we need to fill?

We need to examine both the present duties and skills required and anticipate the changes that will likely occur within the near future. Consult the present job holder, colleagues, the supervisor and any specialists involved.

A vacancy may give an opportunity to revise that job and others within the section in line with new technology or processes.

Job description

This document is the basis of many processes within HRM. It needs to describe tasks and decisions made:

- Main purpose: A single sentence describing the job's objective.
- Main and minor tasks: What is done, including method and equipment used including frequency.
- Scope of authority: Decisions made, and referred.
- Context: Who directly supervises, others reported to during work day and any subordinates.
- Working conditions: Physical, hours of work, shift pattern.
- Special skill requirements.

An example is shown in Figure 2.1.4.

Job title:	Maintenance technician	**Grade:** 5
Reporting to:	Directly to area maintenance engineer	
Relationships:	Co-operate with process supervisors and operators on shifts	
Main job function:	Maintain plant and equipment in optimum condition	
Job content:	Diagnose and rectify faults in electro-mechanical plant and equipment Carry out minor alterations to plant and equipment Carry out specified preventative maintenance tasks Assist production operatives in fine tuning operational control settings Install new plant and equipment	
Qualifications:	NVQ level 3 preferred	
Experience:	Three years minimum in similar job maintaining packaging machinery Experience in microprocessor controls an advantage	
Working conditions:	Often will have to work in cramped and awkward positions Subject to live electricity working, oil spillage and hot equipment Required to work shifts, periods of overtime and during public holidays	
Performance standards:	Plant up-time Number of breakdowns Spares and other material usage	

Figure 2.1.4 *Job description example*

Personal specification

From the job description, we can decide the characteristics of a person who would ideally fit the requirements of the job. As it is not always possible to get the 'ideal', we should also indicate the minimum acceptable, e.g. someone with some of the characteristics needed who may be fully trained and developed. We may also wish to attract people who have the potential to move to other positions within the organization.

We must ensure here that we do not introduce a bias against any particular section of the population by setting any unnecessary requirements. We may have to prove we are operating equal opportunity during recruitment.

There are two commonly used systems for drafting a personnel specification:

Alec Rodger's Seven-point Plan:

- Physical make-up: Health, appearance, bearing, speech.
- Attainments: Education, qualifications, experience.
- General intelligence.
- Special aptitudes: Mechanical, dexterity, words and figures.
- Interests: Intellectual, practical, active, social, artistic.
- Disposition: Acceptability, influence, steadiness, dependability, self-reliance.
- Circumstances: Ability to work unsociable hours, travel, move location.

Munro Fraser's Five-fold Grading System:

- Impact on others: Physical make-up, appearance, speech, manner.
- Acquired qualification: Education, training, experience.
- Innate abilities: Quick comprehension, aptitude for learning.

- Motivation: Goals, consistency, determination, successes.
- Adjustment: Emotional stability, stress handling, ability to get on with people.

Attracting a suitable candidate

We need to select the process that gives the best opportunity of reaching the target group of potential applicants:

- Internal: Often used for promotion, but can also be useful in developing present staff by widening their experience. It can also be useful to move a person into a job that better matches their skill and aptitude. You do have more pertinent information on candidates, but may be accused of favouritism. A point to note here is that being highly skilled at a particular task often has little relationship to an ability to perform satisfactorily in a given one – perhaps ignoring this has led to the Peter Principle that each employee will be promoted up to his level of incompetence.
- Word of mouth of existing employees: Good for strong group feeling amongst workforce but will probably exclude some sections of the local population.
- Local education, schools, colleges, universities and career centres.
- Media:
  - local newspapers, etc. – especially for lower skilled personnel
  - national newspapers, etc. for higher skills
  - professional journals – for specialists.
- Recruitment consultants – temporary, specialists and senior posts.

Any advertisement should be carefully designed to attract mostly those meeting the personnel specification. It is as inefficient to attract the overqualified as it is the underqualified.

Using application forms

Application forms are useful in providing in the same order all the information required to short list applicants. We should, however, use a different form for each type of job, only asking for the pertinent information against the personnel specification for ease of short listing. Unfortunately very few companies do so, which can put off potential candidates.

Selecting the best applicant

The initial stage is to prepare a short list from the applicants of those who appear to best meet the personnel specification. This reduces wasting both candidates' and staff's time and expenses. Where practicable reply to all applicants. You may want some applicants to reapply for other jobs later – if so, keep their particulars on files, and tell them what you have done.

Inviting for interview

Short listed personnel should be invited for an interview/test. The letter requesting attendance should explain to the candidates where and when to attend and also what they will be expected to be subjected to during their selection.

This selection process, unfortunately, is full of opportunities for errors. This is partially due to the methods employed, but also due to the artificial conditions where both sides attempt to match their needs, sometimes concealing the true situation.

Good selection has to be planned. Methods used include:

- Selection tests: Many jobs require skills which can be, and should be, tested to determine competence levels. These can include manual skills, writing and numeric skills, use of IT and even group working.
- Psychological: Although controversial, these are still often used to determine attitudes and mental reasoning by larger organizations.
- Interviewing: The most common technique, but especially prone to snap judgement. May be formal and/or informal. Methods are from interviewing on a one-to-one basis by several people to being interviewed by a panel.
- It is common to have a form (see Figure 2.1.5) completed by everyone involved in the process to ensure all candidates are judged equally against the personnel specification criteria.

Bearing in mind time and cost, the more people who have connections with the post involved the better, to get a wide range of opinions. Make sure that they have received training in the selection process.

It is imperative that the direct supervisor is heavily involved for commitment.

Criteria	Minimum	Candidates					
Impact on others	Confident approach and clear speaking						
Attainment	HND in electronics						
Experience	Two year similar						
Comprehension	Wide range of problem solving						
Assertiveness	Able to argue a point						
Disposition	Able to get people on their side						
Self-motivation	Involvement in study and interests						
Empathy	Ability to see others' point of view						
Circumstances	Flexible						

Figure 2.1.5 *Interview criteria collection*

Job offer

Bearing in mind the importance of this stage, it is necessary that the decision is made to match the job requirement. Everyone involved must accept the decision made.

Remember that the unsuccessful candidates will be disappointed, but you may still want to use them at a later date. Be courteous and inform them gently of your decision.

Make any job offer subject to receiving suitable references, but remember these may not convey the whole truth. They are best for factual information such as job title, length of service, attendance and duties, but a personal phone call may extract more useful background.

When making a job offer, always state any period of probation involved and the stages of review.

Finally, make arrangements for starting the new employee.

Contract of employment

The first legal step is taken here. Under the Trade Union Reform and Employment Rights (TURER) Act 1993, employees must have within eight weeks of starting work an express statement detailing:

- Place of work.
- Date of start.
- Method of pay calculation.
- Hours of work.
- Holiday entitlement (and pay).
- Sickness arrangements.
- Period of notice.
- Discipline and grievance handling procedures.

The Employment Protection (Consolidated) Act (EPCA) 1978 states that employees must receive a wage slip detailing deductions made. It also confers rights against unfair dismissal after set time periods, which are adjusted from time to time by different governments.

Discrimination

At this stage make sure that you have covered yourself legally. The pertinent Acts are:

- Equal Pay Act 1970: Equal pay for work of equal value.
- Sex Discrimination Act 1975: No discrimination on grounds of sex or married status.
- Race Relations Act 1976: No discrimination on grounds of race, colour and nationality.

Under these acts, it is not illegal to take positive action to encourage applications from under-represented groups. It is illegal, however, to make positive discrimination during the selection process. There are exceptions but they must be 'genuine occupational qualified', i.e. actors, care assistants, social workers dealing with particular ethnic groups.

- Disability Discrimination Act 1995: Grounds of physical or mental impairment.

Under this act, you must justify any decision not to employ a disabled person and must make 'reasonable adjustments' to the workplace if this would aid the practical effects of the disability. Remember many disabled people do make excellent employees.

Much of the disparity in pay is not due to the rate for the job itself, but could be due to a restriction in entry to the higher paying jobs. Equal opportunity policies are not yet required by law but are recommended. The Equal Opportunity Commission (EOC) have produced a model policy which organizations are strongly recommended to mirror.

Summary of the EOC's model Equal Opportunity Policy

(1) Introduction: Desirability of policy and need to be strictly adhered to.
(2) Definitions: Direct and indirect discrimination defined.
(3) General statement of policy: A commitment to equal treatment and the belief that this is also in the interests of the organization. Staff should be made aware of this policy and key personnel trained in the policy.
(4) Possible preconceptions: Examples of preconceptions that may be erroneously held about individuals due to their sex or marital status.
(5) Recruitment and promotion: Care to be taken that recruitment information has an equal chance of reaching both sexes and does not indicate a preference for one group of applicants. Care that job requirements are justifiable and that interviews are conducted on an objective basis. An intention not to discriminate in promotion.
(6) Training: An intention to not discriminate with further details.
(7) Terms and conditions of service and facilities: An intention not to discriminate.
(8) Monitoring: Nomination of a person responsible for monitoring the effectiveness of this policy and with overall authority for implementation. An intention to review the policy and procedures. Intention to rectify any areas where employees/applicants are found to not be receiving equal treatment.
(9) Grievances and victimization: An intention to deal effectively with grievances and a note of the victimization clauses in the act.

There are other areas, such as age, where some discrimination appears to take place, which are not at present covered by law. These are under discussion, but no legislation has emerged yet. Governments, however, are showing concern in these areas by requesting organizations to include discrimination practices such as ageism in their equal opportunity policies.

A body which has had some impact in employment matters is the European Court of Justice. The EC is increasingly determining employee rights under legislation such as the Social Chapter with its Working Time Directive and Freedom of Movement for Workers within the EC. (Note that work permits are required for non-EC nationals, which can be a long and difficult process to gain.)

Terminating employee contracts

Eventually all employees will leave the organization – either voluntarily or involuntarily. Sometimes this is because the employment contract has come to a natural end – retirement or the end of a fixed-period contract. Sometimes it will be prematurely terminated by the employee or the organization.

Whatever the reason, this process has to be managed to ensure both that the employee is treated correctly and the organization does not suffer.

Employee resignation

An organization has committed resources into selecting and then training and developing each employee. When an experienced employee leaves, this investment goes with them and must be re-incurred. It will take time and money to build up an equivalent competence level in a new employee – even when the selection process is efficient in finding a replacement.

In addition extra duties and responsibilities will probably have to be undertaken by the remaining staff. This can be difficult to do if staff numbers are tight – for example, in a small company, and may affect the efficient operation of a section.

It is important that the organization establishes why the employee leaves:

- The resignation may be due to factors unconnected with the job such as family member moving to another location, looking after children or old/sick relatives, returning to education, etc.

  Examining these may lead to changes in conditions such as childcare provision, job sharing, flexible time keeping, etc. to retain highly competent staff.
- It may be a sign of problems in the job itself or the management of it. If so then it may require input to prevent, or reduce the impact of the problem.
- It may be due to favouritism or harassment by other employees or managers which may come under headings such as sexual, racialism or bullying.
- It may be a sign that the recruitment process has:
  - incorrectly completed the job requirements, or personnel specification
  - failed to select a suitable person.
- There may have been a failure in the training process, especially when the job has changed and training needs have not been identified.
- It may be that the organization has failed to recognize and make provision for the aspirations of the leaver.

- It may be that the job has changed resulting in a mismatch between the holder and the new job requirements.
- Stress in the job.

It can be difficult to establish the full reasons why an employee decides to resign, but careful questioning can prevent a costly repeat of a solvable problem. It will also limit the organization's exposure to legal action at a later date under unfair dismissal, or discrimination.

Note that when an employee has been 'requested' to resign and does so, a court or tribunal may decide this was under duress and is in effect a dismissal. In addition a resignation gives the organization an opportunity to look again at its management of people and their tasks and could lead to improvements.

Natural ends of a contract of employment

Even where there is a natural end of contract, this requires management from both the employee point of view and the organization.

Retirement

Normal retirement comes with plenty of advance notice giving the opportunity for succession planning and re-equipping the retiree for the new phase of his life.

Some people will be looking forward to their retirement, but others will feel that it is the end of their usefulness. The latter will need careful counselling. The treatment of the retirees will be noted by the remaining personnel and may affect their commitment to the organization.

It may be that retirees will have an opportunity to assist in peak times, holidays or even maternity leave for a remaining member of staff. Some will welcome short period or part-time work afterwards within the organization. Their experience can often be useful.

Early retirement is also used for organizational and/or employee benefit:

- It can ease redundancy situations.
- It can reduce 'log-jams' in promotion.
- It enables staff to pursue other interests, or start up their own business.

The conditions attached to the pension scheme and any enhancement offered by the organization will be critical here.

End of a fixed period contract

Because of the growth of this type of contract, the managing of its completion is important. The reasons include:

- You will need the holder to:
  - complete the time period
  - finish the work satisfactorily
  - hand over correctly to remaining staff
  - not take away important information such as customer details.
- You may require the leaver again in the future.
- Your actions will be noted by any others on the same type of contract.

It is no surprise that the construction and other cyclic industries such as ship-building, suffer from a deluge of industrial action towards the completion of contracts.

You should treat end-of-fixed-contract leavers in a similar manner to those leaving under redundancy. A court may even consider that because of the actual stay of an employee through renewal of short-term contracts this constitutes actual normal employment and the leaver is entitled to redundancy terms and conditions.

Dismissal

Dismissal comes about by action from the organization to unilaterally terminate the contract of employment.

It is another area where the law has introduced constraints upon organizations to ensure that all dismissals are fair. Employees have the right to periods of notice, written reasons for dismissal and not to be unfairly dismissed under the Employment Protection Act 1978. This right is normally tied to length of service which is set, and changed, by government.

A dismissal can be considered fair on the grounds of:

- Lack of capability: Case needs to be shown of the lack in:
  - Skill or aptitude: Attitude may come under this heading. Should have been addressed at the recruitment stage, but mistakes can be made there. A gross shortage needs to be demonstrated that has not been remedied after repeated warnings and attempts at remedial action. Long periods of ignoring the shortfall by management will weaken the organization's case.
  - Qualifications: Simple misrepresentation is easy and may come under misconduct. Sometimes, however, the employment contract may require the holder to attain certain qualifications during service. A driver losing his licence would come under this heading.
  - Ill-health: Providing the circumstances are such that frequent absence, or state of health, prevents the employee carrying out his duties. Alternative methods of carrying out the job and alternative posts need to have been considered.
- Misconduct: Very broad category, includes:
  - disobedience of a reasonable instruction
  - persistent absence or lateness
  - rudeness – especially to customers
  - criminal action
  - harassment of other staff.
- Redundancy: Can be under two circumstances:
  - employer ceasing to trade
  - work of a particular kind is no longer required.

  Note that redundancies also have to be discussed with the employee and there is an obligation to inform the Department of Industry.

What courts and tribunals look at in these cases are:

- What is reasonable in the circumstances?
- What procedure followed (not having suitable procedures is not a defence)
- Are all employees treated the same under the circumstances?

Awards may be financial compensation or an order to re-employ.

Where the behaviour of management contributes towards an employee leaving, the court may agree that there has been constructed dismissal. The employee may then have a case under unfair dismissal.

Redundancy

Although redundancy can be legally fair, is not usually due to the employees' direct actions and needs to be treated carefully to minimize the effect on those leaving and those remaining.

It must be continually stressed that it is the job which is redundant and not the person as such. It can be fairly traumatic for an employee, after many years' faithful service, to be coldly told that he/she is no longer required by the organization. Stress counselling will probably be required and assistance in finding new employment.

There is no legally recognized method of selecting those to be made redundant. It must be shown to be on a reasonable basis. The convention of last-in-first-out has no legal standing and often organizations make the choice on a different basis – such as skill, competence or absences to ensure they keep the employees they feel have most to contribute.

Often organizations employ means to encourage people to accept voluntary redundancy by increasing their entitlement or granting enhancement to retirement schemes. The non-replacement of leavers is another technique to allow natural wastage to reduce numbers.

Care should be taken that accepting volunteers does not result in some sections, or skills, undermanned and others with a surplus. Retraining and redeployment may need to be carried out to rebalance workloads.

Problems 2.1.1

(1) Why do you think some people stay longer than others in a job?
(2) How would you describe the ideal person to carry out your own tasks, using any of the plans.
(3) Look at the job adverts in local papers and professional journals. Why are some more attractive to you than others?
(4) Think about any interview you have experienced. How were you greeted, put at your ease and questioned?
(5) Think about your place of work or study. What practical problems are there in accessing and working for a person in a wheelchair?
(6) Why have you left any position?
(7) Under what conditions do you think both the employee and the employer will feel comfortable in being re-instated after a tribunal decision?

2.2 Making the most of people – work design, training and development

For any organization to be fully effective, it must make full use of the potential of its workforce. One route is motivation, which is a complex mixture, as this section demonstrates. Another is to train and develop the present employees to their maximum skill and potential.

The rise of scientific management

Previous to the rise of the human behaviour movement, management's actions tended to reflect Theory 'X' of Douglas McGregor (1906–1964), that states that management saw workers as:

- By nature lazy and avoiding work if possible.
- Lacking ambition and disliking responsibility.
- Being self-centred and indifferent to organizational goals.
- Resistant to change.

Much of the labour force were uneducated and in a different class structure, which reinforced this view.

With the rise of the industrial revolution, traditional skills were no longer in demand. For example, studies by Adam Smith, the eighteenth-century economist, into pin production demonstrated that the division of jobs into small, highly specialized areas gave rise to large increases in productivity. This division of labour gave an added advantage that selection and training could more easily be carried out because of the limited range of skills needed by the job holder.

In this climate, Fredrick W. Taylor's (1856–1917) investigations at the Bethlehem Steel Company gave rise to the birth of scientific management. He proposed that:

- Work content (time) could be measured to determine a fair day's work.
- Pay can be linked to this measurement.
- All work could be studied to develop better ways of carrying out tasks.
- Workers should be trained in the best methods.
- All planning should be a function of management.

Although Taylor's theories came under severe attack, this has partially been because of misapplication of his techniques by untrained, or even unscrupulous, practitioners. His ideas and method of enquiry have given birth to work study and much of modern management thinking and specialization's – including HRM.

Henry Ford (1863–1947), amongst others, took these ideas further by developing the system of mass production. This again substantially improved output at that time, but led to severe deskilling coupled to close control of work pattern and rate of working.

Motivation theories

Managers, and others, have always been interested in ways of motivating (and sometimes manipulating) the people under them. If an organization can get its people involved, then hopefully they will be more productive, make fewer mistakes and stay with the firm. There has been an input into understanding what makes people function effectively. The driving force is:

UNDERSTAND >> PREDICT >> INFLUENCE

The leading theories put forward are:

Incentive theory

Here it is assumed that given the correct rewards (or punishment), the employees will work harder. This lies behind the concept of performance related pay (see Section 2.3). The misapplication of work study has brought this into disrepute for hourly paid personnel but surprisingly perhaps it is still used for senior executives today.

This theory states that an individual will increase his/her effort in order to obtain a desired reward if:

- He/she has unfulfilled needs.
- The reward meets this need.
- The extra benefits are considered to be worth the extra effort.

Abraham Maslow (1908–1970) identified several layers of human needs:

- Physical: Air; food; water; warmth; sex; sleep.
- Security: Safety; shelter; savings; no threats; familiarity.
- Social: Human contact; belonging; affection; friendship.
- Esteem: Self-respect; others' respect; status; power.
- Self-actualization: Fulfilment; realization of potential; doing and enjoying what one does best.
- Transcendence Spirituality.

Each layer is always in existence but one layer normally dominates an individual at any one time (see Figure 2.2.1). Changes in personal circumstances, such as losing a job, can change the dominating factor.

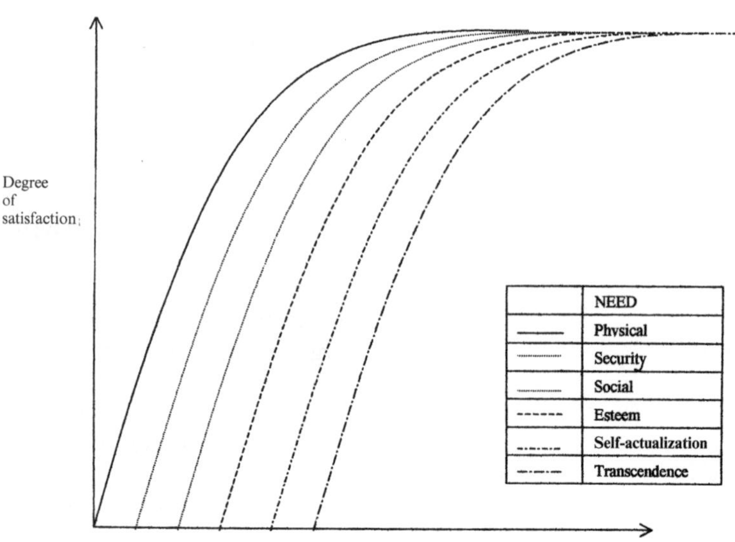

Figure 2.2.1 *Maslow's layers of human needs*

Satisfaction theories

Fredrick Herzberg (1923–) identified that in most situations there are environmental factors that will motivate, or demotivate a person. He termed them motivators and hygiene factors. The factors are:

Motivators (satisfiers)

- Achievement
- Recognition
- Work content
- Responsibility
- Advancement

Hygiene factors (dissatisfiers)

- Company policy
- Supervision
- Salary
- Relationships
- Working conditions

The point he stresses is that should a motivator factor be considered present it will normally motivate a person; however, if it is seen as missing it will not necessarily demotivate that person. Similarly for the hygiene factor its presence, or perceived presence, demotivates but its absence will not produce motivation. Table 2.2.1 demonstrates this concept.

Table 2.2.1 Herzberg's link to satisfaction/dissatisfaction

	Existing	*Missing*
Motivator	Satisfied	Not dissatisfied
Hygiene Factor	Dissatisfied	Not satisfied

There is little evidence that satisfaction does in fact substantially increase productivity. It undoubtedly does, however, appear to reduce job related stress and absenteeism and leads to long service.

The motivators link in with the incentive theory, if they are identified as rewards which meet some of Maslow's hierarchy of needs.

Intrinsic theories

McGregor also offered an alternative Theory 'Y' to describe the average worker:

- People are not by nature passive or resistant to organizational needs. They become so through experience within an organization.
- Work is a natural activity.
- People will exercise self-control to meet objectives to which they are committed.
- People can work towards group goals, if they agree with them.
- People can come to seek responsibility.
- People are able to contribute creatively to workplace problems.

This, if true, presents a challenge to management to capture this natural inclination and direct it towards organizational goals.

Human relations movement

Elton Mayo and his team carried out experiments initially to determine the effect of rest and other improvements in conditions on accidents, absenteeism and labour turnover. Later studies, – the Hawthorne Experiment – established that work groups, and modes of supervision, do significantly control the rate of output from a section. This happened both in positive and negative ways.

The problem is that man is a complex animal whose needs depend on his inborn nature, his upbringing and education and how he perceives his environment. Man is difficult to predict and cannot be generalized.

In addition, money is a general exchange good that can be given in return for the inconvenience of working and which can be exchanged for things needed. This ability of workers to accept money and use it to meet needs outside of the workplace clouds the whole area of motivation.

Job design

Early in the twentieth century, job design was all about work simplification and cost minimization under three criteria:

- Breaking the work down into simple separate elements – deskilling.
- Closely specifying the work to be done – no decision making.
- Closely controlling the work rate – often using machinery to pace actions.

This meant that jobs became extremely limited and did not require the employee to make good use of their abilities. This resulted in alienation of much of the workforce from the goals of the organization. One manifestation of alienation is absenteeism, another is a tendency for industrial action to take place.

As more evidence became available, some organizations realized that changes in the working environment could lead to improvements in quality and output and reduce many of the negative incidents.

Early efforts centred around:

- Job rotation – moving operators around the simple tasks over the working period.
- Job enlargement – increasing the number of simple tasks done by an operative.

These did have some success in alleviating boredom, but still did not make full use of the employee's potential. Efforts since have moved onto:

- Job enrichment – both widening tasks done and increasing decision making and creativity.
- Self-directed teams – giving a group substantial control over the tasks, including administrative tasks such as planning and communication with other groups, in effect doing their own supervision.

The latter two have been shown to give better and longer lasting effects. Both have been used with substantial effect in both manufacturing and administrative processes leading to reductions in technical support and managerial posts. A technique similar to work study has developed called Business Process Re-engineering (BPR).

Business Process Re-engineering (BPR)

The term was coined by Michael Hammer and James Champy in their book *Reengineering the Corporation* (1993).

BPR examines the core processes carried out within an organization.

It is based on using work-flow analysis to identify how an organization produces its product, or service. This highlights the key operations and decision-making points, especially where they are done and who carries them out.

By using IT and different ways of organizing the workforce, substantial reductions in staff can be made whilst reducing the calendar time taken.

An example would be the changes in a bank's procedure for granting personal loans. This used to take several weeks with the application form going to a variety of specialist departments and passing though a number of levels of authority regarding the amount. Now it is normally handled in minutes by one person using IT to access rules and check pertinent information such as home address and credit rating.

This has led to substantial restructuring and reductions in the labour force – especially amongst specialists and management.

BPR, like work study, has had some success and some failure. The probability is that the failures have been caused by mishandling in implementation or resistance from corporate, or workforce, culture.

Improving employees – training and development

Assuming we have the right people working on correctly designed jobs, we have to ensure the holders receive sufficient and effective training to carry them out.

There has been continual, but unfortunately continually changing, government interest in vocational training. This started back in 1964 with the formation of Industrial Training Boards through to the Training and Enterprise Councils (TEC) in the 1990s and the introduction of work-based NVQs.

In 1991, the Investor in People (IIP) initiative was launched based on the need to maintain and increase the UK's competitive position in world markets.

Investor in people

An Investor in People makes a public commitment from the top to develop all employees to achieve its business objectives.

- Every employer should have a written, but flexible, plan which sets out business goals and targets, considers how employees will contribute to achieving the plan and specifies how development needs in particular will be assessed and met.
- Management should develop and communicate to all employees a vision of where the organization is going and the contribution employees will make to its success, involving employee representatives as appropriate.

An Investor in People regularly reviews the training and development of all employees.

- The resources for training and development should be clearly identified in the business plan.
- Managers should be responsible for regularly agreeing training and development needs with each employee in the context of business objectives, setting targets and standards linked, where appropriate, to the achievement of National, or Scottish, Vocational Qualifications (or relevant units).

An Investor in People takes action to train and develop individuals on recruitment and throughout their employment.

- Action should focus on the training needs of all new recruits and continually develop and improve the skills of existing employees.
- All employees should be encouraged to contribute to identifying and meeting their own job related needs.

An Investor in People evaluates the investment in training and development to assess achievement and improve future effectiveness.

- The investment, the competence and commitment of employees, and the use made of skills learned, should be reviewed at all levels against business goals and targets.
- The effectiveness of training and development should be reviewed at the top level and lead to renewed commitment and target setting.

Investors in People Charter from the Employment Department Group's 1990 Brochure.

The Investor in People initiative is interesting not only because of its high aims but also because of its tie-in of training and development to the business plan of the organization.

This is where we will start – the business needs.

Identifying training needs

The events that trigger training needs include:

- New employee.
- New technology, e.g. change from electromechanical design to microprocessor control.
- A change in working methods.
- Increased flexibility – multi-skilling.
- Promotion or transfer to another post.
- Improvement of skills, e.g. to reduce quality problems.
- Job enlargement/enrichment.
- Changes in structure, e.g. a move towards self-managing groups.

Care is required that the identified need can be addressed by training and does not reflect some organizational or equipment problem.

Once we identify the actual need we need to break it down into what we are trying to install. The types of skills are:

- Cognitive skill: Basically thinking process – making decisions, analysing faults.
- Perceptual skill: Seeing and interpreting what we see – e.g. scanning an aircraft's control panel.
- Motor skill: Controlling human physical movement; co-ordination of limbs with senses such as sight and hearing often have to be made.

Many jobs require a blend of each of these which is built up by training and requires continual practice to maintain at peak performance. Ability in training terms is the level of competence in using a skill.

People do have varying degrees of ability in the different skills. Some do appear innate, such as colour discrimination and spacial awareness. Most, however, can be developed, but not necessarily to the same degree in everyone. For example, not everyone has the inherent ability to develop to a world class athlete, but everyone's performance can be improved. Abilities are often transferable between jobs.

Where the required ability is difficult to develop, then the selection processes must identify those who will struggle to attain them.

People also require core knowledge to be able to use skills effectively by having a background against which they can compare what is happening. Knowledge required includes:

- Basic knowledge, which a trainee is expected to have before training commences.
- Background knowledge about the company – especially so when inducting a new employee.
- Knowledge specific to the job: Reporting relationships, procedures, equipment, materials, fault recognition and diagnosis, etc.
- Knowledge of Standards: Quality, output, attitude, etc.

We need to determine the blend of knowledge, skill and competence required to carry out the job. In other words we again have to carry out a job analysis – this time looking at these aspects (see Figure 2.2.2).

This information will come from a blend of observation and questioning of employees, supervisors and sometimes suppliers (in the case of new equipment or computer programmers).

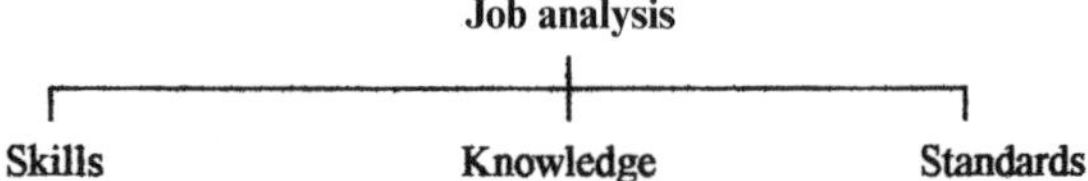

Figure 2.2.2 *Breakdown of job requirements into component parts*

It is the task then of how to install and develop this blend of knowledge, skills and attitudes.

Designing training programmes

Approaches to operative and administration training:

- Learning-by-doing (sitting with Nellie): Often just informal by sitting with an experienced operator and observing and then copying. Can be effective for low skilled tasks or where the experienced operator is a trained instructor.
- On-the-job: Very common. Normally done by experienced operator/instructor or the supervisor of the post. Mainly on a one-to-one basis but can be used for small groups.
- Off-the-job: Enables trainee to be introduced to new concepts and theoretical information. Can suffer if relevance to job not seen.

Methods of training:

- Passive – lecture/demonstration: Passing of information from instructor. Useful for blocks of new information.
- Active: Basically a learning-by-doing situation.
- Workshops to discuss and develop skills.
- Case study to analyse situations and approaches.
- Simulation to practise skills and interpretations.
- Computer-aided to learn and develop skills.
- Distance learning.

Remember that learning is an inductive process – people need time to assimilate what they are learning. At the beginning they will require a lot of guidance, so that they know what to do and this reduces the trial-and-error sequence.

Feedback during the learning process is critical to attain a skill, knowledge or attitude. Constant repetition is required to attain full mastery as most tasks move through the learning curve (see Figure 2.2.3)

The induction of a new employee into an organization is normally a case of too much information too soon. This should initially be limited to the bare essentials of what they need right away with small feeder doses over the next few weeks – even using packages of information to read.

Developing people

People need developing in many ways rather than through pure training. Many of the abilities can be shown but they need to be assimilated in the trainee's own way – especially in the field of management.

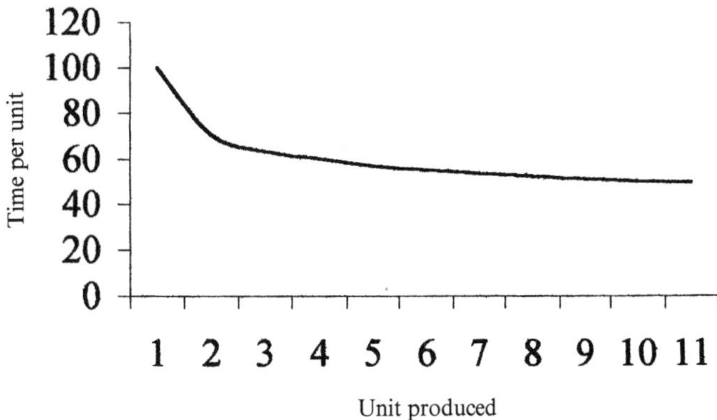

Figure 2.2.3 *Learning curve showing reduction in time taken per unit as numbers completed increase*

The development process can arise in two ways:

- Following appraisal of employees highlighting extra skill or competence needing development.
- Identified future needs of the organization.

Kolb's Model of Learning (*Organizational Psychology*, 1984) demonstrates that a four-stage iterative experiential cycle is required for full learning:

(a) Concrete experience: Describe what is happening.
(b) Reflective observation: Think about what cause-and-effect relationships are involved.
(c) Abstract conceptualization: Think what other relationships could be applied which may change what is happening.
(d) Active experimentation: Try out these other factors. Return to (a).

There is a variety of methods using a similar process to Kolb's cycle.

NVQs. A National Vocational Qualification (NVQ) is built around participants building up a dossier, or portfolio to demonstrate that they not only can perform competently various tasks in a range of situations, but that they understand the reasoning behind the actions taken. The standards for the competencies have been set by lead bodies made up from industry and trade unions.

Shadowing: Basically involves following round an experienced employee to observe how he does his tasks. If the trainee has been training experientially learning can be effective. Often not structured enough to give more than a flavour.

Mentoring: Here a senior experienced manager, who is not usually the direct supervisor, takes charge of the training and development. The mentor acts as an adviser and protector to the trainee. The trainee has the mentor to discuss what is happening and different ways he can tackle them.

Coaching: Normally carried out by the direct supervisor. Similar to a football, or athletics coach, who takes a competent person and improves their performance by constructive critical comment and guidance. Regular feedback must take place for effectiveness.

Action learning: A similar concept to mentoring, except this time the mentor's role is undertaken by a group of the trainee's peers. Basically applies Kolb's cycle. Being used significantly in post-graduate management studies and amongst self-development groups.

Evaluating the training

All training costs money and time. It should therefore be evaluated against its aims of meeting the organizational and the individual needs. The evaluation should also check actual costs involved and benefits received.

Records should be kept of what training staff have received and the levels of skills attained. These should be periodically updated to reflect the present competence level.

Methods of evaluating the training method include:

- Questionnaires: Immediate feedback from trainees – this evaluates the course more than the learning.
- Tests: Common on certificate courses. Can be used to test skills and knowledge competence. Can take the form of practical work such as a welding test.
- Pre- and post-questionnaires: To determine difference in knowledge. The post-questioning can be taken some time afterwards to show retention.
- Post-training appraisal: Of the trainee by his supervisor after training.

Problems 2.2.1

(1) Do you agree with McGregor's Theory 'X' description of the average worker?
(2) What has personally motivated, or demotivated, you recently?
(3) How would you reorganize the tasks in a McDonald's restaurant?
(4) Why does training receive a low priority in many organizations?
(5) Think about a child learning to walk. How long does it take before he/she becomes fully competent?
(6) Do you think constant practice alone is sufficient to enable you to master any task?
(7) Try to apply Kolb's cycle to an activity which has been giving you some problems in getting to grips with.

2.3 Rewarding employees – pay and benefits

Salary, especially when compared to others' within the same organization, can be a source of discontentment as Herzberg stated (see p. 75). People often use salary, and other benefits, as an indication to their relative status within the organization.

This section examines different methods of payment and other benefits in use.

Objectives of a payment scheme

It is important that organizations design an effective payment system to avoid discontentment among staff. Such a system would have as its objectives:

External compatibility. Salaries and benefits should compare to the pertinent job market to attract new recruits and prevent losing existing staff:

- Clerical and manual – local comparison.
- Managerial and professional – national comparison.

Internal equity. Staff should recognize the fairness of:

- Equally graded jobs.
- Differentials between grades.

Easily administered. To reduce cost and ensure the system is error free.

Easily understood. People should understand why their job is a certain grade and why the salary for that grade is what it is.

Management controls. To ensure the best use of money spent on salaries, there needs to be means of analysing and controlling the wage bill on basis of:

- Trends in numbers and various payment.
- Controllability through standards and budgets.
- Predictability over the short term.
- Preventing drifting of earnings irrespective of output.
- Ensuring match to skill needed and responsibility.
- Retaining ability to be flexible and reactive to changes.

Participative. If the employees themselves are involved in the design, not only will their knowledge and judgement be involved, but it will probably be more acceptable.

There are a variety of payment schemes which can be used alone or in combinations to match the needs of the organization and the employees.

Payment schemes

Time rate

This is simply a set payment per hour, which is multiplied by the number of hours worked to reach due wage. There may be premiums for unsociable hours, overtime and even certain poor working conditions.

This is easily understood and aside from the differential issue seldom results in disputes.

However, productivity is not usually measured and hence it is difficult to control, except by budgets and close management. There is no direct incentive to improve output or reduce inputs.

Payment by results (PBR)

This is an effort to relate output/input usage directly to pay, which started formally following Taylor's work. It is best used where personal performance greatly influences the output results.

Performance and PBR for an hourly paid operator

The incentive paid is calculated from the operative's overall performance which is calculated out:

$$\text{Performance (\%)} = \frac{\text{units produced} \times \text{standard time/unit} \times 100}{\text{time on standard work}}$$

Example:
An operative works an 8 hour day, during which 160 units are produced. Each unit had a work study issued standard time of 3 min. During the day the operator attended a meeting lasting 30 min and suffered from a machine breakdown lasting 45 min.

Time on standard work = attendance time – time not on bonus work
= 480 (8 hours) – (30 + 45) min
= 405 min

Standard min. produced = 160 units @ 3 min per unit
= 480 min

$$\text{Performance} = \frac{\text{units produced} \times \text{standard time/unit} \times 100}{\text{time on standard work}}$$

$$= \frac{160 \times 3 \times 100}{405}$$

$$= \frac{48\,000}{405}$$

$$= 119\%$$

If the operative was on a straight proportional scheme which pays a time rate at 75% performance and time + third at 100% performance, then the pay would be 1.59 times time

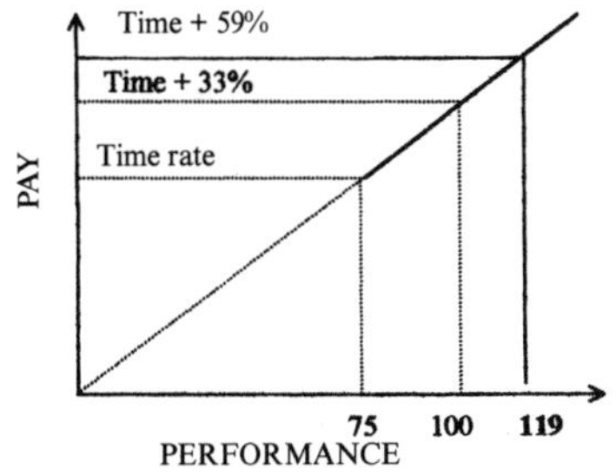

Figure 2.3.1 *Straight proportion payment-by-results scheme (showing relationship performance and pay)*

rate – see Figure 2.3.1. This is calculated from time rate × 1.33 × (119 ÷ 100).

Note: if the operator persuaded the supervisor to allow 90 min instead of the actual 75 min incurred, then his time booked on the work would decrease. In turn his performance would increase to 123% and his pay to 1.64 times time rate.

Performance should be recorded for each individual over an extended period to give a history.

Performance can be used to judge the effectiveness of a group of individuals, such as a section or a department. Where used for payment to large groups, a direct relationship between individual effort and pay can be difficult to discern. This overall performance could be used for a group bonus for supporting staff who are not on measured work.

The relationship between performance and pay can be geared to increase incentive or reduce reward where varying conditions can affect output performance as in Figure 2.3.2.

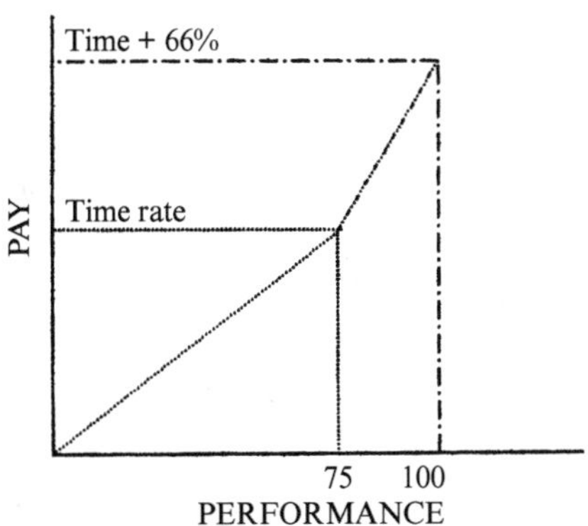

(a) Geared to increase incentive as performance rises above 75 (used when work is highly manual)

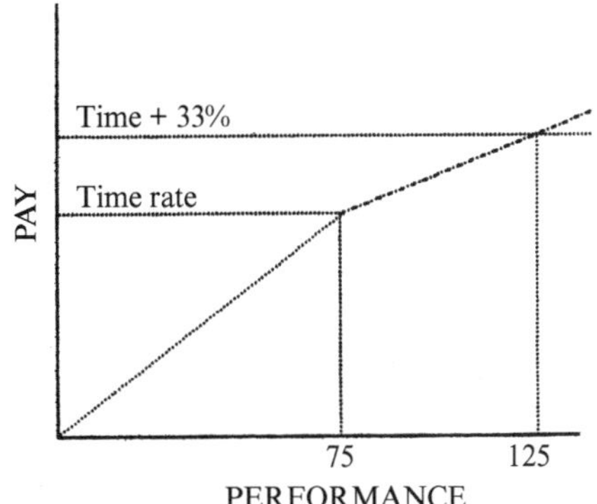

(b) Geared to decrease payment as performance rises above 75 (used when work is highly variable in work involved)

Figure 2.3.2 *Geared payment-by-result schemes*

PBR is expensive to install but initial gains in productivity to the organization can be considerable. However, it must be well maintained to hold accuracy of times set and prevent drift through ongoing improvements by the operator.

PBR often results in friction between different workers when work is allocated due to perceived differences in ease of attaining targets. This can be especially so when the process itself is highly automated and gives little opportunity for the operator to affect output – except in ensuring the unit is constantly working.

PBR schemes will suffer credibility if management try to change targets without justifiable reasons – such as a new method.

By far the main problems arise during periods of fluctuation in work load which could affect time on PBR, and hence earnings. This arises from the frames of reference used by workers being different from that of management:

- Workers' view: Over a period of time, they get used to receiving a set income under a bonus scheme. It becomes their norm. If for a reason beyond the operators' control, their performance drops, then their wage is decreased from this norm.
- Management view: Basic wage is the norm. Bonus payment under PBR is always an extra reward because of harder working. If for a reason beyond the operators' control, their performance drops, then they have not earned the portion of their wage related to performance. Pay then moves towards the management norm.

Therefore we have a failure of minds meeting. This failure in common thinking makes it easy for disputes to arise.

A linked problem is how to pay personnel who contribute towards individuals on PBR indirectly such as material handlers, inspectors and even their supervisors. It is common practice to introduce a payment linked to the bonus received by PBR workers. This often does not go down well with those whose have to expend extra effort to 'earn' their pay.

In addition to the normal measured PBR, there are other schemes which relate payment to performance:

- Sales commission: These are not based on the same detailed measurement of the work involved and are often based on contribution analysis coupled to personal opinion. Few details are available on their make-up and effect.
- Management-by-target meeting: Some staff, mainly managerial, have pay directly linked to measurable output or to attaining set objectives. Again few details are available on their effectiveness.

Measured day work

Because of the difficulty with PBR, and increased automation, many organizations have introduced this hybrid between time rate and PBR.

In this system the work content is still measured and targets set, but short-term changes in individuals' performance has no immediate effect on their wage. Some control is still obtained, and attitudes to changes tend to be more flexible. The employee cannot directly influence his wage by varying work rate.

However, in the short term, this tends to be more expensive than PBR because lower performance (whatever the reason) is not reflected in a drop in labour cost. Remedial action becomes a question of negotiation.

Plant and company schemes

These range from profit sharing to a value-added basis and increasingly through share options. These can catch employee interest because it reinforces the message 'we are all in the same boat'.

It does have problems:

- Often payments are small in relation to the main payment received.
- Because the efforts of many get amalgamated, the more contributing and harder working end up with the same benefit as less performing ones.

- Because of the large time gap between an individual's specific input and payment the relationship to individual performance is muted.
- Payments can vary substantially, as much of the results are determined by market forces.
- Organizations' fortunes depend on decisions contributed to by only a few people – the top management.
- It is especially difficult to operate in adverse trading conditions where complex, uncontrollable external forces exist – often the time when employees have to work extremely hard and monetary rewards are low.

Merit rating

This is a differential time-rate payment made to workers on the basis of certain attributes or skills. It often results in suspicion of favouritism. Considerable variation can occur between sections in a large organization. Over time there is a tendency for everyone to rise towards the higher grades.

Appraisal related pay

More specific and open than merit rating, although it does share many features. Specially useful where individuals can be set targets and

Figure 2.3.3 *Factors under appraisal related pay and associated range of points against performance criteria (suggested by ACAS)*

Factor/Score	0 1 2	3 4 5	6 7	8 9 10
Timekeeping and attendance	Frequently late/poor attendance	Periodically late/ occasional absences	Seldom late or absent	Excellent timekeeper and attendance
Job knowledge	Needs constant guidance	Needs regular guidance	Good knowledge of job	Excellent knowledge of wide range of jobs
Quantity of work	Unacceptable	Occasionally fails to reach target	Targets regularly met	Outstanding productivity
Quality of work	Requires constant checking	Sometimes below standard	Satisfactory – usually accurate	Consistent high quality
Relationships with others	Does not get on well with colleagues	Co-operative, but passive	Gets on with colleagues	Very active team member
Communications	Poor communicator often misunderstood	Not always understood	Satisfactory	Thorough, accurate and clear
Problem solving	Shows little initiative	Able to solve minor problems only	Good problem solver	Shows flair and initiative
Safety awareness	Disregards own and others' safety	Often needs to be reminded re safety	Conscious of importance of safety	Sets example of safe practice
Acceptance of responsibility	Avoids responsibility	Accepts some responsibility	Accepts within own job responsibility	Frequently seeks extra duties
Forward planning	Reacts only to events	Limited to own task only	Good forward planner	Highly proactive

judged on their achievements. The award can be an extra percentage or a step up within grading. Differs from merit pay as the criteria are more explicit (see Figure 2.3.3) but scheme can suffer from similar suspicion and problems.

Performance distributions, e.g. a typical proportioning which can be used for setting payment differentials, are:

Exceptional	5% of staff paid + 10%
Highly effective	15% of staff paid + 7.5%
Effective	60% of staff paid + 5%
Less than effective	18% of staff paid + 2.5%
Unacceptable	2% of staff paid + 0%

Benefits

Pensions

There are three types of pension schemes.

State

There are two schemes administered by the UK government:

- Basic pension: This is contributed to and paid out from National Insurance contributions. It entitles all citizens to a basic pension. In recent years this basic state pension has been effectively reduced in comparison to the national average wage by successive governments.
- State Earnings Related Pension Scheme (SERPS) was an additional payment related to wage which built up credits towards additional pension. It has been phased out.

Government pensions are funded out of present government income. The growth of people being entitled to the basic pension though increased life expectancy has made government increased pressure towards company and private schemes.

Company and industry schemes

There has always been advantages to both organizations and their employees from company pension schemes:

- Helps to recruit and retain employees.
- Improves industrial relations.
- Gives a mechanism for early retirement as part of redundancies or long-term sickness.

Pension schemes vary considerably but normal provisions are:

- Condition of service for all staff.
- Employee contribution based on salary.
- Employee contributions may be 'topped-up' to increase entitlement.
- Employer matches employee's contribution.
- Contributions go into a separately managed fund.

- Pension is made based on final salary and years of service, although some schemes work on average salary.
- Part of benefit may be taken as a lump sum.

Although men and women must have equal access to company schemes, different retirement ages and survivor payments are allowed.

> **Teachers' pension scheme**
>
> Employee's contribution 6% of salary
> Employer's contribution 7.4% of salary
> Normally employees contracted out of SERPS
>
> Payment based on final salary multiplied by years of service (up to a maximum of 40 years) which is divided by a factor of 80 to give a maximum pension of half salary. Extra years can be purchased. Early retirement (below 60 years of age) results in decreased pension payments.
>
> Example:
> A lecturer retiring from service in a further education college.
>
> Age 63 – no reduction in due amount
> Final salary £26 000
> Years of service 25 years
>
> Annual pension entitlement $= \dfrac{25 \times £26\,000}{80}$
>
> $= £8125$
>
> Lump sum entitlement $= 3 \times$ annual pension
> $= £24\,375$

The government is keen that all employees become members of company schemes as they are self-financing and reduce dependence on the basic state scheme and back-up of social security payment.

Personal schemes

Although mainly for the self-employed, other people can take out their own pension schemes or 'top-up' their company schemes.

Sick pay

As with pensions, sick pay is a mixture of state and employer funded.

Statutory maternity pay (SMP)

An employee who is pregnant is entitled to Statutory Maternity Pay. This is for a period of 18 weeks, 90% for the first six weeks and 30%

for the remaining 12 weeks. These periods and payments may be extended in the near future. There are also rights for unpaid leave for family reasons for both mothers and fathers. This may become paid leave within the next few years.

Payment made is mainly recoverable from the Department of Social Security (DSS). Many companies pay in excess of the minimum, but this extra is not recoverable.

In addition the employee has the right to return to work for a period of twenty-six weeks from the date of birth, and again this may be lengthened.

Statutory sick pay (SSP)

Administered by the organization, the employer pays out a set payment when the employee is off due to illness. This is later reclaimed from the state. This is built up from:

- Qualifying days sick.
- Waiting days – these are normally three days unpaid, but are linked to any absence over the previous eight weeks.
- Certification. A doctor's certificate is required for over seven days, but for less than this the employee provides self-certification.
- Records must be kept for three years.

These payments can be transferred to the Department of Social Security (DSS) after twenty-eight weeks.

Company schemes

These can vary from the basic SSP to non-contributing schemes.

These schemes normally have a qualifying period of service and pay out full pay for a limited period followed by a reduced payment.

Absence and sick pay monitoring

Like any other benefits, this can be open to abuse by some employees. Most employees record low absences with small periods off. Others consider the scheme to be a right to take additional leave periods.

Careful monitoring is required – again bearing in mind equality and fairness in treatment.

Other benefits

There is a large range of extra benefits which organizations can offer their staff:

- Reduced or non-contribution to pension schemes.
- Private medical insurance.
- Company car, or mileage, allowance.
- Mortgage facilities and personal loans.
- Subsidised meals.
- Subsidised travel to work.
- Clothing allowances.
- Relocation allowances.
- Discounts on products.
- Education vouchers.
- Childcare: Crèches, nursery cost assistance, etc.

Initially these benefits arose as a cheaper alternative to making payment directly to staff, therefore gaining an advantage through group schemes and a reduction in personal taxation. However, the tax authorities are gradually catching up demanding that tax be paid on the value of the benefit received.

Some benefits are from time to time actively encouraged by government, although sometimes different government regulations appear to contradict this encouragement.

Job evaluation schemes

It is very unusual for everyone in an organization to be paid the same wage. This is understandably so, as different jobs require different skills and have different responsibilities.

How does an organization decide on the 'rate for the job'?

The first way is to look at what the market pays and set the organization's rate around that. However, two factors cause problems here:

- Local factors can often distort the rate for a particular job.
- Similar names can be used for substantially different jobs.

Therefore if these rates are used they may be out of line with others in the organization, which can cause feelings of unfairness.

Another way is to look internally at job rates – at least that is what the present employees are used to. This will result in perhaps in a large number of different rates, which will need consolidating to around four or five rates.

Reducing the number of job rates

If we take all the existing job rates, rank them numerically and then draw a scatter diagram (see Figure 2.3.4), we should find:

- There is a considerable range.
- There is a pattern – a trend upwards as the rates are pre-ranked.
- There could be natural groupings.

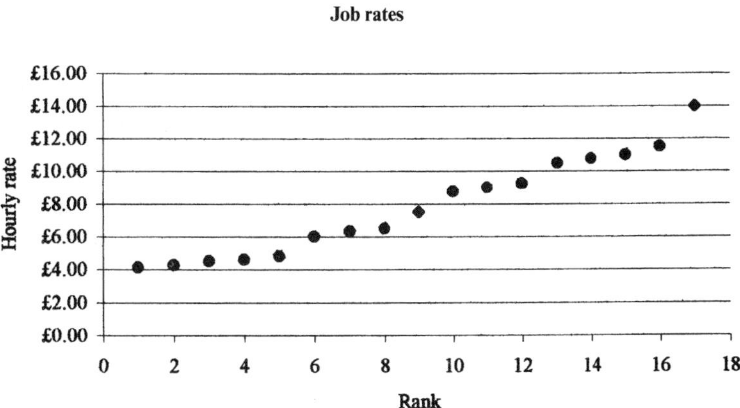

Figure 2.3.4 *Scatter diagram of existing hourly pay rates*

Job rates	Group	Mid-point value	Selected value	Difference	Number in Group Males	Females
£4.10	D	£4.45	£4.50	£0.40	2	3
£4.25	D	£4.45	£4.50	£0.25	2	12
£4.50	D	£4.45	£4.50	£0.00	0	8
£4.60	D	£4.45	£4.50	–£0.10	3	24
£4.80	D	£4.45	£4.50	–£0.30	5	3
£6.00	C	£6.28	£6.50	£0.50	5	15
£6.33	C	£6.28	£6.50	£0.17	3	2
£6.50	C	£6.28	£6.50	£0.00	20	44
£7.50	*				2	5
£8.75	B	£9.00	£9.00	£0.25	15	5
£9.00	B	£9.00	£9.00	£0.00	1	2
£9.25	B	£9.00	£9.00	–£0.25	7	6
£10.50	A	£10.94	£11.00	£0.50	5	0
£10.75	A	£10.94	£11.00	£0.25	15	3
£11.00	A	£10.94	£11.00	£0.00	12	2
£11.50	A	£10.94	£11.00	–£0.50	8	0
£14.00	*				5	0

Figure 2.3.5 *Analysis of existing wage structure showing recommended grouping*

If we carry out an exercise on this information (see Figure 2.3.5) we find:

- There are four natural groups.
- Two job rates do not fall into a particular group. These were found to contain a pay element for supervision for leading a small team.
- If we take an average for each group, this can enable us to select a simple representative rate for that group.
- When we compare the actual rate to that selected value, some rates are higher and some lower. Therefore if we paid at the selected rate, there would be gainers and losers (see page 96 for benefits and problems in job evaluation).
- When looking at the sex in each group, we find in the lower rate groups, it is predominately female and the reverse in the higher groups. This will need investigating to discover the reason – is it due to real differences in value, or does it indicate a lack of either equal pay for equal work or equal opportunity in selection processes? This needs investigating.

The basic remaining questions to be answered are:

- Is the rate a good indication of the value of the job to the organization?
- Are the jobs on similar rates, close enough in characteristics to be classified as having the same value?

If the answer to the latter two questions is yes, then it would be possible to change into a reduced number of job rates.

However, changing technology and processes can often change the original basis for differentials. We therefore need a method that can fairly differentiate between jobs, which is capable of slotting new jobs into, or revising the rate for a changed job.

It must be stressed here that we are looking at the characteristics only of the job, not at how the present job holder performs in it.

Simple comparative methods

Job ranking

This is a simple method – often sufficient for a small organization. It basically means making a list of jobs in order of pay worth to the organization. It follows a simple procedure:

- Make a list of all jobs within the organization – without details of their pay rate.
- Go through that list and rank the jobs in order of the perceived importance to the operation of the organization – remember to look at the job, not how the person in it is performing.
- Compare this to the jobs ranked by pay – can give an opportunity to reduce the number of rates.
- Address any apparent anomalies.

You now have a ranked list – any new jobs can be slotted into this list by examination. Similarly any changes in the job which change its importance can be refitted in.

Paired comparison

The paired comparison method is another simple method. It eases decisions by comparing every job with every other one – one at a time. This can mean a considerable number of comparisons have to be made.

- The basic procedure is to draw up a simple table with jobs heading each row and each column.
- You then move along the row comparing the row job against the column job.
- Award (place in intersection box):
  - two for when the row job is considered of higher value
  - one if both jobs are judged equal value
  - zero if the column job is judged of higher value
    - ♦ Note each job is compared against another on two occasions – once using the row position and the other using the column position – these must agree (i.e. the box contents add up to two).
- Once all jobs have been compared, total the values in each row.
- Check the total of all the jobs' values by comparing it to the multiple of (jobs × (jobs – 1)).
- The resultant total gives the relative value of that job.

The final run of a paired comparison is shown in Figure 2.3.6. The highest score was gained by the maintenance technician – the lowest by the janitor.

Job classification

This involves selecting one particular job description to be representative of all other jobs in its grade. A job is then compared with each of these prime examples to denote which it is nearest to.

Examples of this are shown in Figure 2.3.7.

	Janitor	Material handler	Assembly worker	Tool setter	Jog borer	Maintenance technician	Total
Janitor	X	0	0	0	0	0	0
Material Handler	2	X	1	0	0	0	3
Assembly worker	2	1	X	0	0	0	3
Tool setter	2	2	2	X	1	0	7
Jog borer	2	2	2	1	X	0	7
Maintenance technician	2	2	2	2	2	X	10
						Total:	30

Figure 2.3.6 *Paired comparison method of selecting job importance*

CLERICAL JOB CLASSIFICATION

Grade	Description
A	Tasks which involve a high degree of judgement and responsibility, e.g. secretary to an executive.
B	Complex tasks which require considerable experience, but need limited initiative as detailed procedures/rules exist, e.g. collating technical information and laying out drafts.
C	Tasks of a routine nature following well-defined procedures, e.g. collating simple information before typing.
D	Simple clerical tasks requiring limited skills with detailed rules, e.g. simple copy typing.
E	Simple tasks which require no clerical skills, e.g. mail and document sorting.

Figure 2.3.7 *An example of a job classification scheme for clerical staff*

Factor assessment

This widely used scheme attempts to value manual and clerical jobs through their characteristics, termed factors. The scheme normally uses four to five main factor headings, each of which can be broken down into several sub-factors. The International Labour Organization list over thirty commonly used factors.

The usual main factors are:

- Skill
- Responsibility
- Mental effort
- Physical effort
- Working conditions

An example of the factor descriptions in a particular manual scheme is shown in Figure 2.3.8. The five main factors have been broken down into a total of 12 sub-factors. The descriptions should be simple and avoid ambiguity.

JOB EVALUATION FACTOR DESCRIPTION

EXPERIENCE & TRAINING

EDUCATION: The minimum educational level necessary for a person to be capable of learning the job and reaching an acceptable standard of performance. Consider any mathematical or other skills required by the job.

EXPERIENCE: The accepted time after basic training to fully understand the job tasks and deal with all but a few really exceptional circumstances which may arise.

TRAINING: The time required to receive basic training sufficient to be able to carry out the job.

PHYSICAL REQUIREMENTS

HANDLING: The exertion required to move materials or equipment during the job. Consider range of weights lifted and frequency involved.

WORKING POSITION: The extent to which discomfort is caused by any awkward stance or position required to do the job.

MENTAL APPLICATION

COMPLEXITY/DEXTERITY: The extent to which the job requires concentration and manipulative skills to execute variable sequences in the process and the frequency of such use.

MONOTONY: The extent to which the job consists of short repetitive actions, or otherwise requiring no scope for selecting alternatives or other freedom of action.

RESPONSIBILITY

USE OF RESOURCES: The extent to which the job necessitates skill and care to improve the utilization of resources or yield from materials used in the process. Consider the cost of a careless day.

JUDGEMENT/INITIATIVE: Responsibility for taking decisions for courses of action to facilitate the flow of work to achieve targets, maintain quality levels, prevent damage to materials or equipment and ensure minimal equipment downtime occurs.

CONTACT RELATIONSHIPS: Regularity of contacts of a routine or complex nature required for giving or receiving instructions to control mechanical or manual operations.

SUPERVISION

ADVISORY: Advises or explains tasks to other employees.

SUPERVISION & DELEGATION: Closely supervises work of other employees and/or delegates' work to others.

ENVIRONMENT

WORKING CONDITIONS: Those conditions in which the work is normally or periodically carried out, e.g. in the presence of noise, dirt, fumes, damp, heat, cold. Consider the degree and combination of these factors.

HAZARDS: The unavoidable exposure to risk of possible injury, etc. after normal safety precautions have been exercised. These apply both to the operator and others nearby who may be subjected to the hazard.

Figure 2.3.8 *Job evaluation factor descriptions. (The comments under each factor must be clear, measurable and supportable)*

The next stage is to decide for your particular organization which factor should contribute most towards people's pay. This is highly subjective and can differ between quite similar organizations. The object is to achieve weightings which are understandable and fair to most staff in the organization.

The main factor weighting is then distributed across its sub-factors. Each sub-factor is then subdivided into 3–5 degrees normally with equal steps as in Figure 2.3.9. Note the points figure for a degree is indicative – actual points awarded can be between the values stated.

FACTOR	SUB-FACTOR	TOTAL POINTS	MAXIMUM PER CENT	POINTS LOW	MEDIUM	HIGH	EXCEPTIONAL
EXPERIENCE AND TRAINING	Education	500	5%	25	50	75	100
	Experience		15%	75	150	225	300
	Training		5%	25	50	75	100
PHYSICAL REQUIREMENT	Handling	400	10%	50	100	150	200
	Working position		10%	50	100	150	200
MENTAL APPLICATION	Complex/dexterity	400	14%	70	140	210	280
	Monotony		6%	30	60	90	120
RESPONSIBILITY	Use of resources	400	10%	50	100	150	200
	Judgement/initiative		5%	25	50	75	100
	Contacts		5%	25	50	75	100
ENVIRONMENT	Work conditions	300	10%	50	100	150	200
	Hazards		5%	25	50	75	100
	TOTALS	2000	100%	500	1000	1500	2000

Figure 2.3.9 *Points awardable against factors*

The reason for having a total number of points available as high as the 2000 shown here is simple. Even a low scoring job will end up with a healthy looking number, hence protecting people's pride.

Jobs have to have a detailed description using the same factors as the scheme before they can be evaluated. This requires training in the technique. Similarly the evaluation is best carried out by an experienced team to prevent bias creeping in.

The designed scheme must be tested against key jobs to ensure that the corresponding totals correlate to these job's rankings. Only after this could the breakpoints between job grades be decided on.

You will find that the grade for some jobs has changed – up or down. When a job rate has reduced, it is common practice to protect that person's salary for a set period – up to three years. During this period, the organization should attempt to find another post paying a similar rate to what the employee was receiving.

It also must be tested for any bias against any group of employees, such as women.

Management

There has been some attempts at evaluating executive positions but these have not been developed to a satisfactory conclusion because of the lack of standardization of roles and the relatively fewer posts involved.

These posts tend to remain outside formal job evaluation schemes and salaries relate more to numbers of staff and turnover.

Problems 2.3.1

(1) What do you think is a fair basis on which to pay for different jobs?

(2) If an operative is paid £6.00 per hour for first 37.5 hours with time + half for overtime and works 48 hours what is his pay before deductions?

(3) Should operators be paid the bonus rate for their full attendance time, or just the time on bonus work?
(4) Which payment scheme would you personally prefer to be paid by? Why?
(5) What percentage of final earnings do you think is fair to pay as a combined state and private pension?
(6) Do you think continual ill health is a fair reason for dismissing an employee?
(7) What should you do if, following job evaluation, the rate for a particular job involving a large number of staff comes out with a substantial increase?

Appendix

Productivity

Any organization depends on producing a profit to survive. Profit is basically the difference between the money earned and the money spent by the organization. Productivity is how well the organization makes use of its resources, be it people, machines or material to produce a high profit.

Basic work content of an operation

Most operation times are extended due to bad engineering or management. Figure 2.3.10 gives an impression of the effect of different factors in determining the actual time taken to carry out an operation.

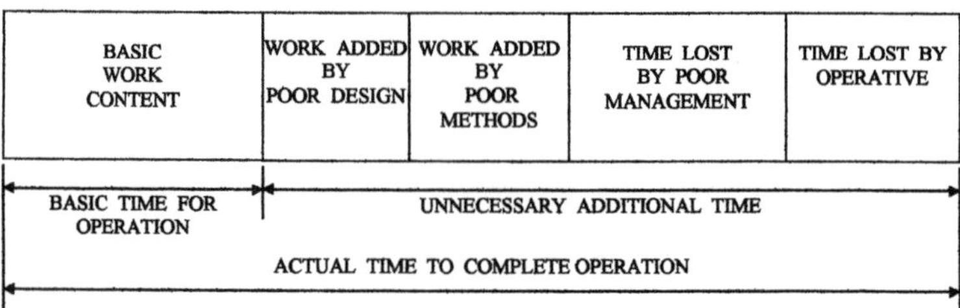

Figure 2.3.10 *Effect of different factors in the actual time taken to carry out an operation*

The basic work content is decided by the function of the part and is the minimum (ideal) required to produce that function. The additional factors increasing the time taken from this basic time up until that actually used are caused by:
Work content added by poor design

- Designer selects material which is difficult to process.
- Designer often selects process, which may not be ideal.
- Design produces variety instead of standardizing hence adding to non-value adding work.
- Designer sets tolerances higher than function requires causing extra processing.
- Designer selects larger starting size than necessary causing extra processing to reduce bulk.

Work added by poor methods

- Wrong process selected by designer.
- Wrong machine selected by production engineering.
- Wrong tooling selected by operator or production engineering.
- Poor plant layout causing unnecessary material movement.
- Process instructions wrong leading to extra reworking.
- Machine operated incorrectly due to lack of training.
- Insufficient handling devices extending process time.

Time lost by poor management

- Excessive variety of products leading to extra non-value adding work.

- Lack of standardization leading to extra non-value adding work.
- Many design changes leading to tool breakdowns and quality problems.
- Poor scheduling where similar work is not scheduled together.
- Poor operational control extending process time.
- Lack of raw materials leading to idle time.
- Plant maintained poorly leading to extended process time.
- Poor working conditions leading to lost time through extra rest periods.
- Poor safety precautions leading to lost time through accidents.
- Poor training leading to inefficient operations and quality problems.

Time lost by operative

- Poor method used different to that trained for.
- Absence and lateness.
- Slow work rate.
- Carelessness.

There are a multitude of techniques which management use to reduce the actual time taken towards the basic required. They all rely on recording and measuring what is happening now, developing better methods of working and then controlling to the new standards produced.

Work study

Work study is the systematic examination of all aspects of the working operation. Although it has been in formal use for almost a century, its application has often been poor and has led to it having negative images.

The modern 'in' technique in management cost reduction – *re-engineering* – basically uses a similar technique. Work Study is useful because of its systematic approach to identifying what is happening and then developing improvements.

The two components of work study are:

Method study

To examine existing work methods and improve the effective use of:

- Material
- Manpower
- Machines
- Movement
- Space

Work measurement

To assess human effectiveness to give improved:

- Performance
- Planning and control
- Cost control
- Incentives

Method study

The starting point for any change should be to examine and improve on what is presently happening – even before actually measuring the time taken. In order of preference the basic objective for each task examined is to:

- Eliminate
- Combine, or
- Make easier

The stages of method study are to select the operation to be studied on basis of perceived importance:

- Frequent/common operation
- High cost operation
- High usage of a scarce resource (operative or machine)
- Bottleneck operation
- High scrap rates
- Large time factor in critical lead time
- New product/process

Record present operations using:

- Process flow charts
- Flow diagrams
- Multiple activity charts
- String diagrams

These should include exact distances moved and the time taken but this is not normally necessary the first time, as movement tends to be the easiest task to alter.

The important tasks are the OPERATION and INSPECTION where a material has work done to it or the material is counted or examined.

Examine the present method, asking the following questions, and then ask 'WHY?' to each answer received to the question:

- What is happening? WHY?
- Where is it happening? WHY?
- Who/what is doing it? WHY?
- How is it being done? WHY?
- What is the required specification? WHY?
- What are the required standards? WHY?

Develop a better method by understanding and critically examining the WHYs collected. Follow the 'WHY?' by 'WHAT ELSE COULD BE DONE?' until 'WHAT SHOULD BE DONE?' is arrived at.

Common examples are moving operations in the sequence that they are done so that the operative is carrying them out whilst the machine is working, as in Figure 2.3.11.

The proposed method should always be tested before finalizing. Define the agreed method so that it can readily be followed and act as a basis for later checking.

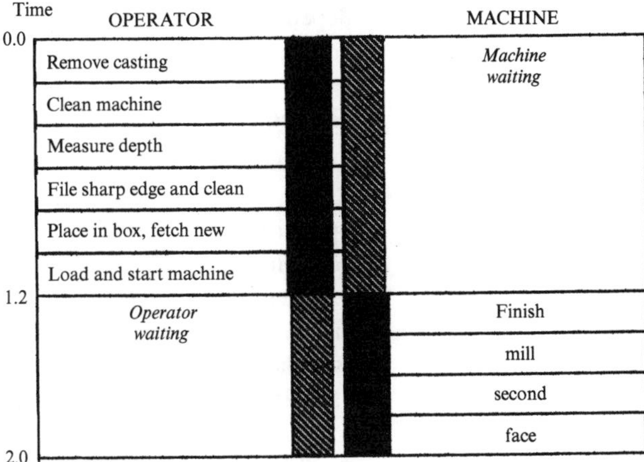

(a) Operation before study carried out

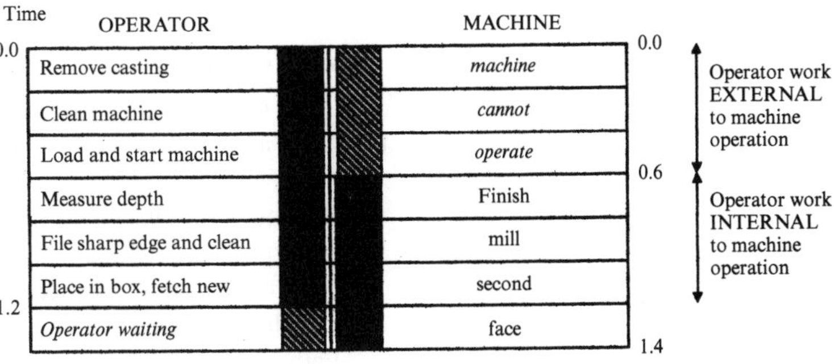

(b) Operation after study carried out

Figure 2.3.11 *Reduction of 30% achieved through method study*

Implement the new method by gaining the acceptance of the operational personnel and their managers and then ensuring the correct training is given.

Maintain the new methods by periodically checking that they are being followed and that any improvements are incorporated into the defined (standard) method.

Work measurement

There is a common saying 'Before anything can be controlled, it must first be measured!'. This is the basic task of work measurement

Accurately measuring the time taken gives a firm basis for planning and control. It also allows management to prepare accurate estimates; determine capacity and its utilization; determine operatives' performance; and give a sound basis for calculating incentive earned in PBR schemes.

Work measurement requires specialized training. It should be carried out only by personnel experienced in the process under study so that optimal conditions are being measured. It is not uncommon for trade union representatives to be trained in work measurement techniques and even be part of the work study team.

Time study is the most common method, although increasingly simulation is being used. A stop watch is used to determine the exact time required for a set operation. The measured time is adjusted depending on the operative's effectiveness to get a basic time. It is then adjusted to give ample allowance for recovery from fatigue.

Rating

To measure work content, the work study observer has not only to record the time taken but also has to apply a judgement on the operative's effectiveness. This judgement is called RATING and is a combination of effectiveness; speed; effort and attention.

The rating is carried out to a scale termed the BSI 0–100 scale. There are two fixed points on this scale – 0 for 'not working' and 100 for 'standard (incentive) performance'. Normally ratings only in the range 75–125 are observed during a study. These can be described as:

- 125: A very high performance requiring a high degree of effort, skill and co-ordination. (Walking at 8 km per hour.)
- 100: A brisk performance. (Walking at 6.4 km per hour.)
- 75: A steady, deliberate performance. (Walking at 4.8 km/hour.)

Taking the study

Just as in method study, there is a set procedure to ensure that the study produces an accurate and usable time:

- Set method: The study should only be carried out on an agreed method. This includes all data such as machine speeds and feeds, fixtures required, position of all tools and components.
- Divide complete job into elements: In order to ensure ease of taking both elapsed time and rating, the entire job should be broken down into easily recognizable small elements of between 0.1 and 1.0 min. The breakpoints between elements should be made easily recognized – e.g. laying down a tool, pressing a button, etc.
- Carry out study: Record start and stop times. Every event during the study should be noted.
- Work out study:
  - Standardize times: Convert every observed time into a normalized time by multiplying by the observed rating:

    Observed time = 0.45 min
    Observed rating = 90
    Normalized time = (0.45 × 90/100) = 0.405 min

  - Select representative time: Table all standardized times for each element and select a representative time for that element:

    ELEMENT STANDARDIZED TIMES

Load machine	0.40	0.43	0.42	0.39	~~0.55~~
	0.39	0.41	~~0.23~~	0.42	0.41
	0.41	0.39	0.40	0.41	0.40

  Selected representative value by discounting any 'odd' values, such as those crossed out, and averaging the remainder:

  This gives a selected time of 0.41 min.

- Apply compensating rest allowance (CRA) to each element to compensate for energy expended carrying it out. This produces a standard time:

Standard time = normalized time × (100 + CRA)/100
If normalized time = 0.41 and CRA is 15% then
Standard time = 0.41 × (100 + 15)/100 = 0.4715

- Decide frequency of element: Where an element occurs every time, its standard time will be allowed once per component.

  Some elements, such as fetching material or tooling may not happen once for each product. The selected time is reduced to that proportionate for one product.

  If an element only occurs once every 10 components then the allowed standard time for that element per component = standard time per occurrence/10.

  Add all allowed element times to give the standard time for the whole operation. This is normally issued in two standard digit formats.
- Issue agreed standard time and output per hour at 100 performance to operative and get their agreement.

Check the accuracy by carrying out production studies at set intervals.

3 Law

Summary

The English legal system has developed over hundreds of years. It forms the model for legal systems in many countries throughout the world and its mix of evolved and prescribed law makes it useful for understanding how other systems in other countries work. Understanding the prescribed legislation relating to the design, production, protection and sale of engineering products and services also provides a basis for understanding compliance of these issues worldwide. In the event that the law has to be called on, there are two areas of action known as civil and criminal law. This chapter introduces these actions, the relevant laws, the legal issues around them and the legal framework in which they sit. An understanding of these is important in reducing risk and optimizing commercial success.

Objectives

By the end of the chapter you should:

- have a grounding how the English legal system developed and why it is different to other countries. You should understand the actions that are involved should the law be breached and the legal system be called upon (Section 3.1);
- be familiar with the legal options available to protect products and services (Section 3.2);
- be familiar with contract law required in the provision and supply of products and services (Section 3.3);
- understand liability issues and legislation in providing goods and services (Section 3.4).

3.1 The English legal system

Evolution and history

The English legal system is a historic one with roots dating back to the localized, tribal Anglo Saxon times of the first millennium. It has been less influenced by the Roman empire than neighbouring countries and is hence different to most European legal systems.

Norman invaders arrived in 1066 and over a 200 year period began to link these regional and custom variations through travelling judges (Assizes) and by establishing central courts. They hence established laws that were common throughout the land ('common law'). Before written

Key point

The procedure for introducing new laws is the circulation of a Green Paper to interested parties followed by a White Paper containing definitive proposals. After approval by the House of Commons, House of Lords and Royal Assent it becomes an Act and is recorded in *Halsbury's Statutes of England*. An Act can only be repealed, amended, consolidated (the grouping of Acts together) or codified (a summarizing of relevant case law and statutes). The Deregulation Act 1994 also allows the Secretary of State to repeal primary business legislation without full parliamentary approval.

Key point

The European Commission formulates the community policy and drafts legislation and treaties which are enabled by administrative Acts. The European Parliament contains 626 MEPs and acts as an advisory/consultative, non-legislative body to the Commission.

Key point

Around 70% of Britain's business law now stems from Europe. The transference of legislative power towards Europe is a concern to people who believe that legislation should remain localized.

'year books' were used, legal issues resolved by one case set an example for other similar cases (the principle of 'precedent'). Special cases were dealt with by the chancery under the principle of 'equity' (fairness) which became precedential too. Common law and equity became combined under 'the Supreme Court of the Judicature' in 1873. Custom is hence now succeeded by precedent but may still be applicable in certain cases, particularly in business. The precedent or case law system is voluminous but is considered to be more flexible and realistic than the alternative Romanized code or list-based systems used elsewhere.

Over the past 150 years, governments have contributed to the legal system by introducing legislation and statutes which overrule common law. A government may also introduce a code (for example, highways, race relations, health and safety) which may not be actual law but may influence a court in reaching a decision.

Legislation is interpreted by courts who must obey the law. Responsibility may, however, be delegated from the courts to subordinate bodies via enabling Acts. A lower court is bound by the decisions of a superior court but not vice versa (although inferior court decisions and even commonwealth court decisions may guide superior courts).

The absence of a written constitution, however, which would allow courts to question legislation from the government, is sometimes considered as being unhelpful to the nation's governance.

The Treaty of Accession with Europe in 1972 led to the European Communities Act which allowed European Community (EC) legislation to be incorporated into United Kingdom law.

Most EC legislation is related to commercial and economic issues surrounding Europe and the Community's budget of around £50 bn. EC legislation comes in a variety of forms:

- Articles, which apply from person to person (horizontal) and person to state (vertical).
- Regulations, which are general and binding to all.
- Directives, which are general but applied to particular states (vertical).
- Decisions, which are specific and have no legislative effect.

Common Market Law Reports, which are in themselves beginning to resemble a list of precedents, detail the legislation. The Council of Ministers are sovereign representatives who act on Commission proposals, and the European Court of Justice, assisted by the Court of Auditors, ensures that the legislation is implemented.

Example 3.1.1

In *R.* v. *Secretary of State for Transport, ex parte* (on behalf of) *Factortame Ltd* (1991), Spanish fishermen were claimed to be catching part of the UK's 'quota' of European fishing stocks by forming British companies and having their boats registered in the United Kingdom. The government tried to prevent this under the Merchant Shipping Act 1988 and regulations made under it. The government lost the case and it was the first time that the UK government itself had been deemed to be acting illegally.

How the system works

Civil law

Civil law concerns the resolution of disputes between individuals. The dispute could be over a variety of issues such as small claims, contract, personal injury, land disputes or insolvency and usually involves an individual seeking redress against another individual or against the crown. Cases are referred to as plaintiff v. defendant.

An act with an element of blame leading to civil wrong is known as a tort (tort comes from the Latin '*tortum*' meaning 'wrong'). Fault can be committed even without intention ('*mens rea*') which is known as strict liability. Buying stolen goods, for example, could make you liable even if you did not know the goods were stolen.

A careless action can lead to the tort of negligence provided that the 'defendant' had a duty of care, that the duty of care was broken and that damage was suffered as a consequence. Designers and engineers have a duty of care towards their customers and hence defective goods can lead to the tort of negligence (it can also lead to claims for a breach of contract). With negligence claims, the onus is often on the designer or manufacturer to prove their innocence. A vicarious tort occurs when a person is held responsible for a tort committed by others. For example, the partners or directors in a firm might be held responsible by plaintiffs for the actions of a company employee which they know nothing about.

Cases of civil law generally start in one of the country's 400 county courts. A county court comprises circuit judges and solicitors in gowns but not usually juries. Cases that are more complex cases, involve a higher magnitude of damages or require juries (such as defamation) go directly to a High Court. There are 26 High Courts organized into three divisions; Queen's Bench, Chancery and Family. It is the job of the District Registrar to determine where the case will start.

County courts particularly are outgrowing the small disputes they were originally set up to handle. Recent changes in proceedings for small claims are intended to make processing quicker and the use of Special courts is also becoming more widespread. Special courts are geared to specific cases such as restrictive trades court or patent disputes. They comprise lawyers and High Court judges and have High Court status. Because they specialize they can be much quicker. Administrative tribunals are also used. These are smaller, less formal bodies that are subordinate to a court. They comprise lay members (that is non-legal members of the public) looking at disputes such as social security, revenue or employment matters. If a tribunal oversteps its powers the decisions can be questioned in the High Court and this is known as a judicial review. A domestic tribunal will relate to matters such as trade unions and trade associations.

Court action takes the following steps:

- A writ (Queen's Bench) or a summons (county court) is issued.
- A preliminary hearing is held by the District Registrar.
- A court appearance follows, usually by a solicitor(s) representing the parties. The plaintiff proceeds by trying to prove guilt and the action is then defended.

A number of steps are, however, advisable before taking civil action:

- Are you in time? There are strict time limits by which time action must be taken and which depend on the type of dispute.
- Can you win your case? The decision will rest on the balance of probabilities. Even though you may be the damaged party you will have to prove it.
- Can you afford to take action? Proceedings are expensive and you may have to pay the other party's costs if you lose. Legal aid covers civil proceedings provided there is a reasonable case and insufficient personal funds although it does not now cover personal injury or administrative tribunals. Insurance may cover the legal costs of losing and in certain cases it is worth discussing with solicitors if they would take a fee only if the case is won (a conditional arrangement).
- Who should you sue? The purpose of suing is normally to extract damages and it is sometimes necessary to claim from the more affluent party rather than the more guilty party. A pedestrian injured by a poorly driven bus might sue the bus company rather than the bus driver, for example, because the bus company is more likely to be able to pay a substantial sum than the driver.
- Is it worth it? How much will you receive? How long and how stressful will it be? Will you get bad publicity even if you win? Will or can the loser pay up?
- Can you negotiate a solution? Consider arbitration to help, particularly for resolving a contract difference. Arbitrators are often based within the county court system and may refer just the legal aspect to a High Court ('case stated').

Criminal law

Criminal law is concerned with conduct and punishment for crimes against the state and public order, crimes against a person or property (a criminal act may therefore also lead to a tort) and others unclassified (such as traffic offences). Liability can again be strict. Cases are referred to as the Crown v. prisoner.

Magistrates' courts contain a minimum of two Justices of the Peace (magistrates) who preside over minor offences, such as motoring, trades descriptions or breaches of the Factory Acts. Magistrates are lay persons appointed by committees and advised on legal issues by the Clerk to the Justices. Powers generally extend up to £5000 fine or 6 months' jail, but they can decide whether remand is necessary or whether the case is reasonable to go forward to crown courts. Crown courts have four levels of seriousness ranging from trial by a High Court judge down to trials with barristers with juries.

In criminal action, the steps are:

- The accused is summonsed. In criminal law there is no time limit to action.
- The accused enters a plea and the prosecution (for example, the Criminal Prosecution Service or Factory Inspectorate) proceeds. Unlike civil cases which rely on the balance of probabilities, there are strict rules and standards of proof which require absolute certainty of guilt.

- If proved guilty, compensation may be awarded to a victim, which avoids a repeat case in the civil court.
- Appeals can be referred to a Divisional Court (two or three High Court judges) ruling on the procedure in questioned cases. The Court of Appeal (Master of the Rolls and Lord Justices) review the case based on court and judges notes. The House of Lords is the highest appeal source in the country and consists of ex-chancellors and life peer Law Lords. Cases can be appealed directly here if parties and the original judge agree. The European Court of Justice is obligatory if no further internal appeal rights exist.

Problems 3.1.1

(1) What is the purpose of a Green Paper?
(2) What is a Consolidation Act?
(3) Explain why in the creation of legislation the European Parliament is different to the British Parliament.
(4) Explain why it is possible to be held responsible for something you did not do?
(5) In the civil case of *Hedley Byrne* v. *Heller and Partners* (1963), a firm of advertising agents asked a bank to give a reference for a customer they were considering giving credit to. The bank provided this but expressly disclaimed liability when giving it. The advertising agency extended credit facilities but when the client failed to repay them the agency took the bank to court arguing that the reference was given carelessly. Who is the plaintiff and who is the defendant? Consider the issues and why the agency is suing the bank and not the customer. Consider whether you think the bank could be held responsible.

Case study

The Gadget Shop v. *The Bug Chain*

The Gadget Shop Ltd ('TGS') ran shops selling novelty items throughout the UK. In November 1999, the first defendant The Bug Chain ('TBC') opened a shop selling a range of similar products and employed former employees of TGS. TGS issued proceedings against TBC for attempting to deceive customers (in breach of the common law of passing off), infringing its shop fitting plans and claiming that the former employees had improperly provided confidential information relating to its product-knowledge file.

A court order was granted against TBC permitting the search for and seizure of its information. The order also restrained the defendants from dealing with, using or disclosing any of the confidential knowledge.

TGS sought a further order that a modified injunction be continued until trial. TBC sought to set aside (annul) the search order on the basis that the legal procedures were not correctly followed and that TGS had failed to make a full and frank disclosure to the court.

The judge held that there had been a number of administrative and procedural errors. There were certain material departures from the standard form of order about which the court was given no opportunity to form or express a view. In particular, the omission to provide for a solicitor partner to be in the search teams and the absence of any undertaking not to inform third parties of the order except for the purposes of the proceedings. TGS also did not disclose in its evidence, or inform the court, that since October 1999, the information contained in its product-knowledge file had been in the public domain via its website. It was improbable that, had the court been aware of the information on the website at the application stage, the order would have been made at all, or at any rate in the form including the product-knowledge file. The order was discharged.

3.2 Product protection

A company's products and services, including its know-how, are sometimes referred to as its intellectual property (IP). It is important that IP is protected to reduce the opportunities for other companies to copy it and to maximize its value.

There are many commercial ways of protecting IP including secrecy, technical complexity, being faster to the market place and branding. However, the law provides statutory methods known as the intellectual property rights. These are explained in the following section.

Key point

Counterfeiting and piracy are estimated to cost companies worldwide up to Euro 300 billion per year, result in 200 000 job losses worldwide and accounting for between 5% and 7% of world trade in value terms. The industries most affected are software, clothing and textiles, toys, music recordings, perfumes, publishing, pharmaceuticals, videos, car parts and sports goods.

Intellectual property rights

The most common form of intellectual property rights (IPR) are patents, design rights, copyright and trade marks. There are different types of right because each protects a different aspect of intellectual property:

- Patents protect technical or functional innovation.
- Design rights protect shapes or configurations.
- Copyright protects the representation of an innovation.
- Trade marks protect a distinctive element.

Note that it is difficult to protect the idea in itself and that there must be a tangible element to the idea. You cannot protect the principle of perpetual motion unless you have proved that it can be done!

Example 3.2.1

Aero by Nestlé ®

Nestlé would have found it difficult to protect the idea of putting bubbles into chocolate *per se*. They can, however, patent the machinery and process that put the bubbles into chocolate, copyright the wrapper, trade mark the name and have design rights in the shape of the bar.

Key point

World trade in knowledge (IP) for the top economies is now considered to be equal to that for goods and services.

Activity 3.2.1

Imagine that the watch you use is a product that is new to the world. Write down as many ways as you think possible to protect it using the intellectual property rights available.

Activity 3.2.2

Form a design group and design a new type of folding bicycle that will fit into the space usually provided in the boot of a car for a spare wheel. Try to design in features that will allow you to protect the bicycle using intellectual property rights.

Patents

Description

Patents are governed by the Copyright, Design and Patents Act (CDPA) 1988 which specifies that a legal monopoly is granted to an inventor for a limited period in return for payment of a small fee and for making the invention known to the public. A patent can provide protection for a technical or functional innovation. This covers the composition, construction, manufacture of a substance article or apparatus of a product or process. Excluded from patenting are artistic creations, mathematical methods, software, schemes, rules, discoveries of nature, biology, methods of treatment, surgery or therapy for humans or animals. A number of exclusions, particularly in software, business methods and biotechnology, are, however, beginning to be accepted.

Benefits in owning a patent are not just in protecting the invention. It includes credibility, marketing tools and expansion through cross-licensing and technology transfer. It is also seen as helping to attract investment capital, particularly to internet companies which otherwise often lack tangible assets.

Key point

In 1998 it is estimated that there were about 130 000 new inventions leading to two million new patent applications worldwide. This translated into 640 000 registrations nearly triple the 220 000 registrations made in the 1960s. Sixty per cent of these were from Japan, 25% from the US and 11% from Europe.

Example 3.2.2

Applications for computer related patents are growing at the rate of 20%. Applications are more likely to be successful if they relate to 'inventive software', i.e. if the program determines the sequence of steps in a process which is to be protected or if the software is embedded within a technical application.

The Patent Office approved 301 patents for internet business methods in 1999 compared to just 39 in 1998. The US Supreme Court (1999) upheld a ruling that an idea, in the form of a business method, was protectable by patent provided that there was a useful result.

This is leading to a growing number of court cases as companies battle over rights in the burgeoning IT world. For example, BT has announced that it holds the patent for hyperlink technology, the basis for surfing and linking the Internet. Internet companies are responding by claiming that BT's patent is invalid by saying that Apple's hypercard system preceded BT by four years and Project Xanadu (a web forerunner) beat BT by 20 years.

In the world of pharmacy and biotechnology, gene sequences have traditionally been considered non patentable. However sequences are being patented provided they fulfil the normal patent criteria, can be isolated, characterised, cloned and their function is known. Arguments over the morality and inventiveness of gene discoveries abound, leading to uncertainty in the law. This is again leading to battles between companies.

Requirements

In order to be patentable an invention must be new, inventive and valuable. To be valuable it must be useful in an industrial or agricultural context. There is no lower limit on the level of value.

Example 3.2.3

Patents have been granted for inventions such as three-legged tights, cheese filtered cigarettes, kissing shields, beer barrel hats and dogs' ear protectors.

Key point

In law, disclosure is often considered a grey area and there is evidence of a movement by courts towards some disclosure being allowed provided that there is no chance of analysis. A six-month allowance is also made for disclosure by theft or release at a trade conference.

To be new the idea must not have been disclosed in any way, whether written, oral or visually. Disclosure includes:

- Simply being made available to the public (it doesn't even actually have to be seen by anybody).
- Using e-mail which is not encrypted.
- Trials to a third party or for profit.
- Being made available to too many people or given to particular key parties even if marked 'confidential'. For example, a report marked 'private and confidential' was disclosed even though it was circulated to just 10 members of a 350 strong group.

Example 3.2.4

Confidential information

Any information is capable of protection if it is imparted in confidence and if it is equitable. Confidential information is therefore information that must be of quality, have been imparted obliging confidence and must have a use for the owner. Novelty is not a prerequisite.

Confidentiality agreement clauses should identify background knowledge, third party acquisition and duration. Courts use a springboard principle (a look at future events) in determining the validity of the duration period.

Agreements do not guarantee against accidental disclosure

Key point

A 'selection' invention is the name given to something that is not new, but is applied in a novel or previously unconsidered way.

Courts use a person who is 'skilled in the art' to determine inventiveness. They are deemed to be experts who are intelligent but unimaginative. If they can generate the same solution given the same problem then the invention is considered obvious.

There is no definition of what comprises an invention but to be considered inventive the idea must not be 'common general knowledge, not be obvious and beyond what is currently "state of the art". The simpler the invention is the harder it is to prove inventiveness. It is also hindered by hindsight. An inventive step can also be destroyed by a combination of documents which pieces together a solution (mosaicing). Success can be improved if the invention goes against generally accepted views or if the idea is commercially successful.

Application and procedure

Ownership

A patent will usually belong to the inventor, except where ownership has been contracted otherwise, for example with consultants. It also includes employment contracts and a company has ownership rights if an invention is made in the course of specified (express) duties or normal duties (expressed or implied). Innovation achieved in an employee's own time is a grey area if it relates to their normal job.

Application

The patent must be applied for. It is a generally recommended rule of thumb not to apply for a patent if work to complete an invention will take more than one year. Time, however, is strictly of the essence because in the UK the first to file a patent is the owner.

Example 3.2.5

Alexander Graham Bell filed the patent for telephones just two hours before Elisha Gray.

Key point

Note that claims can be amended (but not appended) up to 12 months after an application. It is also possible to withdraw and reapply 12 months later. Lapsed applications are still confidential and allow details of inventors and owners to be established and corrected later.

It is hence better to patent little and often rather than waiting for 'the big one'. It is also common practice to use many patents for one device. Kodak's disc camera, for example, has over 100 patents. Some companies such as Canon create a citadel of patents, using many patents surrounding the concept but without actually patenting or revealing the key technology. The procedure for patent applications is straightforward but can be lengthy:

Unofficial search

Undertaking a patent search before applying prevents wasting time and money. You can search for free but there are 36 million patents

Key point

Limited searches can be achieved using the internet. Try:

www.patent.gov.uk
www.patents.ibm.com

and only recent applications are electronically stored. Their organization has also undergone several major revisions and in order to carry out a complete search over the whole period of time, three search classifications with backwards and forwards concordances have to be used. A professional search service is therefore advisable. Patent agents, for example, charge around £100 for overviews and £1000 for summaries (including translated texts). The Patent Office also offers a Patent On Line Search Service.

Filing

Filing an application involves completing Patents Form 1/77 and supplying the following information:

- A bibliography comprising:
  A number
  B name
  C abstract.
  Any named inventor on the patent must have contributed to the intellectual effort.
- A description and title.
  Entries in the following sections:

5–9	technical field
10–31	background
35–50	essential technical features
54–90	list of drawings
91 +	particular description

Careful wording is required. For example:

- 'Display means' for a monitor is far more encompassing than 'screen'.
- 'X for use as' covers X for a wider range of uses than 'the use of X as'.
- Use 'between the range' and 'preferably' when referring to numbers.
- Use 'vertically, or nearly so' in preference to saying just 'vertically'.
- Claims entered in the following sections:

1 and 21	the main claims
2–20 and 22–28	appendants (supplementary built on the main),
29	commercial reasons
31	omnibus (final catch all)

- A search report if applicable.

Careful phrasing is also required. For example:

- It has to be detailed enough so that another person of equal knowledge to the inventor could recreate the invention. 'Real' patents are descriptive and intend to stop use by others. However, whilst these are easy to define in court they are more technically easy to infringe.
- 'Defensive' patents are vague and aim to have the widest scope for an invention so that it can be exploited to the maximum. An

example would be Intel patenting 'microchips', or Henry Ford patenting 'the car'. However, whilst these are harder to infringe they are harder to enforce.

Courts are moving towards a purposive rather than exacting definition of the text, meaning that text is viewed as a signpost rather than a fencepost (*Catnic* case). Software is available to help with composing but professional help is always advised. The charge for wording a simple invention might typically cost around £800.

Example 3.2.6

James Watt was advised to write a vague defensive patent in his application for a patent for the separate condenser principle (which formed the key element of his steam engine innovation) in an attempt to broaden its applications. Watt's original intention of a purposive patent would have subsequently been better in warding off infringers of which there were to be many. Watt particularly hated the Hornblower family who invented the compound engine, referring to them as the 'horned imps of Satan'. The government of the time was not sympathetic to Watt, viewing any monopolistic control of the fledgling power industry as unhealthy for British interests.

Activity 3.2.3

Try drafting a 150 word application for your own product (or for a rotary clothes drier).
Examine somebody else's application and see if you can create a similar invention without infringing the patent.

Filing a UK applications is free and the day of filing establishes a priority date which is a key reference point for UK patents. The information is not released generally but the application is reported in the *Official Journal* (Patents) OJ(P), usually around three months after filing.

An official search

An official search must be started within 12 months of the priority date. This is done by the Patent Office and costs around £200.

Publication

The results of the search are usually published 18–21 months after filing and are reported as the 'A' specification.

Examination

Examination for the Patent/Office to consider whether the invention usually is patentable must be done within six months of publication. This is now 39 months after the filing date.

Grant

The final approval or grant of the patent is known as the 'B' specification. The normal duration for this stage is 51 months after filing. There is also a 'fast track' option for combined filing and searching at the same time. This approach helps to reduce the additional time needed for searching and hence quickens the entire process. Subject to renewal actions every four years, the patent can be held up to a maximum of 20 years.

Activity 3.2.4

Find and comment on the applicability of the international patent application PCT 94/20041 on vacuum assisted wound treatment.

Infringement

Patent infringement is a civil matter. The infringement can be direct or indirect but must be for commercial reasons. You should communicate the problem and the remedy to the infringer before taking legal action. A patent and number stamped on your product improves the argument that infringement has occurred intentionally. With cases of joint ownership, a single owner can sue for infringement. The joint owner can then be sided with the defendants but does not incur those costs. Damages are restricted to an account of profits or a loss of sales.

Example 3.2.7

Pure research and experimentation is not necessarily an infringement but could be if the results have a commercial end (even if the research is early and commercialization is a long way off). Similarly the mere repair of a patented article would not normally amount to infringement of a patent unless it was construed as manufacture.

Key point

Patent agents can now act as litigators in the UK High Court in intellectual property rights cases, following the designation of the Chartered Institute of Patent Agents (CIPA) as an authorized body. This gives companies and individuals a simpler, more open and cheaper way of pursuing intellectual property right claims. Previously patent agents could only act in the county courts.

International

A UK patent can be infringed in other countries unless protection is also sought overseas. Under the Paris Convention of 1886, the priority filing date established in the UK allows a 12 month period between filing in other countries (with the exception of Taiwan). A UK patent must be sought before an overseas patent can be sought and, if necessary, permission must be sought from the Ministry of Defence and the

Key point

A 'first to invent' system requires additional documentation and evidence to corroborate the date that the invention was conceived. It must be shown that diligence was exercised throughout the innovation process and that it was continual and uninterrupted.

Notebooks should therefore be kept that include facts and experimental details, days off and why, be legibly numbered, signed off, have no blank spaces, have additional evidence glued in and excluding opinions, slang or jargon.

Key points

The World Intellectual Property Organization (WIPO) is a specialized United Nations agency with 169 member states whose purpose is the protection of intellectual property rights throughout the world.

The 'utility' model is a European proposal for protecting innovation specifically aimed at SMEs. It will be similar to a patent and aimed at technical inventions, products and processes and excludes discoveries, theories, mathematics, aesthetic creations, schemes, information, biological materials, chemical or pharmaceutical substances/processes but will be industry specific, will require a lower level of inventiveness (not very obvious to someone skilled in the art), simpler description, claims, drawings, abstract, filing fee, have one invention only embodying the underlying concept and provide protection for six years. Combined with a one country does all method the utility model will be easier, quicker and have 50% lower costs than patenting and without the necessity of searching and novelty proof.

Department of Trade and Industry. Patenting overseas is time consuming and expensive. Foreign patents must also be reproduced in the language of the host country (BASF, 1999).

The European Patent Organization (EPO) does co-ordinate a 'European patent' (EPC) to which 19 states subscribe and which from 2002 will include East European countries. The European Commission has also adopted a proposal for the introduction of a single community patent, which should be available by the end of 2001. The World Intellectual Property Organization also administers a Patent Cooperation Treaty (PCT), which covers many of the industrial states but costs upwards of £20 000.

One problem with establishing a common 'worldwide' patent that would reduce the cost is that different countries have different patent rules. The US and Japan operate a first to invent system rather than a first to patent system. Filing can take place up to 12 (six) months after the disclosure of an innovation. Japan also has a less inventive 'utility' measure midway between a patent and a design. The US and Japan also have an 'inventor owns' system rather than a company owns system and assignment is then given to the company by deed or nominal payment.

Design rights

Design right (unregistered)

Design right is concerned with protecting the shape and form of an invention, as detailed in Part III of the CDPA. Work must be original, more than copyright but less than an inventive step and includes microchips but not 2D designs or surface decorations (such as wallpaper).

Design right is granted automatically to the designer or originator, again unless contracted otherwise such as in employment. In a commissioned work it usually belongs to the person who commissioned the work rather than the designer. It is not a monopoly as the same design can be repeated elsewhere if an originator comes up with the same creation independently. It does prevent outright copying or copying by reverse engineering for a period of up to 15 years from the time of the design or 10 years from the first sale. The design should hence be publicized as soon, as loudly and as widely as possible to try and diminish an argument that a similar design was produced independently.

Registered design

An invention that has an aesthetic element as well as shape and form can be registered, as detailed in the Design Act 1949. The act defines aesthetics as features that 'appeal to the eye' and includes patterns and 2D designs. 'Must fit' and 'must match' articles such as functional car body panels are however specifically excluded (*BLMC* v. *Armstrong* (1986)).

Registration is through the Patents Office. An application is around £260 plus any agent fees with subsequent five yearly renewals up to 25 years. In infringement cases, proof is by a consumer's 'imperfect recollection'. In other words would a user confuse a registered design with an infringing design.

Between 8000 and 10 000 designs have been consistently registered throughout the 1990s. Patterns and 2D designs are less frequently protected because they tend to go out of fashion fairly quickly, but high levels of filings are evident in electronic apparatus (walkmans, pagers, radios), containers and measuring instruments, with games, toys and packaging showing the biggest area of growth.

Protection is automatic in some countries, but protection in the UK does not automatically extend abroad as the UK does not subscribe to the international treaties that exist. Applications must therefore be sought in individual countries (at roughly double the UK cost per country). Registering a UK design right does though establish a six month priority date, which can be carried forward into other countries in much the same way as a patent application.

The European Directive 98/71/EC proposes a Community-wide right, similar to the UK's design right but with a shorter term of three years. A Community-registered design will be a single right, which will be effective throughout Europe. Nominated national courts in each member state (for example, the High Court in the UK) have jurisdiction as Community trademark or designs courts in order to allow the enforcement of such rights. The Directive requires member states to implement the new legal framework into their law by October 2001. The Directive agrees to disagree on certain matters, most crucially the 'right to repair', leaving this up to national law for the time being.

Copyright

Description

Copyright is the protection applied to the expression of an idea rather than the idea itself, as described in the CDPA. Originally this protection established the rights of ownership of literary work but now includes artistic works such as sculptures and period furniture, dramatic works, musical works and arts (LDMA) as well as graphics, typography and computer programs. Generally speaking if the expression of an idea is worth copying it is worth protecting and is therefore copyrightable. It requires intellectual effort through skill and labour and not necessarily novelty or innovation. Facts, ideas and names, for example, are not copyrightable but compilations of facts are.

Example 3.2.8

Drawings are considered artistic works and are therefore covered by copyright but not the subject of the drawing unless that is itself artistic. Similarly a film may be copyrighted but there should be skill and care in taking the photograph. Photographing a three-dimensional object for an exhibition involving positioning an object, the angle at which the shot was taken, lighting, focus and choosing the particular object to exhibit, for example, are matters of judgement and skill that could be copyright protected (*Antiquesportfolio.com plc* v. *Rodney Fitch & Co. Ltd*). If the subject of the photograph is human then the photographer could be infringing that person's copyright in their face and their privacy rights particularly if the photograph is commercially exploited.

Application

The originator of a work is the owner of the copyright although employers have an automatic right of assignment if the work is produced as part of the employee's normal duties. Ownership of commissioned, consultative or independent consultant work is less clear but the originator is usually the owner unless agreed otherwise.

Key point

'© Copyright (date) by (author/owner)' is the usually preferred form of notice.

Copyright is granted automatically. It is not necessary to use the symbol ©, which has never officially been given legal status in the UK, but courts may find it helpful in establishing dates of authorship. Unopened, stored documents may also help in this respect. The ©, however, may be needed overseas and exports should always include a copyright warning.

Protection commences when the work is first recorded and lasts 70 years after death. The period of protection is reduced if items (typically more than 50) are sold as they are then to be considered industrially exploited. Protection is also less (50 years) for computer generated works, sound recordings (50 years) and typography (25 years).

Key point

For education purposes, fair dealing allows reporting, research, abstracts, review, a 400 word text, multiple extract text totalling below 800 words, and copying for current events (excluding photographs and filming). Individual establishments may also have their own arrangements.

Copyright is a 'negative right' in that it prevents copying, issuing, performing or adapting the work. Independent creation is not infringement. There are also certain automatic rental and public lending rights. 'Fair dealing' allows for some copying without owner's permission.

Infringement, however, can be primary (for example, by direct copying) or secondary (for example, by using material that has been copied). Therefore assume that all works are copyrighted to someone, somewhere unless the work specifically states that it is for public use domain. Even basing work on someone else's work requires permission, e.g. a story using somebody else's characters (except for parody).

Current legislation imposes the burden of proof on copyright owners. At present, proceedings can fail because a well-advised and publicly funded defendant is able to drag out a good case by pleading not guilty and requiring the copyright owner to prove all aspects of copyright (subsistence of copyright, authorship, chain of title, assignments). This considerably increases the length and the cost of the trial, and often frustrates meritorious cases.

Key point

A lax organizational attitude to copyright infringement makes the managing director liable. Make sure, for example, that photocopiers have adjacent warning notices.

International

Unlike the UK, copyright in the US is not automatic and must be registered. Also, where work in the UK is copyrightable if it is formed with effort (i.e. skill or labour), in the US there must be some creative effort. The US also has a very low threshold of infringement which is deemed a criminal matter.

Generally, however, similar rules for copyright exist as they do for design. These are covered by the Bern Convention, Universal Copyright

Convention, Rome Convention and various EC Directives (including the Protection of Computer Programs, Rental and Lending Rights, Television without Frontiers, Cable and Satellite, Terms of Protection, Protection of Databases).

WIPO attempts to harmonize global copyrights in 1997 led to the subsequent EU Copyright Directive and related Acts (Rights in the Information Society) which may ultimately give rise to changes in UK law. These might include:

- Broader reproduction rights (including temporary copying).
- Rights to communicate works to the public.
- Protection against technical circumvention.
- Copying some text commercially onto paper for fair compensation.
- Copying for private non-commercial in return for fair compensation.
- Copying in libraries and universities including use for teaching or science which is not commercial and which must be acknowledged.

Related rights

Moral rights

Moral rights are related to copyright but are concerned with the right to be recognized as the originator of the work rather than the owner of the work. Moral rights do not apply to computer programs, computer generated works, typefaces, news and current events work, collective reference works. A moral right is an inalienable right (which means that it cannot be transferred unless it is in a will) but it is a right which must be asserted. It covers the following for which infringers can be sued:

- Paternity or the right to be identified and credited as the originator of the work.
- Integrity and the right not to have work subjected to derogatory treatment. Work protected by moral rights should not be intentionally destroyed, damaged, altered, retouched, modified or changed in any way whatsoever without the written permission of the originator.
- The right not to have false or joint work incorrectly attributed to you or by others.

Performing rights

Performing such as street theatre or singing is normally considered as the delivery of a performance which is viewed as the supply of a service rather than a product. Current legislation implies that the free circulation of such services is freely applicable. However, more recently courts have determined that Article 59 of the Copyright Act applies which would exercise the same rights as exists for goods. The re-enactment of a performance may hence be an infringement if the performer is offering the performance in your guise. Rights in performance, however, represent protection against unauthorized recording, broadcasting, copying, issuing, showing of a live or pre-recorded show including the Web. Reproduction, distribution and rental rights are transferable property rights with protection that lasts for 50 years.

Digital copyright

Electronic information receives the same copyright protection as written or printed matter. Unless covered by contract or fair dealing, infringement of copyright includes, scanning, rekeying, downloading or sending copyright material on a LAN or fax. Copyright was, however, created for a paper world and the facility to gather, clone, manipulate and distribute information without detection makes its application in the modern electronic world problematic. The problem is likely to increase as processing, compression and bandwidth technologies improve and as knowledge and 'virtual companies' become more important. Testing the law in the cyber world is also often new and untried.

Database right

Databases qualify for copyright protection under the 'compilation' rule of the CDPA. Hence copyright exists where authors have selected and arranged database contents using intellectual means. This would, for example, exclude a telephone directory. 'Database rights' also exist where a substantial investment has been made in obtaining, verifying, or presenting the contents of a database. It is an automatic right that prevents unauthorized use or extraction from the database. Protection lasts for 15 years but major revisions will continue to generate 15 year extensions.

The Web

Net published work is copyrighted with the exception of URLs and e-mail addresses which are facts and therefore not applicable. Reading a web page automatically involves making a copy of the page and storing it in a computer's RAM which might in itself be viewed as a copying infringement and make both user and internet service provider (ISP) liable. It is, however, generally accepted that there is an implied licence to do this provided the provider has the right to offer the page initially.

In the UK, ISPs are considered publishers in common law because they have the power to store and delete information from authors (*Godfrey* v. *Demon*) and they are therefore liable to claims against infringement. Preventative steps should include limiting the number of newsgroups, restricting links and access to personal sites, publishing a contact's name, observing information regularly and removing contentious information immediately. Beware also of the Obscene Publications Act 1959 which covers simply making such material available. An ISP would be responsible if it failed to remove the material once its obscene nature was determined.

Key point

The global nature of the Web, makes it easy to infringe a variety of laws in other countries including copyright and obscenity.

The Shetland Times case has, however, raised the issue of copyright infringement of one site linking to another. An interactive site can be considered a web, with this implied licence for all to use, but a non-interactive site might be considered a cable television service which would put the onus of infringement on the viewer rather than the service provider.

This perspective stands more akin to US rulings which consider ISP as providers of equipment similar to telephone service providers. The US Online Copyright Infringement Liability Limitation Act gives greater power to copyright owners over digital reproduction but limits the liabilities of ISPs at the expense of users. The ISP must not be aware of infringement, should not receive a direct benefit from the infringing article and acts rapidly when made aware. The Digital Millennium Copyright Act (World Trade Organization) endorses the US stance of user rather than provider liability.

Key point

It is estimated that of the 615 million new business software applications installed worldwide during 1998, 231 million or 38% were pirated.

Key points

Be particularly aware of accidental infringement. Just because work may be offered free, on the Web, for example, does not mean it is not copyright protected and indeed the offer may not come from the originator. Making use of such information is a breach of copyright if the owner loses income or the breacher gains income. Even forwarding material sent to you may be in breach of copyright and it may be advisable to destroy electronic copy once used. Passing on information as to where to find downloadable data is also illegal.

The arguments for and against greater control are wide ranging and include the right to reward for innovating against the right to the freedom of information.

Technological protection may anyhow provide better solutions than the law. 'Digital guardians' include know-how, encryption, electronic locks, 'watermarked' pictures and passwords. Companies are now producing software that can protect either the printing or downloading of a part or a whole document, including the source code to protect the web design itself.

Example 3.2.9

As an example of the dangers to users, a US Federal District Court in Utah has issued a temporary injunction against a Salt Lake City non-profit organization that admittedly publishes 'critical research' on the Church of Jesus Christ of Latter-day Saints, also known as the Mormon Church. According to reports, the ruling was intended to stop the Utah Lighthouse Ministry from posting e-mail messages on its website telling readers where to find online copies of a book published by the Mormon Church. In essence, the court upheld the contention of the Mormon Church that the Lighthouse Ministry was guilty of 'contributory infringement' of the Mormon Church's copyrighted *Handbook of Instruction*. The implication is that simply providing an address or links could be a copyright infringement.

Software

Rules for software are covered by the Copyright (Computer Programs) Regulations 1992 appended to the Copyright Act 1988. Independent creation is allowed and software companies separate departments which analyse competitive and market information from departments which write software, to ensure that independent creation is seen to be done.

The United States Department of Justice, the Federal Bureau of Investigation and the United States Customs Service have established a law enforcement initiative aimed at combating the growing challenge of piracy and counterfeiting both domestically and internationally. The Intellectual Property Rights Initiative will focus on people who steal trade secrets, counterfeit chips and software, and pirate programs over the Internet. The US No Electronic Theft (NET) Act was passed in 1997 specifically to close a loophole in piracy laws, which previously meant that it was illegal only to sell copyrighted software and other electronic media.

Electronic mail

Technically even forwarding an e-mail sent to you requires the originator's permission. E-mail messages sent to a public list are regarded as having an implied licence for publication, which will allow others to keep a copy, quote and forward the piece. Beware of distortion of the message and false attribution which may infringe moral rights, however.

Case studies

(1) The United Kingdom recently upheld the decision to grant a software company only limited injunctive relief against suppliers, who were found to be using counterfeit software and dismissed the appeal. The court found that there was no invariable rule or principle as to the relief to be granted in intellectual property cases. The judge had found that the infringements were relatively

minor and unintended. This fully entitled the judge therefore to restrict the relief, tailoring it to match the wrong that had been committed. The case in point dealt with software, supplied to the company acting on behalf of inquiry agents employed by Microsoft, purporting to be Microsoft's Windows 95 software. They were in fact counterfeit and had been purchased by the company from an existing customer with whom it had longstanding business dealings. The company denied knowing that the supplies were counterfeit. The judge found that the company had infringed Microsoft's intellectual property rights, but only in a relatively minor and unintended way and in circumstances in which there was no evidence of any intention of repeating the infringements.

(2) The US Department of Justice won its first MP3 case against a University of Oregon student for illegally distributing copyright material (MP3, or Mpeg 3, enables digital audio material to be compressed into compact file sizes and transmitted on the Internet or burned onto recordable CDs). A United States Federal Court has also heard a plea of guilty from a counterfeiter and software pirate, who faces charges of distributing more than US$13 million worth of counterfeit compact discs and computer software programs. Using state of the art copying equipment the defendant operated two warehouses, where CDs were copied by the thousands, according to prosecutors. The counterfeit discs included Microsoft Office 97 programs and an assortment of popular musical CDs. The defendant admits to distributing more than 332 000 counterfeit musical CDs and more than 50 000 copies of Microsoft software programs. The musical CDs which would have had a legitimate retail value of about US$16 each, sold for about US$10 on the black market, while the Microsoft programs, costing about US$400 at legitimate retail outlets, had a street value of about US$200 according to prosecutors.

Trade marks

Description

A trade mark can be any representation that differentiates goods and services, goodwill and reputation. The common law of 'passing off' affords some protection against the copying of these trade marks but you must prove your historic ownership, an intention to confuse/ deceive, a tangible loss and demonstrate rigorous active protection. Being proactive in trade mark protection helps, such as indicating your recognition of your trade marks by the use of the symbol 'TM'.

It is, however, possible to afford stronger protection by registering the trade mark. Registration used to be limited to words and logos but following the Trade Mark Act 1994 the scope has widened to include other distinctive features such as three-dimensional shapes, three letter words, acronyms, uncommon words, foreign surnames, small places, foreign places, slogans, a descriptive word with a logo, sounds, colour and even smell. Registration is still excluded for common words, surnames or phonetical similarities which require proof of distinctiveness and where others have 'honest concurrent use, shapes by nature or functions which add value'. Liability for breach is strict but the onus of proof of distinctiveness is with the originator.

Key points

It may be difficult to register your trade mark if it has become a generic name for products of that type throughout industry.

Anybody attaching metatags (hidden lines of code) to their web pages to attract attention, and providing unauthorized linking of your product with other sites/products, may be infringing trade marks. In a move to exercise control over who has the right to offer links to its site, Universal Pictures has threatened a website (www.movie-list.com) with legal action if it did not remove links from its website to movie trailers at Universal's own site (www.universalpictures.com). This action places a question mark over the whole legality of hyperlinking.

Example 3.2.10

Coca Cola has not only registered its name as a trade mark but also the shape of its bottles. Also registered are the shape of Toblerone bars, the colours green for BP petrol forecourts and turquoise for Heinz beans cans, the Direct Line insurance jingle, Mr Kipling's 'exceedingly good cakes' slogan and the smells of Chanel perfume and beer scented darts.

Application

It is generally advisable to conduct a search before applying for a trade mark to assess whether the mark is likely to encounter any difficulties as a result of its use and registration. A self-search using proprietary CDs might cost £15 and the Patent Office may charge £100. The mark must then be registered within one or more of 42 classes of goods or services available. The registration is renewable every 10 years indefinitely costing around £200. Once granted, the ® symbol can be used to show that the mark is registered.

Key point

Trade mark applications are rising at a rate of 5% over previous years.

Example 3.2.11

An educational CD may have two classes:

- Class 9 (CD ROMS and computer software goods)
- Class 41 (educational and training services)

Internationally, the recommended approach is to search in the countries that are of interest. A typical search by patent agents may cost between £250 and £600 per country but it is also possible to do a 'World Wide Identical Screening Search' (WISS). WISS is limited to using identical and phonetically identical trade marks published since January 1976 but at £550 is a cheap and quick guide. Application costs vary per country but the average is around £700 to £800 per country per class.

Key point

An objection to the registrability of the trade mark because it is phonetically identical to a common surname is less likely to arise using the CTM rather than a one country only route.

The 1994 Trade Mark Act harmonized and simplified the rules within the EU enabling a commonly available Community Trade Mark (CTM). The cost of filing the application is approximately £700 but this covers up to three classes. When the mark is accepted there is an additional registration fee also equal to around £700.

Under the Madrid Protocol, an International Registration (IR) can also cover a number of associated countries worldwide. Each application is examined separately by the relevant national trade mark authorities so there is a risk that an objection concerning a common surname may be raised. The cost of filing an IR is about £800 for up to three classes with an additional cost payable for each designated country of approximately £100. The additional cost of a prior UK application should also be taken into consideration. The application can be filed centrally and is worthwhile if protection is sought in approximately three or more of the associated countries.

Domain names

Key point

InterNIC is the American-based body which allocates domains. In the UK it is NOMINET. There is a great deal of activity within the domain naming system. In July 1999 there were over 6.8 million registered domains for the top five top level domains (com, net, org, edu, int).

The registration of trade marks and internet domain names is not linked. Domain names are allocated on a first come first served basis and ownership of a registered trade mark does not automatically entitle you to ownership of the domain site. The disparity has resulted in a great deal of confusion and litigation around this issue, particularly in relation to people owning domain names that trade mark owners would like ('cybersquatters'). Where there is dishonest intention (such as 'passing off') then the trade mark owner is more likely to be successful in claiming against the domain name owner.

A small company with the same name as a large company with a similar name can legitimately register the domain name at the expense of a big company (*Prince* v. *Prince Sportswear*).

The World Wrestling Federation Entertainment claimed that the owner of the domain worldwrestlingfederation.com registered the newly available address and then attempted to sell it to the wrestling promoters for a hefty profit. The name was ordered to be handed over to the WWF.

WIPO has also upheld the claim of US actress Julia Roberts against the domain name owner of juliaroberts.com. The tribunal ruled that the domain name owner had 'no rights or legitimate interest in the domain name' and had registered it in bad faith. The domain name has also now to be transferred to the actress.

In *Marks and Spencers* v. *One in a Million*, One in a Million argued that they were not passing off because they were not intending to deceive, they merely intended to sell the domain name to M & S. The argument was overruled and M & S were given the domain name site and One in a Million (a collection of students) were handed a £65 000 legal bill.

Key point

Registering 'keywords' with gateway sites may also infringe a trade mark.

However, failure to prove dishonest intention (particularly if the domain name owner has not attempted to sell the name) may be difficult. The owner of the domain name may be prevented from using it if it infringes a trade mark but will not necessarily result in the trade mark owner gaining ownership of the domain name (*Shetland Times*).

Problems 3.2.1

(1) In *Evans & Sons Ltd* v. *Spritebrand Ltd* (1985), a company was sued for breach of copyright. Could a director be held personally responsible as well?

(2) Comment on the application of copyright, design right and registered design right protection for computer generated kitchen scales created for the commercial market.

(3) In January 1995, Glaxo announced a bid for the company Wellcome with the intention of forming a combined business called Glaxo Wellcome plc. A day after the announcement, X registered the company name of Glaxowellcome Ltd and offered the name to Glaxo for £100 000. Glaxo took X to court to insist that X did not have the rights to register the name in the first instance. Was Glaxo successful?

(4) If you receive an e-mail from a friend, explain why you might need to be careful if you decide to forward it to someone else.

Case study

Affymatrix is a biotech company that invented the use of photolithography to place biomaterials onto a small chip, a process similar to integrated circuit production. It is the tool that enables others to explore genes. Affymatrix has adopted the following intellectual property strategy:

- Patent protect everything using broad 'defensive' patents that seek to protect all aspects and possibilities for the invention, in this case all dense placement of biomaterials in small areas.
- Offer licences to anyone who asks but avoid arguments over royalties by selling expensive bundled packages, that include knowledge, computers and machinery, and attempting to make the format world standard.
- Fight all infringers.

The problems it faces are being reliant on one key patent which may be liable to attack. The stance also alienates higher education and cash short companies who will find other solutions. Affymatrix faces counter alliances from 3M, Motorola, PE (owner of Celera), Hewlett Packard and Hitachi ready to break into the market. Litigation also ensues with antagonized competition. In this case by compeitors backing the work of Edwin Southern at Oxford University who patented an early gene sequencing technique. Suing and counter suing is prevalent with Incyte and Hyseq.

3.3 Contracts

Agreements are made between suppliers and buyers of goods but can be made between people or organizations for a variety of reasons. Agreements form a regular and important part of business and this section provides an overview of the formal creation of an agreement.

Requirements

The definition of contract from the *Oxford English Dictionary* is that in its simplest form it is 'a mutual agreement between two or more parties that something shall be done or foreborne by one or both'. It is also commonly interpreted as 'an agreement enforceable by law'. A contract is hence a step beyond a promise or an understanding that forms a legally binding agreement. It can be implied, by actions or understandings, or it can be expressed in written terms. Certain essential elements are, however, necessary to form an enforceable contract and these are intention, consideration and agreement.

Intention

Both parties must have an intention to create a legally binding contractual relationship.

The 'offer' is an expression of willingness by one party (usually the seller) to enter into a legally binding relationship. The offer must be communicated to the offeree by the offerer or his agent. It may be expressed or implied and the test of whether a statement is an offer is objective. Its legal validity depends on whether the offerer would reasonably have been understood as making an offer and not whether he thought he was making an offer. The following are normally held to be offers:

Key point

Non compliance with a contracted offer can result in Court action but may not nullify the contract.

- Quotations
- Estimates
- Orders

Distance selling includes offers by e-commerce, mail order and telephone sales and legislation includes the Mail Order Transactions (Information) Order 1976. The legislation excludes one-off transactions. The minimum information required for electronic transactions must include the identity of the sending party. E-mail is now considered to have the same status as letters so that all e-mails must bear the company's place of registration, registration number and registered address (section 351 of the Companies Act 1985). The seller must also provide information on the main characteristics of goods, special conditions, price, payment details and details on the right to withdraw information. Hard copy details must include a seven day right to retract, without reason or penalty starting from the date of goods received. There is also a requirement to identify further junk mail (spam).

Key point

Cyber contracts are contentious because it is not precisely clear where the contract is made which makes it difficult to apply the normal legal framework.

An offer may be made in a conditional form which is divided into two groups. Conditions precedent (suspensive conditions) and conditions subsequent (resolutive conditions). Clauses inserted into offers which are commonly encountered in commercial transactions include the phrase 'subject to contract' (i.e. subject to the making of a formal, written contract) and 'effectiveness clauses', covering matters such as the obtaining of licences, finance and so forth necessary for the performance of the contract. Clauses of this type usually create a condition precedent, but in some cases a party may be under a contractual liability in respect of the condition, for example there may be a duty to obtain an import licence.

An offer may lapse for a number of reasons:

- Because the time specified for its validity expires.
- Because the offer is communicated as being terminated. The offerer can also withdraw the offer at any time, even if he has promised to

keep the offer open, unless he has contracted (for good consideration) to keep it open.
- Because of a counter-offer which kills off the original offer.
- Because the offer is rejected.

An offer is usually contrasted with an invitation to treat which is an invitation to the other party for them to make an offer. The following are usually held to be invitations to treat:

- Advertisements
- Circulars
- Price lists
- Displays of goods in a shop
- Tenders. When a contract is made by tender, the invitation to tender is construed as an invitation to treat, the tenders are construed as offers and it is then up to the invitor to decide which tender (if any) he will accept. Particular care should be exercised in dealing with 'acceptances' in the form of letters of intent. These almost certainly do not create a binding contract, except in so far as they induce the tenderer to do specific things in response to the terms of the letters of intent.

Consideration

A contract must be supported by consideration. This means money to be paid or obligations undertaken in return for the goods or services to be provided. Any party wishing to sue upon the contract must prove that they put something into the bargain struck between the parties. The obligations and contents of the contract are called its terms and consist of the consideration and other promises which the parties have made to one another in respect of their duties under the contract.

Example 3.3.1

Common areas covered by terms include:

the people or organizations involved (these must be legal entities and are called the parties)
introduction (or preamble)
definitions
the work to be undertaken
quality of the work
product variations
defects
intellectual property
price
payment
terms of payment
time for performance
schedules
delays and conditions for extension
storage
inspection, testing and rejection

delivery
acceptance
passing of ownership
passing of risk
liability
insurance
warranties
indemnities
performance defects
contractor's default
bankruptcy
arbitration
exemption
termination
force majeure
exit clauses
contract variations
arbitration
regulatory law

A distinction is drawn between express terms (the things which the parties explicitly agreed) and implied terms (whose existence a court would infer from the circumstances surrounding the making of the contract). Terms may then be discovered from any documents which the parties used to record their agreement and/or any oral communications between them forming their agreement or part of it.

In any disputes, a court will attempt to decide upon the basis of the written and oral evidence what the parties intended to include in the contract. The court will usually imply only such terms as are necessary to give the contract 'business efficacy'. It is particularly influenced by the relative importance of what was said or written in relation to the subject matter of the contract. However, courts do not permit oral evidence to be given which would vary, contradict or add to the written terms if the parties have reduced the whole of their agreement to writing (the parol evidence rule).

Careful wording is required when drafting sentences, otherwise problems can also be encountered in interpreting terms even when they are expressed in writing. Unless undertaken by a qualified legal expert, contracts should be drafted in simple language, avoiding the use of Latin, jargon or slang, and progress in a logical fashion using clear punctuation.

Activity 3.3.1

Consider the construction problems in these contract terms and consider how they might be redrafted.

(a) Acceptance of these terms shall be by signature of a director of the company. The usual conditions of acceptance shall apply.
(b) The price of the goods is £2500 and will be paid on standard hire purchase terms.

(c) We shall meet 12 months after the first product is sold to decide on a suitable royalty.
(d) Disputes between us arising out of this contract shall be referred to two valuers, one to be nominated by each party.
(e) The goods shall be delivered to the buyer.
(f) Tenders are to be submitted on the following terms, namely that tenders are a single offer for all shares held by us. We bind ourselves to accept the highest offer provided that such offer complies with the terms of this telex.
(g) This policy does not cover any consequence of confiscation, nationalization requisition or damage to property by under or through the order rule or regulation of any government public statutory municipal local customs or health authority.
(h) The tenant agrees that the landlord may take the lifts out of service for emergency repairs and maintenance.
(i) The contract between us will run for six months. If we as sellers are satisfied with you as a customer, the contract is automatically extended for six months. All the conditions of the first contract will then continue to apply.
(j) Either party may determine this agreement by notice in writing if the other shall have committed a material breach of its obligations hereunder.

Example 3.3.2

Incoterms as defined by the Institute of Chambers of Commerce are common ways of clarifying the point at which ownership (and risk) are transferred from one party to another. For example, FOB is the delivery of goods by the seller 'Free on Board' to a ship. They become the responsibility of the buyer as they pass the ship's rail (see Fig 3.3.1 overleaf). Other examples include:

- Ex works — delivered to the seller's factory gates.
- C & F — delivery includes cost of transport and freight.
- CIF — delivery includes cost of transport, insurance and freight.
- Ex ship — delivery to the buyer's port.
- Ex quay — delivery to the nominated quayside.
- DDP — delivery to the nominated dock or port storage.
- Franco domicile — delivery to the buyer's premises.

A convenient solution to the problem of ascertaining the terms of the contract is to employ a printed standard form containing terms suitable

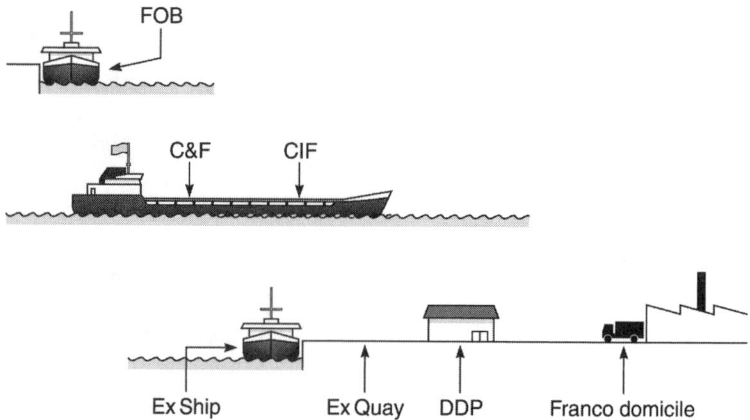

Figure 3.3.1 *Incoterms*

> **Key point**
>
> To avoid accusations of unfair terms in standard forms, it is sometimes advisable to use a neutral parties standard form contract. Institutions and professional bodies often provide model contracts.

> **Key point**
>
> If you are concerned with signing a standard contract you should do nothing to indicate acceptance of the other party's terms.

for a variety of transactions of the same type. These must still be used with caution, however, as they effectively reduce the ability of one party to negotiate and are hence open to criticisms of unfairness. An unfair term is one that has not been individually negotiated and causes an imbalance to the detriment of the consumer. This may render the contracts 'champertous' (illegal). Consumers are protected against unfair terms by the Consumer Contracts Regulations 1999 which have replaced the unfair terms legislation of 1994. The new Act allows Trading Standards to prosecute on behalf of the consumer. Standard forms are also restricted on exemption clauses that they can include.

Making quotations and placing orders with standard conditions attached, usually in small print, can also produce a 'battle of forms' between organizations where each side seeks to impose its terms and conditions over the other. The best solution to this problem is not to get involved in it at all, but to endeavour to reach an agreement with the other party on disputed terms.

Terms of the contract must be distinguished from 'representations' which are statements that do not form part of the contract but which induced the making of the contract. When such a statement is false, this amounts to misrepresentation. No damages for misrepresentation can be made with respect to the contract but tort damages are available for fraudulent or negligent misrepresentation. Any remedy for misrepresentation will depend on factors such as the degree of fault of the maker of the statement:

- Intention of the parties.
- Timing of the statement.
- Incorporated into written contract.
- Indication of importance of issue.
- Special knowledge of statement maker.
- Made on purpose to induce contract.
- The state of mind of the maker of the statement and whether the statement was fraudulent (where the maker of the statement knew it to be false, negligent (where the maker of the statement was careless in determining the truth) or innocent (where the maker of the statement believed it to be true and had reasonable grounds for believing that).

Agreement

In order to complete the agreement the offeree must accept the offer and communicate his acceptance to the offerer. This means that there will be an agreement where one party has made an offer to the other to enter into a contract on certain terms, and the other has accepted on those terms. If the other party fails to accept all of the conditions and terms of the offer, or adds any further conditions and as such supersedes the previous offer, this is in itself a new offer which can be open to acceptance. In order for a party to show that the other party is bound by a particular offer he must show that the other party either (a) signed it, or (b) entered into the contract with notice of it. This notice must be given before the contract was made.

The offerer can specify the way in which the offeree is to signify acceptance. This could be assigning of the contract by hand or under seal but any reasonable means of communication is acceptable, including:

- Word of mouth
- In writing
- Conduct

The acceptance must usually come to the attention of the offerer, but there are some exceptions such as the rule that a postal acceptance is effective as soon as it is posted. Unilateral contracts are made to an unspecified party and also differ in that:

- The offeree does not have to communicate to the offerer the fact that he is accepting the offer.
- (Probably) it is unnecessary for the offeree to know of the offer at the time when he performs the act required in order to bind the offerer.
- The offeree is never bound to perform the act required by the offerer.

In the event that legal action is required to determine whose terms are valid in a battle of forms dispute, the courts usually consider the procedural circumstances of offer, counter-offer, rejection, acceptance and so forth (*Trollope & Coils Ltd* v. *Atomic Power Constructions*). Courts will determine whether they have reached agreement on all material points, even though there may be differences between the forms and conditions printed on the back of them. Applying this guide, it will be found that in most cases when there is a 'battle of forms' there is a contract as soon as the last of the forms is sent and received without objection being taken to it. In some cases the battle is won by the man who fires the last shot. In other cases, however, the battle is won by the man who gets the blow in first. The deciding factor is often the scope of any differences and whether or not these were pointed out.

Case study

The chain tensioning device

On 8 September 1978, the sellers (the defendants) offered to sell the buyers (the plaintiffs) a device for tensioning the chains on a motoring buoy in the Buchan Field in the North Sea. The offer incorporated the defendants' terms and conditions including conditions that any disputes should be settled by arbitration in California and that Californian law should govern the formation, construction and performance of the contract. The price quoted was FOB factory in California and the offer covered the engineering, design and fabrication of the device.

On 29 September the buyers sent the following telex to the defendants, referring to the quotation and stating *inter alia*: 'It is our intention to place an order for one chain tensioner . . . with your goodselves. A purchase order will be prepared in the near future but you are directed to proceed with the tensioner fabrication on the basis of this telex. A purchase order will be issued subject to our usual terms and conditions.'

Three days later the buyers sent the sellers their purchase order dated 5 October which contained the following clauses: All disputes arising in connection with the contract shall be settled by arbitration. The arbitration shall be held in the UK and conducted in accordance with UK law. The written acceptance of this contract, the commencement of performance pursuant hereto . . . by the sellers constitutes an unqualified acceptance by the seller of all of the terms and conditions of this contract. This contract . . . constitutes the entire agreement between the parties, either oral or written.

This purchase order led to a number of telexes between the buyers and the sellers. The sellers commented on the purchase order complaining that they had proceeded with the work on the understanding that their offer was acceptable and were now facing *ex post facto* contract terms but made no objections or comments to either the arbitration clause or acceptance clause. Negotiations continued between the parties and agreement was reached on variations to the plaintiffs' purchase order.

On 20 October the buyers sent a telex to the sellers agreeing the one outstanding point and asking the sellers whether in view of the changes they would prefer that the buyers reissued the purchase order. On 20 December the sellers replied to the effect that they saw no need for the reissue of the order, and enclosed their formal acknowledgement order which contained the following clause. 'Acceptance of buyers' order is conditional and subject to the following conditions . . . unless the buyer shall notify seller in writing to the contrary within 5 days of receipt of this document the buyer shall be deemed conclusively to have accepted the exact terms and conditions hereof. The copy was signed by the buyers and returned on 3 January 1979.

The buyers alleged that defects in the system were discovered when it was being installed in the North Sea in July 1979. On 29 July 1980, the buyers issued a writ in the UK court against the sellers claiming damages for breach of contract. On 7 August 1980, the sellers commenced proceedings in the Californian court alleging *inter alia* that the disputes should be referred to arbitration in California and that Californian law applied to the contract. On 7 November the Californian court having heard argument on the issue, concluded its judgment with the request that the English court should either on its own initiative or on motion stay further proceedings in this action. However, on 11 November, the buyers without notice to the defendants and without informing the court of the request entered judgment in default, and on 21 November the sellers applied to the court for

(a) An order that all further proceedings in the action be stayed pursuant to s. 1 of the Arbitration Act 1975, or the inherent jurisdiction of the court.
(b) An order that the judgment entered on 11 November be set aside. It was accepted by the sellers that the application under the Arbitration Act 1975, s.1 would fail unless they could establish that the contract incorporated the Californian arbitration clause.

It was held by the court that:

(a) The buyers' telex of 29 September was not an acceptance of the sellers' offer of 8 September and its only effect was to enable the sellers to recover on a *quantum merit* for work done pursuant to the direction.
(b) The purchase order by the buyers of 5 October was in English law a counter-offer which destroyed the original offer *in toto* (and was not that form of counter-offer which incorporated the original offer because it so was specifically varied).
(c) On 20 October, in English law, the contract was concluded between the parties and it was well understood by the sellers that it did not incorporate any part of the original offer.
(d) The sellers' submission that the final exchange between the parties on 20 December and 3 January 1979 constituted the contract with the Californian arbitration clause incorporation would be rejected in that the clause in the sellers' letter of 20 December was meaningless since there was nothing left to accept; the contract had already been made and that document was not, and was not intended to be, anything more than a mere formality and the reference to the original offer was for identification only.
(e) In the circumstances the contract did not incorporate the sellers' terms and conditions and thus did not incorporate either the Californian law clause of the Californian arbitration clause; it was a contract upon the terms of the buyers' purchase order as subsequently varied by the telexes up to and including the telex of 20 October.
(f) The sellers' contention that the original offer expressly limited acceptance to the terms of that offer and the telex of 29 September was self-evidently a definite and reasonable expression of acceptance would be rejected in that the telex stated that the buyers intended to place an order, the instruction to proceed was in respect of part only of the goods offered and it was made plain that when the order did come it would be subject to terms and conditions so that the telex of 29 September could not be regarded as a definite expression of acceptance.

(g) In the light of the telexes the intention of the sellers was to accept all those proposals save those to which they had raised objection and specifically negotiated and since no objection had been taken to either there was acceptance of both these clauses.

(h) Even if Californian law was the relevant law for the purpose of determining what was the contract between the parties, under the law the sellers' Californian arbitration clause was not, and the buyers' UK arbitration clause was included; and the sellers' first application to stay the proceedings under the Arbitration Act 1975 failed.

(i) Since none of the sellers' terms and conditions were included as a matter both of English and Californian law there was no basis upon which, under the inherent jurisdiction, there could be a stay to enable the Californian court to decide that matter; and in the event the evidence was that English law was the proper law of the contract.

(j) As the buyers had regularly begun their action here and since the dispute itself both as to liability and damages involved matters occurring in this country and since the defendants could have applied for a stay without ever resorting to the Californian court it would be manifestly unjust to the buyers to grant a stay.

(k) The sellers were unable to rely upon their standard terms and conditions, there was no positive defence shown and as there was nothing left but a mere non-admission of liability this was not sufficient ground upon which to set aside the judgment.

(l) On the facts, the buyers had an early opportunity to inspect the defects on the system and knew what was being alleged and since they did not challenge any of the allegations but only made an effort by resort to the Californian court to include their terms and conditions and sought to rely on such conditions there was in the circumstances no arguable defence on liability and no other sufficient reason for the action on liability to be tried and the application to set aside failed and the sellers' summons would be dismissed.

It has been argued that this traditional analytical approach to battles of forms is out of date. The difficulty is to decide which form, or which part of which form, is a term or condition of the contract. The better way may be to look at all the documents passing between the parties and glean from them, or from the conduct of the parties, the intention (Lord Wilberforce, *New Zealand Shipping Co. Ltd* v. *A.M. Satterthwaite*). In other words extend the evaluation over a wider area and look into the minds of the parties to make certain assumptions. In this way there may be a concluded contract in which the terms and conditions of both parties can be construed together to give a harmonious result. If the differences are irreconcilable, so that they are mutually contradictory, then the conflicting terms may have to be scrapped and replaced with a reasonable implication.

Performing the contract

Once a contract has been concluded between the parties, it should then be clear what each party is obliged to do in order to perform the contract effectively, and the extent to which each party will undertake liability. If these obligations are not met, then an action for breach of contract can be brought leading to damages and other remedies. The general rule is that liability for failure to perform is strict, i.e. it is not a defence to show that you did your best to perform the contract.

The doctrine of 'discharge by frustration' is concerned with the situation which arises when, by reason of some event occurring after the contract was made, it is physically or commercially impossible to perform the contract. The effects of frustration upon the rights of the parties to payment is governed by the Law Reform (Frustrated Contracts) Act 1943, which provides that all sums paid or payable by a party before the frustrating event may be recovered or cease to be payable as the case may be, subject to the right of the other party to claim payment for expenses properly incurred by him before the frustrating event. The kinds of event capable of frustrating a contract include:

- A change in the law making performance of the contract illegal.
- Outbreak of war.
- Destruction of the subject matter of the contract.
- Cancellation of an expected event.
- Requisitioning of the subject matter of the contract.

Because of the very narrow scope of this doctrine it is sound practice to include in contracts a *force majeure* clause which relieves the parties from obligations if the duty to performance becomes onerous or impossible because of some clearly unforeseeable event, such as strikes or delays in transport.

A difficult issue arises when a party wishes to know whether a breach by the other party entitles them to refuse to continue their obligations of the contract. The basic approach of the courts is to draw a distinction between 'conditions' which are more important terms and 'warranties' which are less important terms. A breach of condition will give rise to a right in the innocent party to terminate the contract, whereas a breach of warranty merely gives rise to a right to claim damages. However, recent cases have held that minor breaches of condition may not in themselves entitle the innocent party to terminate.

A failure by a party to perform the contract strictly on time does not usually entitle the innocent party to terminate the contract, unless:

- The parties have expressly agreed that time is to be of the essence.
- Where the contract is of such a nature that a time fixed for performance must be strictly complied with.
- Where time was not originally of the essence, but one party has been guilty of delay and the other has given notice requiring performance within a reasonable time.

Misrepresentation can also render a contract voidable and the injured party may seek to have the contract rescinded (set aside) which puts the parties back into their original positions. No rescission is available

Key point

The Rights of Third Parties Bill has abolished the privity of a contract. This will allow a third party specified in a contract (but not an actual party in the contract) to sue on the contract. It will, for example, allow consortiums to sue sub-contractors or the 'promisor'. Large co-ordinating contractors may be able to download responsibility onto smaller contractors by making them directly liable.

where the injured party delays too long or where third party rights have intervened so that there is no possibility of restoring parties to their original positions. Contracts created by duress may also be voidable.

Licensing

Often it is more preferable for others to exploit your ideas or products on your behalf. In return for a lump sum payment and/or a royalty IP ownership could be transferred to others via an assignment. Ownership can also be temporarily transferred in return for a lump sum and/or a royalty and this form of agreement is detailed in a licence. Most licences are reduced to the written form as per a normal contract. Because they are tailored to a variety of situations there standard forms are less likely. They also have specific terms that relate to the grant of rights, consideration and specific obligations.

Key point

It is estimated that 80% of the inventions developed by corporations worldwide are unused and unexploited with an estimated $115 billion (120 billion euro) of unused technology assets in the US alone. This could be licensed to organizations that would use it. Many 'technology transfer' organizations also now offer services that match potential licensees with licensors.

Example 3.3.3

Licensing is a common form of agreement in business and every day life. By buying this book, for example, the author and publishers have implicitly granted you a licence to read the material in it. They may also have explicitly explained when and how you may reproduce parts of it.

If you have eaten a meal today containing genetically modified food you may not actually own what is inside your stomach. You have, however (probably), been given a licence to place this food there by the biotechnology company that created it.

Grant of rights

The grant of rights establishes the details behind what is being licensed and how:

- Is it the product and/or the 'know-how' behind the product that is being licensed? Is it defined by a patent? Potential licensors should always check to ensure that the licensee actually has the rights to own or license what is being offered ('due diligence').
- Is the product/know-how being transferred in part or in totality?
- Is the transfer a simple once and for all deal or does it include future improvements?
- Are the rights existing or prospective?
- What is the geographical extent to which rights are granted?
- What is the duration of the grant of rights?
- Will the licence convert to an assignment?
- Is the grant exclusive to the licensee in which case only the licensee has the right of use? The licence may otherwise be non-exclusive in which case the licensor may grant rights to other parties. It may also be a sole licence in which case both the licensee and licensor have rights.

In establishing the rights it is important to avoid unfair terms and create monopoly situations. These are prohibited by the Treaty of Rome (Article 85(1)) and enacted in the UK by the Competition Bill 2000. Exceptions to this can be allowed for technical or economic progress (Article 85(3)). For example, price fixing would not be allowed and feature on a 'black list' of unacceptable exclusions. Allowances on a 'white list' would include sole or exclusive licences if granted in specific areas. Any doubts should be referred to the EU for clarification.

Consideration

Consideration in respect of the grant of rights might be a lump sum payment or a royalty. Often it is a combination of both with an upfront lump sum payment followed by a royalty stream. A royalty stream is the licensor's remuneration based on the licensee's use of the transferred goods. It is usually based on a percentage of the profits the licensee can make. A reducing sliding rate is normal and it is often preferred that the return is based on the number and price of trade sales, rather then end-user sales, as this is easier to check. There may also need to be clarifications on whether sales include or exclude any taxes.

It is important not to undervalue the goods and services you are intending to license. IP is traditionally underrated because the technical expert who developed the IP may not have the same commercial expertise. The value should be based not on the cost incurred in making the goods or in replacing the goods but on the sales value to the licensee.

Example 3.3.4

Watt patented the piston effect in 1769 and later its application to wheeled carriages in 1784 but remained actively opposed to the desire amongst his team to develop the steam driven locomotive. He wrote to his partner Matthew Boulton stating 'I wish we have what could be brought to do as we do, to mind the business in hand, and not to hunt shadows' (it was left to Stephenson to invent the train).

Mary Jacobs patented the bra in 1913. The device comprised two handkerchiefs tied together by a ribbon. She sold the idea to Warner brothers for $1500 who made £15 million from it over the next 30 years.

In the 1940s a classification system using four black lines on white card was patented. Today 5 billion bar codes are now scanned each day.

The great inventor Edison actually saw little value in his light bulb and made little money from it. In Germany shortly after, however, Professor Nernst sold his similar idea to AEG for £250 000.

Specific obligations

Because there are now two parties with an interest in the goods or services, i.e. the licensee and the licensor, it is important in the licence agreement to clarify which party has any specific obligations. These might include:

Key points

Many organizations offer services that help potential licensees gain access to copyright material. These include the Licensing and Collection Society which can collect on behalf of authors, Pilot Site Licence Initiatives, National Electronic Site Licence Initiative. The Copyright Licence Agency (CLA) can arrange terms and payment systems for organizations wishing to regularly use copyright material belonging to others. Newspaper Licensing Agreements (NLA) can be arranged for organizations wishing to use newsprint material. It is, for example, illegal for a company to photocopy a business article and distribute this to its employees without permission. Parliamentary Copyright covers work deriving from Parliament and administered by HMSO. It includes Parliamentary reports and papers, Acts and press releases. Use of this material is usually free subject to certain conditions. Guidelines on the use of these and materials from other government agencies can be sought from the Crown Copyright Unit.

Software licences vary but usually invoke some of the following terms:

- One licence for one machine.
- Shareware is copyrighted and not free.
- Public domain is not copyrighted and may be used freely.
- Shrink wrap displays conditions to which you agree when you open the software packaging.
- Click wrap displays conditions which you must agree to before you can enter or download software.
- Escrow is a deposit (software) placed in third party organizations in case a small company servicing a large company goes bankrupt.

Many of these licences are contentious, sometimes discriminating against the user and other times against the supplier. Under the terms of any licence, however, you may back up, observe, test or study, to operate with other programs.

- What happens if the licensee does not meet a minimum level of performance?
- Accountability and penalties for late payments.
- Who will maintain any intellectual property rights such as patents and who will confront any infringers?
- What happens if either party makes any improvements to the goods?
- Does the licensee have any right to sub-license?
- The confidentiality of knowledge and trade secrets.

The changes in the privity law also affect licences. For example, a software developer may enforce licence terms on a user even if a contract only exists between the user and the selling organization.

Problems 3.3.1

(1) What is a letter of intent?

(2) What are the three components necessary to form a legally binding contract?

(3) Decide if there is a valid contract and if so, is it subject to the buyer's terms or the seller's terms?

   (a) Buyer sends an enquiry to seller (no condition of contract proposed).

   (b) Seller submits a quotation together with his standard conditions of sale which include a Contract Price Adjustment formula. The quotation also includes the sentence 'Our conditions of sale attached override any conditions incorporated in any resulting purchase order'.

   (c) Buyer sends purchase order with his Standard Conditions of Purchase which state that contract price is fixed. The purchase order has a tear-off acknowledgement slip.

   (d) Seller signs and returns the acknowledgement with a letter in which it is stated that the agreement is in the seller's terms.

(4) Consider the following cases:

   (a) In *Neale* v. *Merrett*, the seller offered to sell a piece of land to a buyer for £280. The buyer 'accepted' this offer and sent a cheque for £80 with a promise to pay the remainder by instalments of £50. Was there a contract or not?

   (b) In *Victoria Laundry (Windsor) Ltd* v. *Newman Industries Ltd*, the Newman laundry firm ordered a new boiler which arrived late causing Newman to consequently lose a large new contract. Were they entitled to recover damages for normal loss of profits from the supplier?

   (c) In the *Heron II* case, a ship was late in delivering a cargo of sugar to the busy port of Basra, and by the time the delivery was actually made the market price had fallen. Could the normal loss of profits be recovered from the ship owner?

   (d) In *Tulk* v. *Moxhay*, the owner of land which included Leicester Square in London sold the Square itself but retained some land round it. The buyer contracted not to build on the Square. He later sold the land and it then passed through several hands. Ultimately the new owner proposed to build on it. Could they build or not?

Activity 3.3.2

The following Patent Licensing Agreement has been drafted between two parties. Examine the agreement and consider where the content is weak and where future problems may lie. Suggest changes that would overcome these difficulties:

Parties
1. Portsea Ltd located at Port Road, Portsmouth ('Port')
2. EMCA Sales Division located at London Road, Crawley ('EMCA')

Preamble
i) Port is prepared to grant a licence to EMCA and TJ

1. Definitions
   Product any commercial commodity containing the innovation detailed in the Patent
   Patent shall be as Patent number 1,XXX,XXX
2. Grant
   2.1 Port grants to EMCA a sole and exclusive licence to
      i) import, manufacture and sell products using technology identified in the Patent
      ii) adapt the technology for future development
   2.2 This licence shall not confer on any third party the rights to use the technology
   2.3 A shall procure that any customers purchasing the Product shall be enjoined from resale of the commodity outside the territory
3. Consideration
   3.1 On execution of the agreement EMCA shall pay Port a one sum payment of £10 000 Plus £3.00 for every product sold
   3.2 EMCA shall sell the products at £15.00 per unit
4. Payment
   4.1 Payment shall be made by the due date into the Bank of Brighton
   4.2 Port shall have the right to inspect EMCA's financial ledgers to determine that all Transactions and Payments are in order
5. Obligations
   5.1 EMCA shall not be permitted to decompile the inherent software
   5.2 Any improvements by EMCA shall be disclosed to Port who shall thereon become the exclusive owner
6. General
   6.1 Failure by EMCA to adhere to the terms of this agreement shall allow Port to terminate the agreement

Signed by J. Blake ....
(personal assistant to the director)

3.4 Liability

Liability for the production and sale of products has emerged through a series of legislation over a number of years. This section provides a summary of the relevant legislation and the key issues that they cover.

Key point

'Unless otherwise agreed, the goods remain at the seller's risk until the property in them is transferred to the buyer. Whereby the goods are at the buyer's risk whether delivery has been made or not.' (Sale of Goods Act 1979)

Sale of Goods Acts 1893 and 1979

The Sale of Goods Act was created to protect the interests of both businesses and individuals with respect to both buying and selling (excluding bartering and hiring). The legislation clarifies and defines terms and makes it easier for one party to take action against another in the event of any irregularities.

A summary of rights in relation to selling is:

- To be able to sell goods.
- To be free to sell goods without interference.
- To have agreements (which may be oral with the exception of credit arrangements).
- To claim for damages (the difference between contracted price and subsequent resale), withhold delivery and reclaim goods where title as been reserved for non-payment or acceptance of an agreed sale.
- To quantify the point at which ownership transfers from the seller to the buyer.

A summary of rights in relation to buying is:

- To be recognized as an outright owner (quiet possession).
- To buy goods that match their description.
- To buy goods that are of merchantable quality (unless defects are highlighted).
- Buy goods that are reasonable for the purpose required.
- Have terms of sale created by implication in the way the goods are used.

If any of the buyers rights are broken, the conditions of contract are broken ('breach of contract') and damages for injury and loss of profit may be claimed from the seller. Liability is strict. Note that the contract itself is not invalidated if the buyer has paid or accepted the goods.

Privity has traditionally allowed only the contracting parties to take action against each other, hence the buyer must sue the immediate seller. If goods have passed through a series of sellers this may become a chain of actions, but time, or one good argument, may break the chain, protecting the first seller (e.g. the manufacturer). The changes in privity law may reduce the protection afforded by privity. A collateral contract (e.g. manufacturer's guarantee, manufacturer's advertised claims) may pass on to subsequent buyers and hence may also render a manufacturer liable.

A tort action, such as negligence, will also overcome the issue of privity and allow the buyer to take action beyond the immediate seller, e.g. to the manufacturer directly. It also allows people affected by but did not buy the product to take action (i.e. not the buyer of a car but a passenger). Negligence occurs when the common law duty of care is not applied creating physical or (more rarely) economic harm. It is not 'strict' but would include design or manufacturing faults, inadequate testing or poor quality control. The burden of proof of innocence may fall on the producer.

Where it is financially prohibitive for a buyer to take a private tort action against suppliers of faulty products then they may be able to claim breach of the Trades Descriptions Act to enable criminal liability. Similarly a breach of warranty may provide an avenue for damages if advertising promotes an image of a product that does not bear witness in reality.

Consumer Protection Act 1987

As a result of the European Directive on Product Liability, safety regulations after 1987 became proactive rather than reactive. The Consumer Protection Act (CPA) provides civil remedies for injuries or damage to property to individual consumers by defective goods which are devoid of contract or negligence issues. The Act includes non-consumer goods including raw materials and components but not buildings, printed matter, stated poor quality or older products (superseded by newer, safer products).

Defective is defined as 'there is a defect in a product . . . if the safety of the product is not such as persons generally are entitled to expect, and for those purposes "safety" in relation to a product shall include safety with respect to products comprised in that product'. There may be three types of defect – design (generic), manufacturing (one off) or marketing (poor instructions).

Liability on the supplier or manufacturer is strict and all parties may be sued. Claims must begin within three years. Defences include contributory negligence by user, that goods were not faulty at the time of sale, that the product conformed to the state of knowledge at that time, that the component was placed later into someone else's system which was faulty (or vice versa) or that the component supplied was requested with poor specifications.

Example 3.4.1

In January 2000, a woman sued LRC Ltd (formerly the London Rubber Company) because she became pregnant after a condom she was using split. Damages included trauma and the cost of bringing up the child and the claim was based on the negligent manufacture by LRC. LRC claimed that the product must have been made faulty after being sold. The judge dismissed the case arguing that there was no case to answer under the CPA.

Key point

Surveys show that more customers are complaining about faulty goods and services. Those that do not complain are often not satisfied but believe complaining is a waste of time. They may have a point since fewer companies are addressing fewer complaints. Most complainers, however, aren't looking for recompense – only that things are put right.

Further protection is afforded to the consumer by the Data Protection Act 1998 which seeks to prevent misuse of data. The Act has established a Data Protection Commissioner, ensures enhanced privacy and increases access to health, education and credit files. It includes manual registers as well as computer processed records.

Key point

A safe product is defined as 'any product which, under normal or reasonably foreseeable conditions of use, including duration, does not present any risk or only the minimum risks compatible with the product's use, considered as acceptable and consistent with a high level of protection' taking into account the product's characteristics, effect on other products, presentation and categories of users.

General Product Safety Regulations 1994

The General Product Safety Regulations (GPSR) introduced the concept of a 'safe product' with a duty not to knowingly distribute, supply or intend to supply unsafe products. It applies to new, used and reconditioned consumer products (except antiques) that are bought for commercial purposes but includes industrial goods that may end up in the consumer market. The regulations are intended to support rather than supersede specific protective legislation and are enforceable by the Trading Standards Office.

The duties of anybody supplying goods are:

- Make products safe.
- Supply appropriate information including the risks involved using the product and the precautions needed.
- Batch mark products.
- Sample test products.
- Investigate complaints.
- Monitor product performance.

Liability is strict with designer, senior managers and/or the corporation all liable. Infringement is a criminal offence. The degree of safety is conjunctive and assessed on:

- Product characteristics.
- Packaging.
- Instructions for assembly, use and disposal.
- Effect on other products.
- Labelling.
- Those at risk (e.g. children).
- Conformity to standards – voluntary, EU, community technical specifications, industry codes of practice.

Defence includes demonstrating that all reasonable steps have been taken, that due diligence had been exercised and that relying on the information of others had been reasonable to do (which might then make these people liable).

Some products carry further legislative consumer protection, particularly, for example, food and vehicles. The European New Approach Directives detail requirements for specific products such as toys and personal protective equipment.

Key point

In the US, following the investigation into the dominant position of Microsoft, the second largest anti-trust (anti-monopoly) action has been bought against a number of biotechnology companies including Novartis and Monsanto by US farmers protesting against the licensing and ownership terms behind the buying and selling of genetically modified foods.

The Competition Act 2000

Free trade is a cornerstone of westernized market principles and in Europe is defined by Articles 81 and 82 of the European Union Treaty which seeks to restrict anti-competitive practices.

The Competition Act represents the incorporation into English law of the European legislation. It focuses on effects rather than the purposes of actions formerly detailed by the Restrictive Trades Act 1976, the Resale Prices Act 1980 and the Fair Trading Act 1973 which it has replaced. Uncompetitive undertakings and practices (such as price fixing) which restrict or distort competition or trade are illegal and the prevention of unfair dominant trading positions is sought. The legislation comes in two parts:

Prohibition I
There must be no agreements or practices which prevent, restrict or distort competition between organizations. This includes fixing prices or trading conditions, controlling technology, production, marketing or investment, market or supply sharing, and discriminatory obligations against businesses.

Exemptions from this can be sought from the Office of Fair Trading (OFT) if the benefits to the public can be demonstrated to outweigh any anti-competitiveness. The EU can also create block exemptions. Tests for anti-competitive behaviour are based on an analysis of the public interest and the degree of competition. The prohibition is unlikely to effect businesses with less than a 25% share of a market.

Prohibition II
Prohibits abusive behaviour by any business in a position to dominate a market, either by nature of the product they are selling or by geography. This includes, for example, actions which may inhibit new entrants to a market such as loss leaders, seasonal exploitation, technical innovation without purpose, or holding technology back. Consideration of whether the action is illegal includes the actual likelihood of new competition arising, the prices charged, conduct, vertical restraints (that is the number of suppliers and buyers in the supply chain) and refusals to supply. It may not apply if the market share of the dominant company is less than 40%.

The Act is concerned mostly with large organizations and the 'Small Agreements' regulations will exempt companies with a turnover under £20 million. All companies, however, are advised to take basic precautions such as:

- Checking management procedures
- Checking specific regulations
- Installing quality assurance systems
- Maintaining business insurance
- Monitoring and record keeping
- Undertaking risk assessments
- Providing continuing Professional Development (CPD) for staff

Breaches of the Act are investigated by the OFT. The OFT powers are strong and include seizure and 'dawn raids'. Penalties for infringement are based on turnover (nominally around 10%) but do not preclude the termination of agreements and being sued by damaged parties. Guidance on proposed actions can be obtained in advance from the OFT and usually cost between £4000 and £10 000.

Health and safety

In addition to protecting the consumer, there is a great deal of legislation framed to protect employees. There are broad Acts including, for example; the Management of Health and Safety 1992, the Provision and Use of Weights (PUWER II), Workplace Health and Safety Welfare, and there are specific regulations that relate to focused areas such as dealing with asbestos or use of compressed air. Legislation used to be prescriptive but is now wider and more vague placing the onus on employers to ensure that they are acting reasonably.

For any employer:

- The Health and Safety at Work Act (HSWA) 1974 requires that employees have a safe place to work.
- The Factories Act 1961 and the Offices, Shops and Railway Premises Act 1963 have been largely superseded by Work Place Health, Safety and Welfare Regulations 1992, and the Management of Health and Safety and Welfare at Work Regulations 1992 still applies requiring care on temperature, conditions etc. Registration of offices and shops is required.
- Good practice requires a written statement on safety covering hazards, evacuation, medical emergencies etc.
- Employers Liability Insurance certificates must be displayed.

Employing more than five staff requires:

- A written statement on safety covering hazards, evacuation, medical emergencies must be published and available.
- There is a duty to ensure adequate staff training and a safety committee if requested by the unions. More stringent requirements on fire precautions are required by the Fire Precautions Act 1971.

- The Health and Safety (First Aid) Regulations 1981 advise one first aider per 50 staff and a first aid room for staff numbers over 400.
- Injuries must be recorded and reported to the Health and Safety Executive if serious (Records and Notification of Injuries, Diseases and Dangerous Occurrences Regulations 1985).
- The Management of Health and Safety Regulations 1992 requires that health and safety risks are assessed against industry standards and accident prevention measures are taken.

Tort actions may also be taken against employees who feel that employers are liable for injuries they have received.

The Health and Safety Executive (HSE) frames the legislation behind employee protection. Its mission is 'to ensure that risks to people's health and safety from work activities are properly controlled'. The Health and Safety Commission ensures that the legislation is complied with.

Problems 3.4.1

(1) A consumer cannot sue a manufacturer if they bought a defective product through a retailer. Discuss.
(2) All companies who engage in anti-competitive actions are liable to closure. Discuss.
(3) What defences might a designer use against a claim for negligence?

Chapter review questions

(1) Comment on the application of copyright, design right and registered design right protection in the following inventions:
  (a) a new car exhaust system;
  (b) a specially made suit from a new design (comment also on the issue of ownership between designer and wearer).

(2) X registered the name Harrods.com long before the Harrods store wanted to claim it. Harrods took the case to interNIC who suspended the name but X refused to give the name up. Harrods took X to court claiming trade mark infringement, passing off and conspiracy to injure. Was Harrods successful?

(3) The trefoil rotary shaver head was registered by Philips as a trade mark but the legitmacy of this was challenged by Remmington. Was Remmington's challenge successful?

(4) In the US, X conceived an idea but although being diligent in 'reducing the invention to practise' Y was the first to actually complete the product. X filed for a patent in the US but Y objected. Who should win the patent rights?

(5) X commissions Y to paint a portrait but is so incensed with the painting that he destroys it. Y sues claiming under the integrity of moral ownership issue, X has treated his painting in a derogatory manner. Is Y successful?

(6) (a) ® is the symbol for a registered design – true or false?
  (b) What are the three requirements of a patent?

(7) The small son of an inventor accidentally picks up some of his father's papers which includes a draft patent application. He then leaves the papers at school on his desk overnight. In the morning, the inventor rushes to the school and collects the application which is undisturbed on the desk. Can he still pursue the patent?

(8) A plaintiff made a short film, *Joy*, which he sent on a show reel to the first defendant, an advertising agency. It was a film of one man dancing to music. The plaintiff had made extensive use of jump cutting, an editing process whereby pieces of the film within a sequence of movements by the actor were excised. The result was that the actor appeared on the edited version to have performed successively and without an interval movement, which in reality could not have succeeded each other. The first defendant, wishing to make an advertisement based on *Joy*, produced *Anticipation*, a film with a different actor and different story but with extensive use being made of the plaintiff's jump cutting technique. *Anticipation* was subsequently used as a Guinness advertisement. Was the plaintiff successful in suing the defendant for copyright infringement?

(9) (a) In *Mutual Life Assurance* v. *Evatt* (1971), an insurance company made negligent statements to Mr Evatt about the financial state of an associated company. In reliance on this, Evatt invested money in the other company and, as a result, suffered financial loss. Could he recover damages from the insurance company?
  (b) In *Sayers* v. *Harlow UDC* (1958), Mrs Sayers found herself locked in a public lavatory. Unable to summon help, she tried to climb out over the top of the door. She found this impossible and, when climbing back down, allowed her

weight to rest on the toilet roll which 'true to its mechanical requirement, rotated'. Mrs Sayers fell and was injured. Could she sue the council?

Activities

(10) ***Machine Tool Company* v. *Corporation Ltd***
A Corporation (the buyer) made an enquiry to a Machine Tool Company (the seller) concerning the purchase of a machine tool. In response, the sellers made a quotation offering to sell a machine tool to the buyers for the price indicated with delivery in ten months' time. The offer was stated to be subject to certain terms and conditions which 'shall prevail over any terms and conditions in the Buyer's order'. These conditions included a price variation clause providing for the goods to be charged at the price ruling on the date of delivery.

The buyers replied by placing an order for the machine. The order was stated to be subject to certain terms and conditions, which were materially different from those put forward by the sellers and which in particular, made no provision for a variation in price. At the foot of the buyers' order was a tear-off acknowledgment of receipt of the order stating that 'we accept your order on the Terms and Conditions stated thereon'.

The sellers completed and signed the acknowledgment and returned it to the buyers with a letter stating that the buyers' order was being entered in accordance with the sellers' quotation.

When the sellers came to deliver the machine costs had increased so much that the sellers claimed an additional sum due to them under their price variation clause. The buyers refused to pay the increase in price and the sellers brought an action claiming that they were entitled to increase the price under the price variation clause contained in their offer.

The sellers argued that they had stipulated there would be an increase in the price if there was an increase in costs and so forth in the terms and conditions on the back of their quotation form. The buyers rejected the excess charge, relying on their own terms and conditions. They said: 'We did not accept the sellers' quotation as it was. We gave an order for the self-same machine at the self-same price, but on the back of our order we had our own terms and conditions. Our terms and conditions did not contain any price variation clause'.

Who is correct, the buyers or the sellers? Discuss the issues in determining this case. What would your judgment be?

(11) Produce a product protection plan for a product of your choice. A product protection plan in the main should be applied to the actual protection of a particular product in question within the context of the real world. It should hence show a consideration of the competition, evidence of commercial understanding and a logical application of techniques and ideas.

(12) By following and reviewing the history of an innovation which has ended up in litigation, your role is to provide an appraisal of 'what went wrong'. You should be able to broaden your analysis

so that it applies to all new innovation. You should also familiarize yourself with the workings of the legal system.

(13) List the points you would need to agree on if you were negotiating with a colleague who wanted to borrow your computer or car for the week. A list of points is referred to as 'heads of agreement'. Using the heads of agreement draft a formal licence.

Further reading

Barker, D. & Padfield, C. (1996). *Law.* Butterworth Heinemann.

Clayton, J. (1999). *Law for Small Business.* Kogan Page.

Dworking, G. (1996). *Copyright, Designs & Patents.* Blackstone.

Marsh & Soulsby (1996). *Business Law.* Stanley Thomas.

Smith, J. and Thomas, ?. (1992). *Cases on Contract.* Sweet and Maxwell.

Whincup, Michael (1999). *Sales Law and Product Liability.* Gower.

4 Project management

Summary

We spend much of our life, both in work and at home, being involved in projects of various size and complexity. Some projects appear quite simple and we think we do not need to plan them, but just rush in and start them. Often, as a result, these are the ones that drag on and never get finished satisfactorily.

Most projects, even the simpler ones, must be carefully planned to make sure they are successful. This requires that the project be tackled in phases, breaking it down into various interdependent activities and then allocating people and other resources to carry these out. This chapter takes you through these important phases.

Objectives

By the end of this chapter, the reader should:

- understand the importance of the initiation phases in projects, where the aims and objectives are formed and how to carry out the initial planning stages (Section 4.1);
- be able to detail plan a project using network diagrams, Gantt charts and allocate the required resources (Section 4.2);
- use Goldratt's Critical Chain methodology to allow for uncertainties in activity times by applying project and feeder buffers at critical points (Section 4.3);
- be able to control, and report on, progress, including managing change and analysing variances and then conducting an end audit (Section 4.4).

4.1 What is project management?

This section introduces the topic of project management and the importance of the activity. It briefly describes the initial stages of a project such as the setting of objectives, the team selection and the feasibility study. The section concludes with the start of the planning operation where the activities are quantified and dependencies derived.

As discussed in Chapter 1, organizations exist to provide a product or a service and that is where their main day-to-day effort is directed. However, not all the activities within the organization are concerned with these repetitive operational issues. Some activities are part of a one-off effort directed towards a major change within the organization – these are termed projects. They have to be carried out alongside the normal activities with minimum disruption.

Project management deals with the organization and control of the activities involved in planning and carrying out these projects. Projects come in all sizes and shapes from relatively short, one-man affairs to large multi-million pound projects involving hundreds of people and even several organizations.

Examples of projects vary from highly technical ones, e.g. the design of a new aircraft, to organizational ones such as discussed in Chapter 2. Although each project tends to be unique from previous projects, there are sufficient similarities in the approach needed, and the problems encountered, that we can examine and learn from.

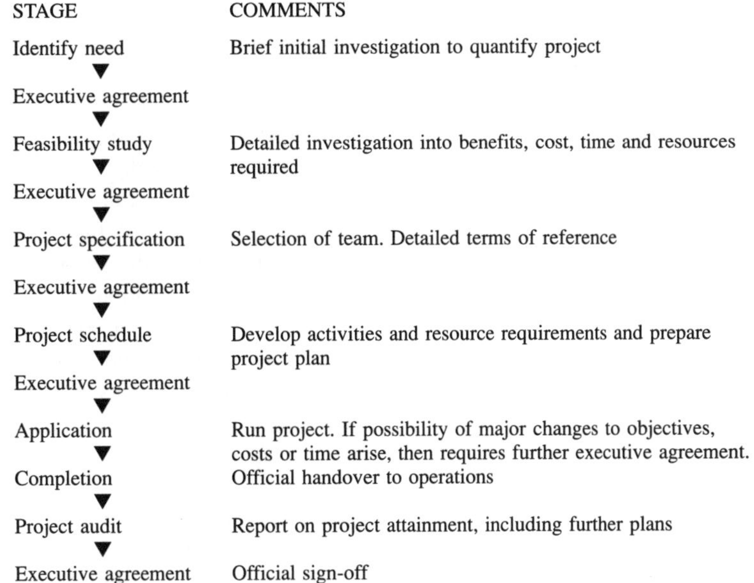

Figure 4.1.1 *Flow chart of the stages of a major project. (Note the need for executive agreement at key stages)*

A major project goes through several stages:

- Concept: Here the initial need for the project evolves. This stage identifies an opportunity of some importance to the organization to justify the diversion of resources from the normal day-to-day activities.
- Feasibility study: This is where the project gets firmed up, or discarded. An initial examination is needed to check the technical feasibility, the time and cost involved and the resources required. Many companies fail to invest sufficient effort here. A project manager is often selected at this stage, although this may be left until the next stage.
- Project specification and initiation: This is where the project starts to fill out with clear goals and constraints laid out in the terms of reference. The project manager and the initial team are confirmed. The team could change through the project stages, but the project manager should stay with it.

- Schedule: The detailed planning now takes place. The activities are developed showing their sequencing and the time and resources required. The latter need checking out for availability constraints at this stage. This finalizes in the detailed project plan.
- Application: This is the 'doing' part of the project where the plan gets put into operation.
- Completion: The penultimate stage is when the project is completed and handed over to the operational personnel, who may be the project team themselves.
- Evaluation: This final stage is too often not carried out properly. It is the stage where lessons for future projects are gathered and the life plan for the project is completed.

At each of these stages there should be considerable discussion and negotiation with senior staff and other stakeholders to ensure that the whole organization understands the project and agrees with the direction it is taking. As shown in Figure 4.1.1, there are key points during the process when a formal commitment to proceed is required from senior management.

Unfortunately all too often a project does not achieve full success and is judged to be a partial, or complete, failure. Projects can be late, fail to be within their budget, fail to meet all their aims or a combination of these. Some of these failures are due to technical problems but the main reason for failure is the human factor – the least of which is overconfidence in the organization's technical or managerial ability which leads to little input at the initial stages. Table 4.1.1 demonstrates some major projects of recent years, and their outcome.

Because projects tend to be a unique one-off task, many of the activities or technologies involved are unfamiliar to the people involved. This leads to a degree of uncertainty in them. This uncertainty shows up in various matters:

- Technological uncertainty: This form of uncertainty is present in most product and process development – especially where new technologies are blended with existing technologies.

Table 4.1.1 Recent major projects and their outcomes

Project	*Criteria*			*Success*
	Quality	*Time*	*Cost*	
Concorde aircraft	Technical success	7.5 years late	Seven times budget	In operation over 25 years
Advanced Passenger Train	Specification not met	2.5 years' delay	Within budget	Cancelled
Thames barrier	Technical success	3.5 years late	Four times budget	In operation
Nimrod early warning system	Specification not met	Incomplete	£2 Billion spent	Cancelled
London Air Traffic Control	Specification changing	5/6 years late	? over budget	?

- Scheduling uncertainty: When people are doing something new, they will be unable to predict accurately how long it will take them, or others, to complete their tasks. There may even be doubts about their release from other ongoing tasks.
- Cost uncertainty: The uncertainty in the time factor means it is difficult to anticipate associated spending. In addition the full cost of equipment suffers from a similar lack of identification.
- Organizational/Political: This is where many projects get into real difficulties. Part of this is the struggle for scarce resources to meet personal targets and part is the rearward action of the potential losers in any change.
- Staff turnover: The longer a project runs the more chance that key staff involved will change. This leads to loss of time as the replacements come up to speed and develop the required knowledge.
- Environment: New legislation and competitors' actions may affect the marketplace.

Any project plan therefore must anticipate these problems and seek to eliminate, or minimize, them by careful contingency planning and management to reduce the risks involved.

Project initiation

A project is started with the identification of a need for a change. This need can arise from many reasons some of which will be proactive, i.e. an attempt to evolve, or use, a new technology, and others will be in reaction to events in the organization's marketplace. The needs can come from a variety of sources, such as a SWOT analysis (see page 305). The following give an indication of some project aims:

- Developing a new product.
- Entering a new marketplace.
- Installing a new computer system.
- Selecting and installing a new process.
- Retraining in new technologies or techniques.
- Improving quality levels.
- Reducing operating costs.
- Refocusing managerial priorities.
- Meeting new legislation.
- Solving a major environmental problem.
- Amalgamating with another organization.

Whatever the need identified, for it to be accepted as a project it must primarily contribute to the business objectives of the organization, and be formally recognized by senior management.

At this stage the full solution will probably not be known. In fact too often an intuitive jump to a solution is made without fully examining the implications or the alternatives. Instant answers and quick commitments may look good, but quite often they contribute to the lack of success as organizations become committed to one solution. It then becomes emotionally difficult, and costly, to change later as shown in Figure 4.1.2.

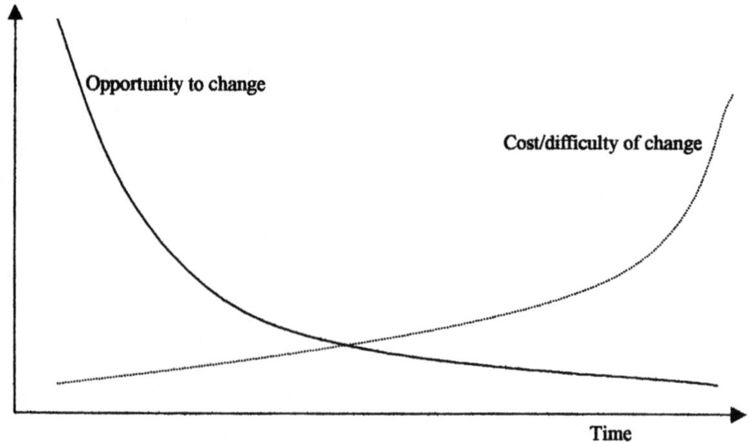

Figure 4.1.2 *Problems in changing project details as time progresses*

The project team

Early on in the project life, a team must start to be formed to carry out the detailed planning and control activities.

The first appointee is not the project leader, but the project sponsor or champion. This is a member of senior management who is responsible for the progress and success of the project. This gives the project status and top level access and commitment. The sponsor may delegate most of the planning and control of the project to a team, but must retain a full comprehensive involvement in the progress.

The next decision is how to manage the project. Depending on its size and complexity, it may be manageable within existing structures with one manager, or technical expert, in control. It may require a team which could be a task force made up of managers and others involved on a part-time basis, or it may require a full-time focused team.

The important thing is the composition of skills, attitudes and commitment required to fully understand and manage all the different aspects of the project. These cannot all be learned during the project so the team needs to bring most of these with them. The team therefore needs both technical knowledge and the ability to carry out the following management tasks:

- Leading
- Planning
- Organizing
- Co-ordinating
- Budgetary control
- Communicating
- Negotiating
- Managing user's expectations

The key person is the project manager, who is usually full time on the project from start to finish, although other members of the team may change. This individual could be someone from the area most involved in the project, but choice may be limited. The person requires a broad range of experience, ability and organizational standing which may not be available, therefore the appointment of someone from outside the immediate area often has to be made.

The team itself needs a blend of personalities as well as skills and experience. They must be able to work together openly. This does not mean submerging into one homogeneous group but a willingness to openly respect and discuss points of difference and come to a group consensus which is acceptable to everyone. To aid this blending, the group size must be limited and made up of people without too wide a status within the organization.

Terms of reference

This is the important authority to proceed document. It sets the objectives, the scope and constraints of the project. This can be brief initially, i.e. prior to the feasibility study, but must be firmed up before the main project goes ahead. Areas needing detailing are:

- Authority and sponsorship – identification of project sponsor.
- Contact with users.
- Objectives: few in number:
  - Alignment with business objectives.
  - Measurable in quantity, quality, time, cost and end product.
  - Consistent, understandable and achievable.
- Scope – departments, people, products, processes, etc.
- Constraints.
- Budget and timings of cash flows.
- Resources required.
- Interim deliverables.
- Timetable showing phases and involvement.
- Risks foreseeable and factors outside influence:
  - Identified assumptions including probability of occurrence.
  - Impact assessment and contingency plans.
- Team roles and responsibilities.

This requires firm agreement at top level.

The feasibility study

A comprehensive feasibility study is not just directed to answering the question 'Can it be done?' It is better that this stage be carried out by the project team, although some organizations use consultants for initial confidentiality. Whoever does it, they must address all the questions surrounding the issue such as:

- What are the alternatives?
- How technically feasible are the alternatives?
- What is involved in doing it?
- Do we have the internal skill and experience to carry it out?
- Who needs to be involved during the process?
- What are the potential problems that may arise?
- What training will be required?
- When can it be done?
- What disruptions will it cause?
- Areas of potential controversy and disagreement?
- What will be the effect on people directly, and indirectly, involved?

- What is needed in resources?
- How much will it cost?
- How long will it take?
- What are the dependency chains?
- What are all the benefits?
- Are there existing benchmarks available to use to set targets?
- Is there legislation that needs consideration?

For these questions to be fully answered will take a fair degree of investigation to gather facts and opinions. This may mean visiting other establishments and suppliers, use of consultants, etc. During this investigation there will have to be considerable discussion with those directly, and indirectly, affected to determine their support, or resistance.

There will be resistance to spending a considerable amount of time and money at this stage and impatience to get on with the job. Although it is important to be cost effective and efficient in carrying out the feasibility study, stinting it too severely can easily lead to problems later on in the main project itself.

The final stage of the feasibility study should be the selection from the available alternatives.

Selection process

The final point of the feasibility study is where the alternatives are evaluated, compared and one selected as the preferred option.

Each of the alternatives can be compared under as many aspects as deemed desirable, perhaps by ranking them using the paired comparison technique (see Chapter 2) to reduce them to the one or two main contenders. These remaining contenders then need to be put through the organization's economic appraisal system (see Chapter 4) to determine their economic standing.

Another way of selecting can be by a similar process to that used in design, i.e. by ranking each alternative's match to a set of key criteria. This can be carried out by producing a matrix, in which each alternative is scored on a scale from a relative value, or cost, viewpoint, i.e. a low of one up to a high of ten under a range of significance criteria. Depending on the circumstances, one set of criteria may be:

- Operational capability: This judges the degree of internal skill and experience available to operate and maintain an alternative. A score of one means that there is very little internal ability and a score of ten means that internal abilities are more than sufficient.
- Technical capability: This is looking at the match between the project team's technical competence to design and evaluate an alternative. It may show that the project will need outside support of consultants.
- Financial investment: The full cost-benefit analysis will show up the total amount of investment required. Here a very high capital investment scores just one – with little or no investment scoring ten.
- Financial return: The time to payback and total return can be judged here to determine future financial benefits.
- Time scale: The ability to complete the project within a certain time frame which can be set by the organization. The longer the timescale, the lower the points gained. It is often easier and better to have several limited connected projects than to try and have them rolled into one big project.

Criteria	Weighting	Alternative A		Alternative B		Alternative C	
		Raw score	Weighted score	Raw score	Weighted score	Raw score	Weighted score
Operational Capability	2	7	14	6	12	5	10
Technical Capability	1	5	5	3	3	7	7
Financial Investment	2	4	8	3	6	8	16
Financial Return	1	4	4	9	9	6	6
Time scale involved	3	2	6	7	21	4	12
Disruptions involved	2	8	16	3	6	6	12
Meeting aims	3	9	27	9	27	6	18
TOTALS		39	80	40	84	42	81

Figure 4.1.3 *Decision matrix comparing alternative solutions to meeting project aims*

- Disruptions to present activities: This judges the disruptions to present activities that an alternative will involve.
- Ability of the alternative to meet aims such as quality, delivery speed, etc.

All the criteria need not carry the same weighting. A survey of top management may decide at a particular time that the criteria may be weighted as follows, with three being the most important aspects in present circumstances:

- Operational capability: 2
- Technical capability: 1
- Financial investment: 2
- Financial return: 1
- Timescale: 3
- Disruptions: 1
- Meeting aims (each aim): 3

At another time they may increase the weighting of the financial investment to 3 because of plenty of available cash funds, or reduce it to 1 where investment needs to be curtailed due to lack of funds.

For the scoring we can produce a decision grid as in Figure 4.1.3. The columns show the total score against each alternative under both weighted and non-weighted conditions. From these, the alternative with the highest weighted score would be selected for first choice – in this case project B.

The result of the feasibility study and the basis of selecting a particular alternative would be presented to the management committee for approval before proceeding to the detailed planning stage. It is at this stage that the main direction and cost of the project is decided. It will be difficult, but not impossible to change these later.

Initial project planning

As with any job, the key to a successful completion lies in the preparatory work carried out beforehand, in this case the detail project plan. Planning projects involves several stages:

- Logical analysis of the many different activities to be done.
- Estimating the resources and time required for each activity.

- Calculating the costs involved.
- Setting these activities in a sequence of dependency.
- Scheduling the activities.
- Preparing a cash flow requirement.
- Setting milestones to denote important stage completion.
- Identifying potential problems and making contingency plans.

Quantifying the activities

Use people familiar with the work involved to gradually break down the main activities into smaller component activities at lower levels. Initially this should stop at a department or section level. The more detailed breakdown can be done by the managers within these areas.

The main characteristics of an activity at this level should be that it has a clear start and end point which can be measured by an end product. Once this is established, then the activity can be quantified in cost, effort, resources and time. This should be under the clear responsibility of one person, not necessarily the one who will be actually carrying it out.

The Information required is:

- Description
- Inputs and preconditions
- Deliverables
- Resource requirement
- Skill requirement
- Estimated cost and time
- Assumptions made
- Degree of variability
- Causes of variability

Because of the uncertainty that is often involved at the planning stage, there may be difficulty in arriving at an estimate of time and cost. What is required is a reasonable, honest figure which is based on using committed, skilled staff. Depending on personalities involved and the culture of the organizations there will be a tendency to over- or underestimate the task needs:

- Overestimation occurs when there is a blame culture and people are wary of committing themselves when the consequences of doing so may rebound on them, this leads to extra padding being inserted as a safety margin. Overestimation may also be based on concerns about the complexity of the task, or may be based on overengineering.
- Underestimation occurs when people are overconfident and take risks or make false assumptions about what is required to be done, perhaps because they do not realize the complexity, or what is required of them.

Both need guarding against because of the overall picture.

Dependency

Alongside the breakdown of the work involved into smaller activities, we need to determine the sequence in which these must be carried out.

This is arrived at by looking at what needs to be complete before each activity can start. Sometimes this will be information but at other times there will need to be a physical entity available such as an object or perhaps an area being made available or prepared.

When this condition is arrived at in another activity, then we say there is dependency on that activity. The point need not be at the end of an activity, but needs to be clearly identified.

Where there is no dependency, it may be possible to carry out several activities at the same time, i.e. in parallel.

However, when there is dependency this controls the sequence that the activities have to be scheduled. The normal type of dependency is that of waiting until the end of the preceding activity, but the start signal can arise at any point – even at the beginning where some degree of overlap can take place. In a few cases, the dependency can come from the ending of two separate activities, i.e. we may need them both to be completed before we can start the next activity.

Once we have all the information about the activities, we can then schedule them together to make up the overall project plan, and then get on with the project.

Problems 4.1.1

(1) Think of all the different types of projects you have been involved in – what made you think of them as a project?

(2) Why do you think the terms of reference for a project require top level agreement?

(3) Under what circumstances do you think that financial considerations will play a minor part in deciding a project has top priority?

(4) How long would you say it takes to travel from your home to your place of work or study?

(5) Can you think of an activity in your daily life that can start only after another activity has started?

4.2 Project scheduling – the critical path

This section looks in detail at scheduling activities within a project to find project completion dates by using dependencies between activities and establishing float (spare) times associated with these activities. This is examined first in small projects and then the effect of combining these small projects into one larger project is investigated. Methods of changing activities and resource allocation to alter the overall completion times are also addressed. Network diagrams and Gantt charts are introduced using Microsoft™ Project to provide the illustrations.

Assuming that all the activities and dependencies have been correctly identified, we now need to complete a schedule showing when each activity should start (and finish), the use of resources especially when money needs to be made available and of course when the project will be complete. This will identify those activities which must be completed on time and those where some spare time exists.

There are dozens of different software packages available to assist project scheduling. In this textbook Microsoft™ Project is used for illustrating software usage. This is merely for convenience – if you wish

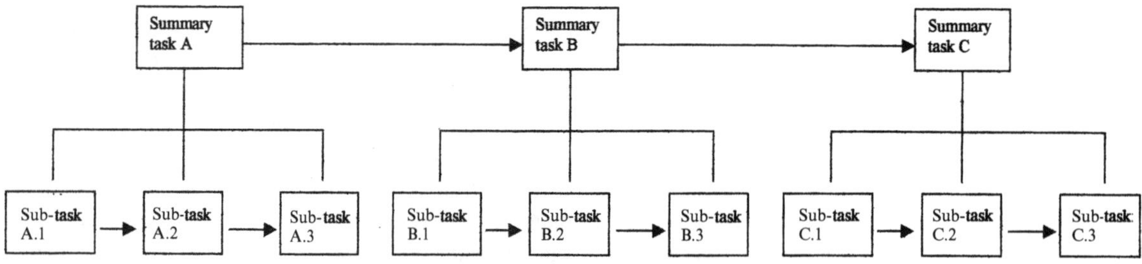

Figure 4.2.1 *Breakdown of main (summary) tasks into sub-tasks*

to purchase a suitable package, you are recommended to conduct your own investigation to meet your particular requirements. A good place to start is the annual *Source Book* published by *Project Manager Today.*

To explain the stages in scheduling we will use three interconnected projects:

- The construction of a new building.
- The design and prototyping of a new product.
- Installing the manufacturing process for the new product in the new building.

Each will be used to explain a principal technique and then they will be amalgamated to show the problem in scheduling multi-projects.

When scheduling projects we need to have a manageable number of activities, say a maximum of twenty. As most projects will consist of a large number of small activities, it is best to group these logically into higher level groups as in Figure 4.2.1. The lower level activities can be scheduled interdependently within Microsoft™ Project. In this software we refer to the upper level activities as summary tasks and the lower level activities as sub-tasks.

Project 1 – the new building: Gantt charts and cash flow

The higher level activities involved in this project are shown in Table 4.2.1. The times and costs are just for demonstration purposes. For the present, let us ignore the resources required, assume that the durations

Table 4.2.1 Activities involved in constructing the new building

Ref.	*Activity*	*Duration weeks*	*Immediate predecessors*	*Cost £*
B.1	Design	6	–	12 500
B.2	Site acquisition	6	–	75 000
B.3	Planning application	12	B.1 and B.2	7000
B.4	Site preparation	3	B.1 and B.3	9000
B.5	Order materials	5	B.1	15 000
B.6	Install drainage	3	B.4 and B.5	5500
B.7	Erect structure	2	B.4 and B.5	7500
B.8	Clad structure	3	B.7 and B.5	6000
B.9	Install flooring	2	B.6 and B.8	4500
B.10	Install services	2	B.9	8000
B.11	Finish interior	3	B.10	3500

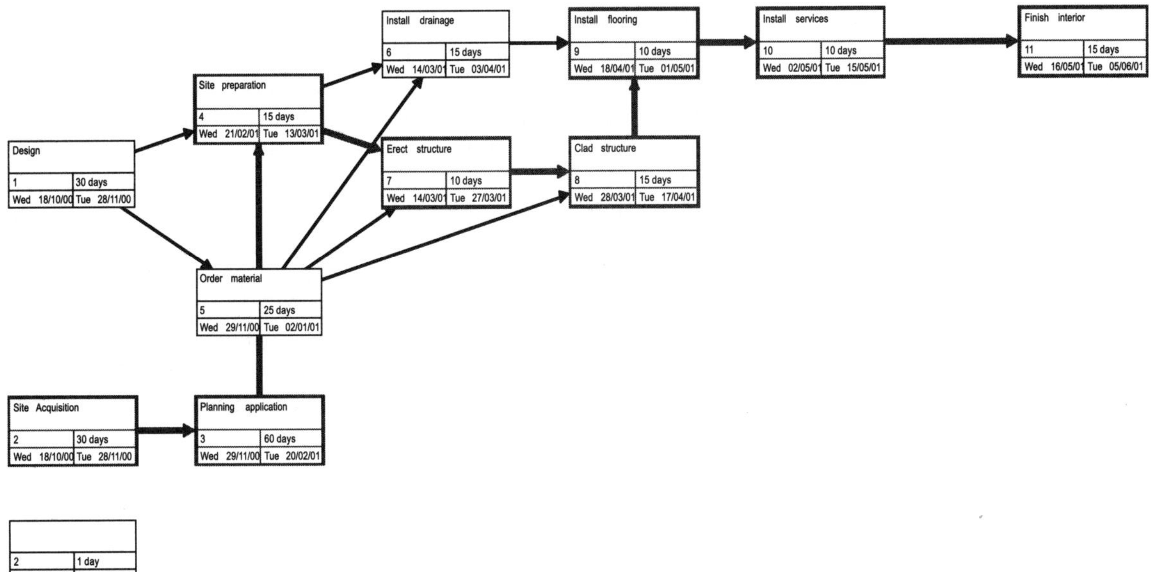

Figure 4.2.2 *Network diagram for building project*

and costs are fixed and that all preceding activities must be complete before the following activities can start.

If we needed to know when the building will be completed, we do not just add up the weeks required for each activity – this would come to 46 weeks. Some activities may be able to be undertaken simultaneously and reduce the overall time needed.

We need to calculate the dependency effect. First we shall manually work this out and then produce a graphic illustration of the plan. The first stage is to draw a network diagram as in Figure 4.2.2. This shows the linkage between activities which eases preparation of the schedule.

We could try to schedule in a longhand layout working through each activity in turn, but it is easier to carry it out using a matrix as in Table 4.2.2. The column headings we are initially concerned with are:

ES = Early Start, i.e. the earliest time an activity can start

EF = Early Finish, i.e. ES + Duration

Table 4.2.2 Calculation of early start and finish times

Activity Ref.	*Duration weeks*	*ES*	*EF = ES + Duration*	*LS*	*LF*	*Float*
B.1	6	0	6			
B.2	6	0	6			
B.3	12	7	18			
B.4	3	19	21			
B.5	5	7	11			
B.6	3	~~12~~/22	24			
B.7	2	~~12~~/22	23			
B.8	3	~~12~~/22	26			
B.9	2	~~25~~/27	28			
B.10	2	29	30			
B.11	3	31	33			

We start with identifying which activities can start at any time, i.e. they have no predecessors. In this case this only applies to activities B.1 and B.2, the design and site acquisition activities, and they therefore have an ES of zero. We then proceed to start any follow-on activities when its predecessor ends, i.e. the week following their EF.

Where there is more than one predecessor, we take the *latest* EF value by which all the predecessors finish. This is shown in the matrix by scoring out the ES values which are not selected.

By looking for the highest EF time we can see how long the project will take to complete. In this case it is the EF for activity B.11, i.e. 33 weeks, assuming all the activities are completed on time. However, do we need to start every activity at its earliest start time? The way to determine this is to complete the next two columns in the matrix as in Table 4.2.3. These are:

LF = Latest Finish, i.e. the latest time an activity can finish to prevent extending the overall project completion time

LS = Latest Start, i.e. LF – Duration

This time we take the final activity, i.e. B.11, and put its EF value into its LF box and then work backwards to see when all predecessors have to be completed by. This is found by transferring the LS value less one week into all immediate predecessors' LF boxes. If an activity has no follow-on activities, it would share the LF time with the final activity.

Where an activity is directly preceded by more than one activity, we use the *earliest* LS value by which all the follow-on activities start. This is shown in the matrix by scoring out the LF values which are not selected.

The float, or slack as it is also called, value is the time that an activity's EF time can be delayed without affecting the end date of the project. It is shown in Table 4.2.4 and calculated from:

Float = LF – EF

The only activities with a float are B.5 and B.6. This means that activity B.5 could either start up to ten weeks after its ES time, or take ten weeks longer than estimated, and not affect the overall project time. Similarly activity B.6 has two weeks' float available.

Table 4.2.3 Calculation of latest start and finish times

Activity Ref.	*Duration weeks*	*ES*	*EF*	*LS = LF – Duration*	*LF*	*Float*
B.1	6	0	6	0	6/~~18~~	
B.2	6	0	6	0	6	
B.3	12	7	18	7	18	
B.4	3	19	21	19	21/~~28~~	
B.5	5	7	11	17	~~26~~/21/~~28~~	
B.6	3	22	24	24	26	
B.7	2	22	23	22	23	
B.8	3	24	26	24	26	
B.9	2	27	28	27	28	
B.10	2	29	30	28	30	
B.11	3	31	33	31	33	

Table 4.2.4 Calculation of floats

Activity Ref.	*Duration weeks*	*ES*	*EF*	*LS*	*LF*	*Float = LF − EF*
B.1	6	0	6	0	6	0
B.2	6	0	6	0	6	0
B.3	12	7	18	7	18	0
B.4	3	19	21	19	21	0
B.5	5	7	11	17	21	10
B.6	3	22	24	24	26	2
B.7	2	22	23	22	23	0
B.8	3	24	26	24	26	0
B.9	2	27	28	27	28	0
B.10	2	29	30	28	30	0
B.11	3	31	33	31	33	0

The other activities are critical to the project finishing in 33 weeks, i.e. they must be started at their ES time and take no longer than their estimated duration. These lie along what is termed the *critical path*, i.e. the path along which all activities must adhere to their expected time so that the project is completed in the minimum possible time. This critical path is indicated as a heavy line in the network diagram in Figure 4.2.2. Any activity that takes longer than its estimated duration plus any associated float will increase the overall project completion time.

To ease communication of the sequence in which the tasks require completion, tasks can be portrayed in a Gantt, or bar, chart as shown in

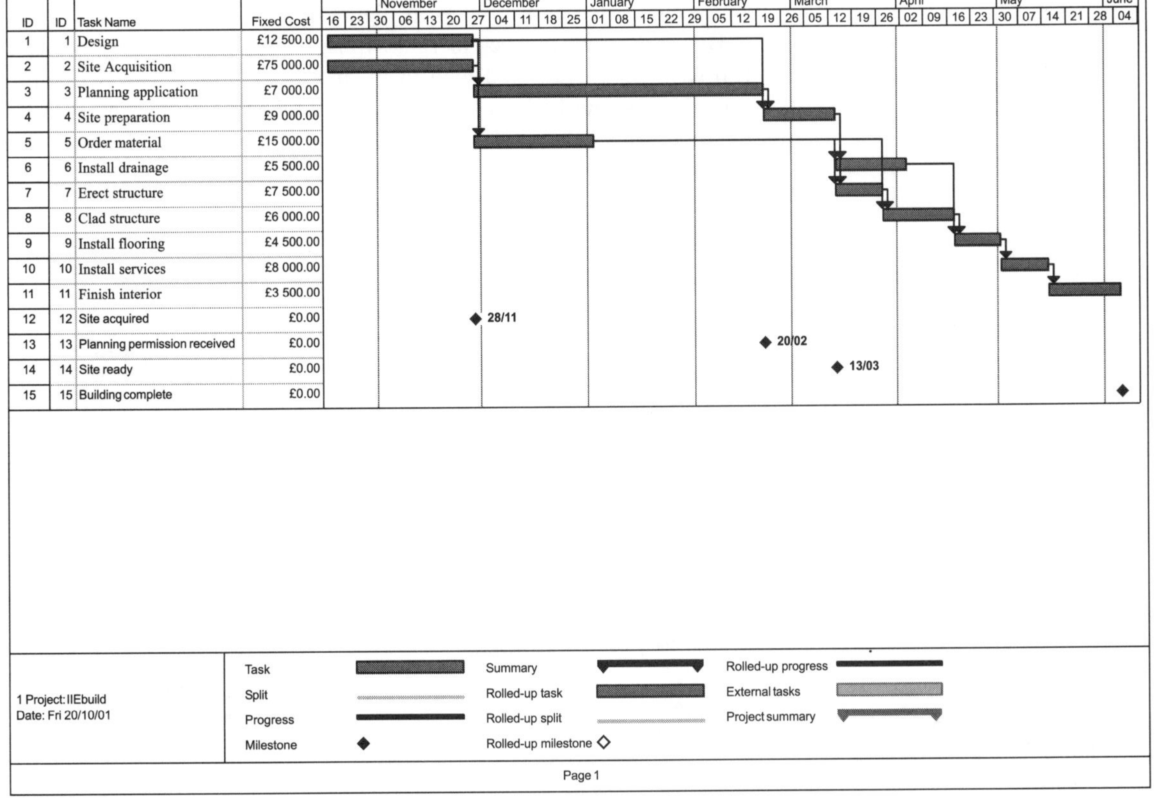

Figure 4.2.3 *Gantt chart for building project*

Figure 4.2.3. The Gantt chart shows the activities on a timescale and therefore is easy to understand. The linkages and floats also stand out. This chart shows the ES values of each activity and certain key events, e.g. planning permission received, have been selected as milestones, i.e. important points during the project's life.

Having scheduled the activities, we now need a schedule of cash flow so that the accounting department can ensure money is available when required. To do this, we need to determine the cash flow for each activity as in Table 4.2.5.

Table 4.2.5 Cash flow per activity

Ref.	*Activity*	*Cost £*	*Cash flow pattern*
B.1	Design	12 000	Even spread
B.2	Site acquisition	75 000	£5000 week 1 and balance at end
B.3	Planning application	7000	£2000 weeks 1–3, balance at end
B.4	Site preparation	9000	Even spread
B.5	Order materials	15 000	One month after delivery
B.6	Install drainage	5500	Even spread
B.7	Erect structure	7500	Even spread
B.8	Clad structure	6000	Even spread
B.9	Install flooring	4500	Even spread
B.10	Install services	8000	Even spread
B.11	Finish interior	3500	Even spread

Using this and the information on the activity's ES and EF times from Table 4.2.3, we can complete a spreadsheet showing the cash outflow on each week, as in Table 4.2.6. From this we can show a graph of weekly and accumulated cash flow as in Figure 4.2.4.

However, if we go back to Table 4.2.4, we see that activities B.5 and B.6 have float associated with them. If we delay these two activities' start dates to the LS times, we also delay when their costs need to be incurred. This delay can earn interest on that money for this time. The difference in cash flow is shown in Figure 4.2.5.

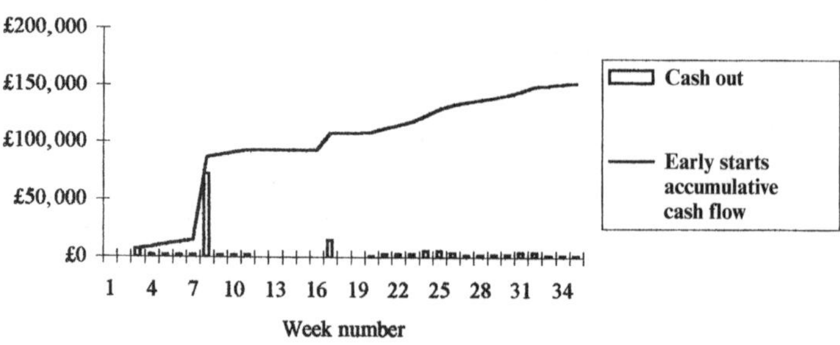

Figure 4.2.4 *Cash flow for early start times*

Figure 4.2.5 *Difference in cash flows for early start and late start times*

Table 4.2.6 Cash flow for build project

Week	*Activities funded*				*Cash out*	*Early starts accumulative cash flow*
	Activity	*Amount*	*Activity*	*Amount*		
0						
1	B.1	£2 000	B.2	£5 000	£7 000	£7 000
2	B.1	£2 000			£2 000	£9 000
3	B.1	£2 000			£2 000	£11 000
4	B.1	£2 000			£2 000	£13 000
5	B.1	£2 000			£2 000	£15 000
6	B.1	£2 000	B.2	£70 000	£72 000	£87 000
7	B.3	£2 000			£2 000	£89 000
8	B.3	£2 000			£2 000	£91 000
9	B.3	£2 000			£2 000	£93 000
10					£0	£93 000
11					£0	£93 000
12					£0	£93 000
13					£0	£93 000
14					£0	£93 000
15			B.5	£15 000	£15 000	£108 000
16					£0	£108 000
17					£0	£108 000
18	B.3	£1 000			£1 000	£109 000
19	B.4	£3 000			£3 000	£112 000
20	B.4	£3 000			£3 000	£115 000
21	B.4	£3 000			£3 000	£118 000
22	B.7	£3 750	B.6	£1 833	£5 583	£123 583
23	B.7	£3 750	B.6	£1 833	£5 583	£129 167
24	B.8	£2 000	B.6	£1 833	£3 833	£133 000
25	B.8	£2 000			£2 000	£135 000
26	B.8	£2 000			£2 000	£137 000
27	B.9	£2 250			£2 250	£139 250
28	B.9	£2 250			£2 250	£141 500
29	B.10	£4 000			£4 000	£145 500
30	B.10	£4 000			£4 000	£149 500
31	B.11	£1 167			£1 167	£150 667
32	B.11	£1 167			£1 167	£151 833
33	B.11	£1 167			£1 167	£153 000

Table 4.2.7 Activities involved in new product design

Activity Ref.	*Activity*	*Precede by*	*Estimated durations (weeks)*			*Expected duration*	*Std Dev.*	*Variance*
			Optimist	*Likely*	*Pessimist*	$= \frac{t_o + 4\,t_l + t_p}{6}$	$\frac{t_o + t_p}{6}$	
			t_o	t_l	t_p	t_e	δ	δ^2
D.1	Market research	–		25 fixed time				
D.2	Concept design	D.1	5	10	25			
D.3	Detailed design	D.2	25	30	35			
D.4	Produce prototype	D.3	10	10	15			
D.5	Engineering tests	D.4		5 fixed time				
D.6	Test market	D.5		5 fixed time				
D.7	Finalize design	D.4	10	15	15			

Project 2 – new product design: variable estimates

In the previous project, we assumed that all the activity durations were fixed. However, in practice, we do not always know exactly in advance how long an activity will take – especially if we have not done that exact task before. What we often have to do in that situation is seek a range of possible estimates and use these to calculate an expected duration.

Let us assume the second project is such a case and we have three estimates for the durations for some activities as shown in Table 4.2.7. The others still have a fixed estimate for their duration. The three estimated durations are based on:

- Optimist: Everything goes exactly right – no problems.
- Likely: Normal activity – some minor problems.
- Pessimist: Everything goes wrong – many problems.

Where we have a fixed duration activity, this will become the expected duration. However, to calculate an expected duration for the variable activities, we apply the formula:

$$\text{Expected duration} = (\text{optimist} + (4 \times \text{likely}) + \text{pessimist}) \div 6$$

i.e. $t_e = (t_o + (4 \times t_l) + t_p) \div 6$

If we take as an example the estimated durations for D.2, the concept design activity, then the formula fills out to:

$$\begin{aligned} t_e &= (5 + (4 \times 10) + 25) \div 6 \\ &= (5 + 40 + 25) \div 6 \\ &= (70) \div 6 \\ &= 11.67 \text{ weeks} \end{aligned}$$

Activity 4.2.1

Use a spreadsheet and calculate the expected duration for activities D.3, D.4 and D.7.

You can also have software calculate the expected time, e.g. Microsoft™ Project using its PA_PERT Entry Sheet view, but this does not calculate out a standard deviation for the activity (see later). Once we have all expected durations calculated, we can complete the matrix as in Table 4.2.8.

Note that the expected durations are not the same as the likely durations, but are altered by expectations about how the activity may vary.

We can now schedule this project using the expected durations and a network diagram as per Figure 4.2.6 to produce the project Gantt chart as per Figure 4.2.7.

Table 4.2.8 Expected times for activities

Activity Ref.	*Activity*	*Precede by*	*Estimated durations (weeks)*			*Expected duration*	*Std Dev.*	*Variance*
			Optimist	*Likely*	*Pessimist*	$= \frac{t_o + 4\,t_l + t_p}{6}$	$\frac{t_o + t_p}{6}$	
			t_o	t_l	t_p	t_e	δ	δ^2
D.1	Market research	–	25 fixed time			25		
D.2	Concept design	D.1	5	10	25	11.67		
D.3	Detailed design	D.2	25	30	35	30		
D.4	Produce prototype	D.3	10	10	15	10.83		
D.5	Engineering tests	D.4	5 fixed time			5		
D.6	Test market	D.5	5 fixed time			5		
D.7	Finalize design	D.4	10	15	15	14.17		

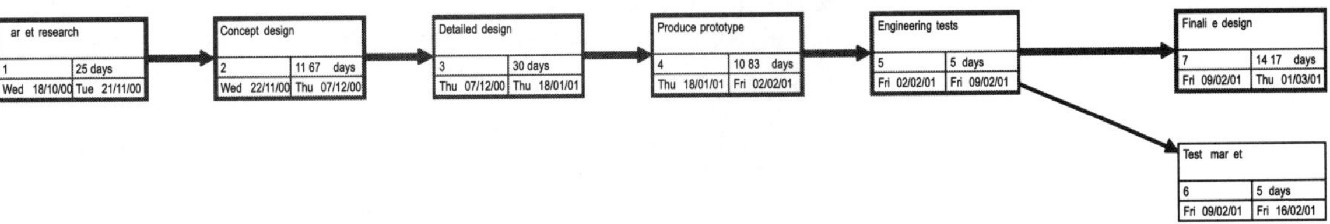

Figure 4.2.6 *Network diagram for design project*

ID	BS		Task name	Duration	Start	Finish
1	D.1		Market research	25 days	ed 18/10/00	Tue 21/11/00
2	D.2		Concept design	11.67 days	ed 22/11/00	Thu 07/12/00
3	D.3		Detailed design	30 days	Thu 07/12/00	Thu 18/01/01
4	D.4		Produce prototype	10.83 days	Thu 18/01/01	Fri 02/02/01
5	D.5		Engineering tests	5 days	Fri 02/02/01	Fri 09/02/01
6	D.6		Test market	5 days	Fri 09/02/01	Fri 16/02/01
7	D.7		Finali e design	14.17 days	Fri 09/02/01	Thu 01/03/01
8	8		MR report due	0 days	Tue 21/11/00	Tue 21/11/00
9	9		Initial design complete	0 days	Thu 18/01/01	Thu 18/01/01
10	10		Final design complete	0 days	Thu 01/03/01	Thu 01/03/01

November December January February M
16/10 30/10 13/11 27/11 11/12 25/12 08/01 22/01 05/02 19/02
21/11
18/01

Project: Project1
Date: Fri 22/10/00

Task — Split — Progress — Milestone — Summary — Rolled-up task — Rolled-up split — Rolled-up milestone — Rolled-up progress — External tasks — Project summary

Page 1

Figure 4.2.7 *Gantt chart for design project*

Activity 4.2.2

Complete a matrix similar to Table 4.2.4 for this project using the expected durations for each activity. Calculate the ES, EF, LS, LF and floats values.

Figure 4.2.6 shows the critical activities to be D.1, D.2, D.3, D.4 D.5 then D.7. These total to 96.17 weeks based on the expected times. Activity D.6 alone has a float – equal to 9.17 weeks.

Maths in action

Patterns in an activity's duration

If we drew graphs of the possible durations for each of the variable activities, we would get distinctly different patterns as shown in Figure 4.2.8. These show:

(a) Long tail, e.g. activity D.2. On occasions may possibly suffer a major problem, or need an extended examination and therefore take a much longer time.
(b) Normal bell-shaped distribution, e.g. activity D.3.
(c) Most likely to be short, a few may have problems and take longer, e.g. activity D.4.
(d) Most likely to be long, a few may go well and take shorter, e.g. activity D.7.

The expected duration is only a calculated value – this means that 50% of the time it will be less than this, and 50% of the time it will be more. Many organizations are unhappy with scheduling a project with only a 50% probability that some activities' duration will be within that used for planning. They may wish to schedule a longer duration than that calculated as expected to ensure a higher probability of the scheduled duration being met.

The extra time added will depend on the desired probability and the distribution pattern of the activity. In a normal distribution, a good value to take is that of the standard deviation (SD). Unfortunately we do not have sufficient values to calculate out a true SD – the practice in project planning is to calculate an approximate SD, as follows:

$$\text{SD} \cong (\text{pessimist value} - \text{optimist value}) \div 6$$

If we take activity D.2 as an example, this works out to:

$$\begin{aligned}\text{SD} &\cong (25 - 5) \div 6\\ &\cong (20) \div 6\\ &\cong 3.33\end{aligned}$$

Note: use this approximation with care as it assumes a normal distribution which is often not the case, e.g. the pattern for D.2 shows a long tail rather than the normal bell shape.

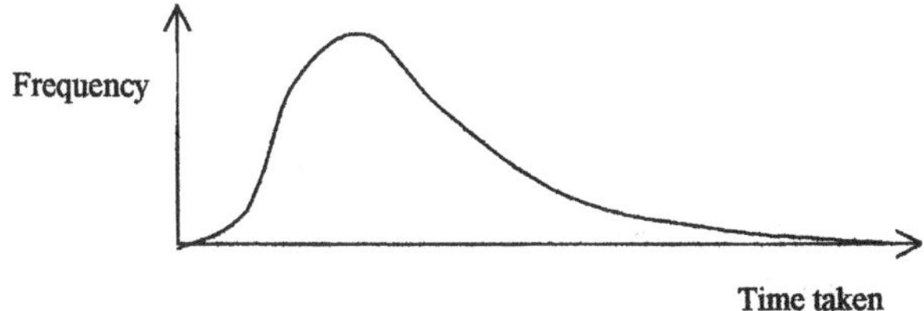

(a) Long tail distribution

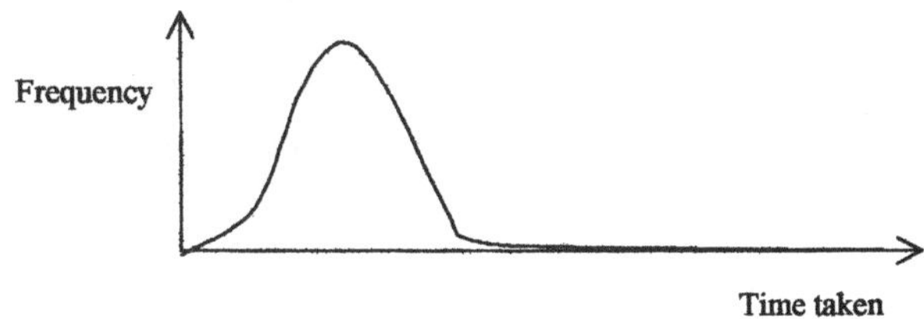

(b) Normal distribution

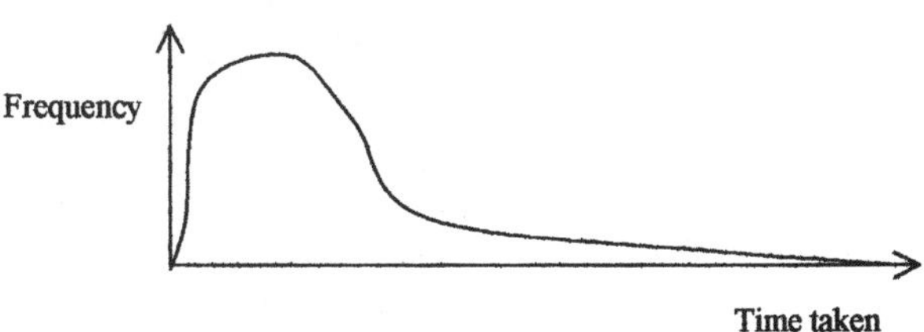

(c) Main short times, but a few extended times

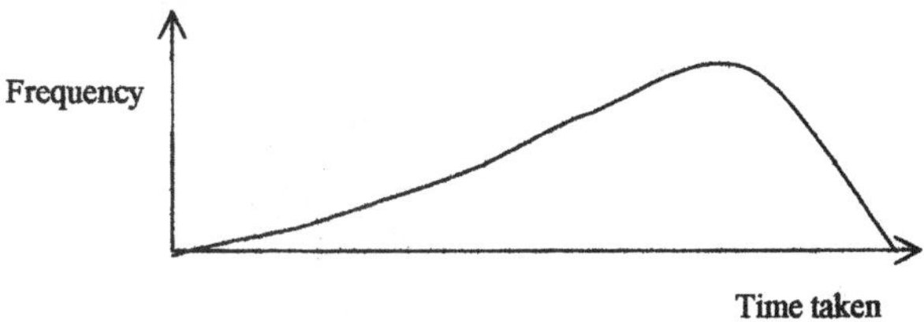

(d) Mainly long times, with a wide dispersion

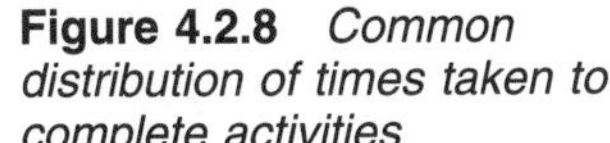

Figure 4.2.8 *Common distribution of times taken to complete activities*

Activity 4.2.3

Calculate out the approximate standard deviation (SD) values for activities D.3, D.4 and D.7. Also calculate the variance which is the square of the SD – we'll come to why we calculate these variances later.

Putting the approximate standard deviations into the matrix, we have Table 4.2.9. We enter the variations and then sum these variations on the critical path activities.

Table 4.2.9 Standard deviations for activities

Activity Ref.	*Activity*	*Precede by*	*Estimated durations (weeks)*			*Expected duration*	*Std Dev.*	*Variance*
			Optimist	*Likely*	*Pessimist*	$= \frac{t_o + 4\,t_l + t_p}{6}$	$\frac{t_o + t_p}{6}$	
			t_o	t_l	t_p	t_e	δ	δ^2
D.1	Market research	–		25 fixed time		25	0	0
D.2	Concept design	D.1	5	10	25	11.67	3.33	11.11
D.3	Detailed design	D.2	25	30	35	30	1.67	2.78
D.4	Produce prototype	D.3	10	10	15	10.83	0.83	0.69
D.5	Engineering tests	D.4		5 fixed time		5	0	0
D.6	Test market	D.5		5 fixed time		5	0	0
D.7	Finalize design	D.4	10	15	15	14.17	0.83	0.69
						Total of variances on critical path		15.28

Maths in action

Normal distribution

In any situation where we have a range of variables, we normally expect a completely random variation created by repeating common causes. A histogram of the measurement of the output would then create a *normal* distribution curve, as shown in Figure 4.2.9.

With a normal distribution an indicator for the probability of being within a set value from the mean value is the standard deviation. Table 4.2.10 gives a sample of these probabilities:

Table 4.2.10 Relationship between probabilities and standard deviations

Desired probability of meeting value	50%	75%	90%	95%	99%
Standard deviation from mean	0	0.67	1.28	1.65	2.33

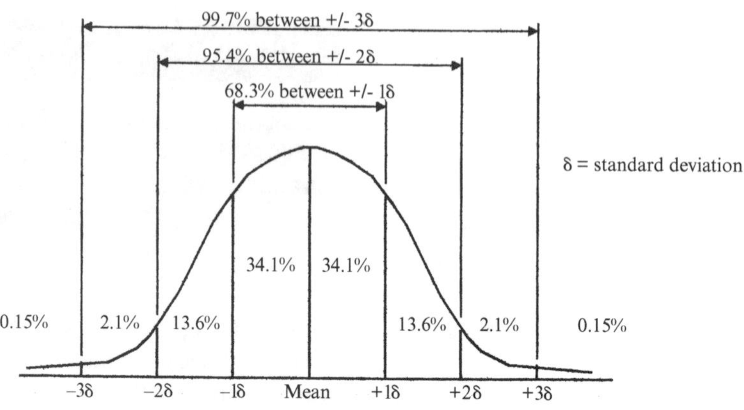

Figure 4.2.9 *Normal distribution curve showing percentage values within multiples of the standard deviation*

This means that if we wish activity D.2 to be met on 90% of occasions, then we take the expected duration and add to it (1.28 times the SD), viz:

D.2's new duration = expected duration + (1.67 times SD)
= 11.67 + (1.67 × 3.33)
= 11.67 + 5.56
= 17.23

This is quite a large increase – imagine what it would do if we had to apply the same calculation to each activity on the critical path to arrive at a value by which we have a 90% probability of completing the whole project. Fortunately we do not.

As the critical path is formed from a string of activities, we can assume that some of these activities will take longer and some will take less than their expected values. These ups and downs will offset each other to some degree and we can allow for that by calculating the overall SD for the project using the variation values for each activity on the critical path. We then find the overall SD by summing the critical path variances and finding the square root of this sum. Note: only these on the critical path are used.

Project SD = square root of (sum of variances on critical path)

Therefore, for this project,

Project SD = $\sqrt{(15.28)}$ = 3.91

If we wish to calculate for the project a new overall completion time with a set probability, we use the overall expected completion time and add to it a multiple of the project's SD value. For example, if we wish to find the overall completion time with a 90% probability of being met, we have:

Project's new duration = expected duration + (1.67 times SD)
= 11.67 + (1.67 × 3.33)
= 96.17 + (1.67 × 3.91)
= 96.17 + 6.53
= 102.80 weeks

We can set the project end date in Microsoft™ Project based to this new value and it will show up an overall project float of 6.53 weeks. We can then use this overall float as a buffer in managing the project. Microsoft™ Project cannot accommodate variable activities except in this way. In Section 4.3, however, we introduce add-in software which does allow variable activities to be coped with and makes use of buffers throughout the project.

Project 3 – new plant installation: changing the overall time

The activities for this project are given in Table 4.2.11. and the network in Figure 4.2.10. Let us assume that the durations are fixed for these activities. Resources will be needed for all the activities, but this time

Table 4.2.11 Activities involved in plant installation

Activity Ref.	*Activity*	*Duration (weeks)*	*Precede by*	*Resource used*
P.1	Procure plant	12	–	Engineers
P.2	Procure tooling	4	P.1	Engineers
P.3	Install plant	1	P.2	Fitters
P.4	Plant trials	2	P.3	3 plant operatives
P.5	Operator training	1	–	3 plant operatives
P.6	Build up inventory	4	P.4, P.5	3 plant operatives

we will look only at the plant operatives which are required on activities P.4, P.5 and P.6.

Activity 4.2.4

Complete for this project a matrix similar to Table 4.2.4 using these expected durations for each activity. Calculate the ES, EF, LS, LF and floats values.

The Gantt chart for the project is shown in Figure 4.2.11. This has been prepared using the latest start times for activities P.2 and P.5 as both have a considerable degree of float available.

The resource information for the plant operatives is shown in Figure 4.2.12. In here the details entered are:

- Max. units – express as 100% = 1 unit, i.e. 300% = 3 operatives.
- Std rate – enter full cost rate (see pages xxx–xxx).
- Ovt. rate – enter overtime full cost rate. Unfortunately only one rate can be entered in Microsoft™ Project which will need consideration if several different rates are paid, e.g. different rates at different times of the week.

Figure 4.2.10 *Network diagram for production project*

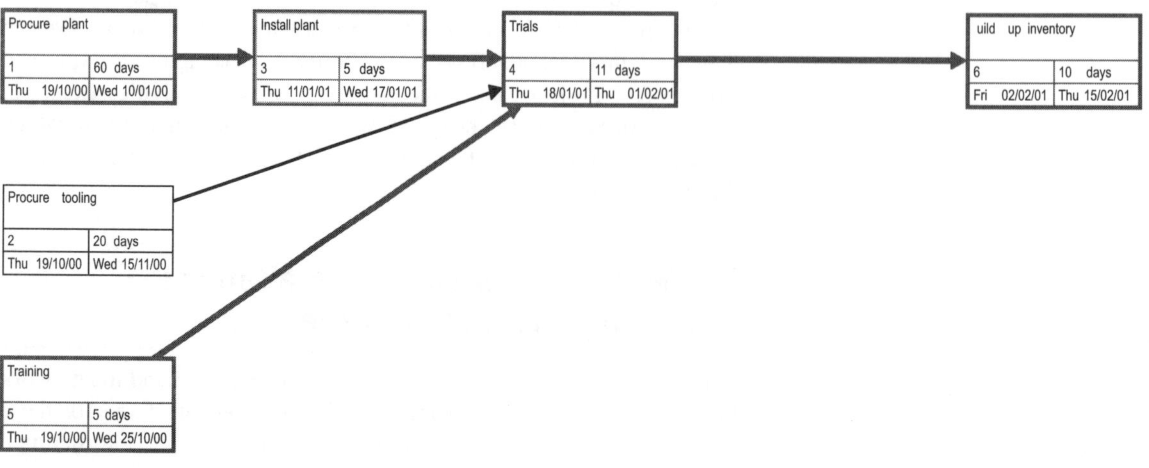

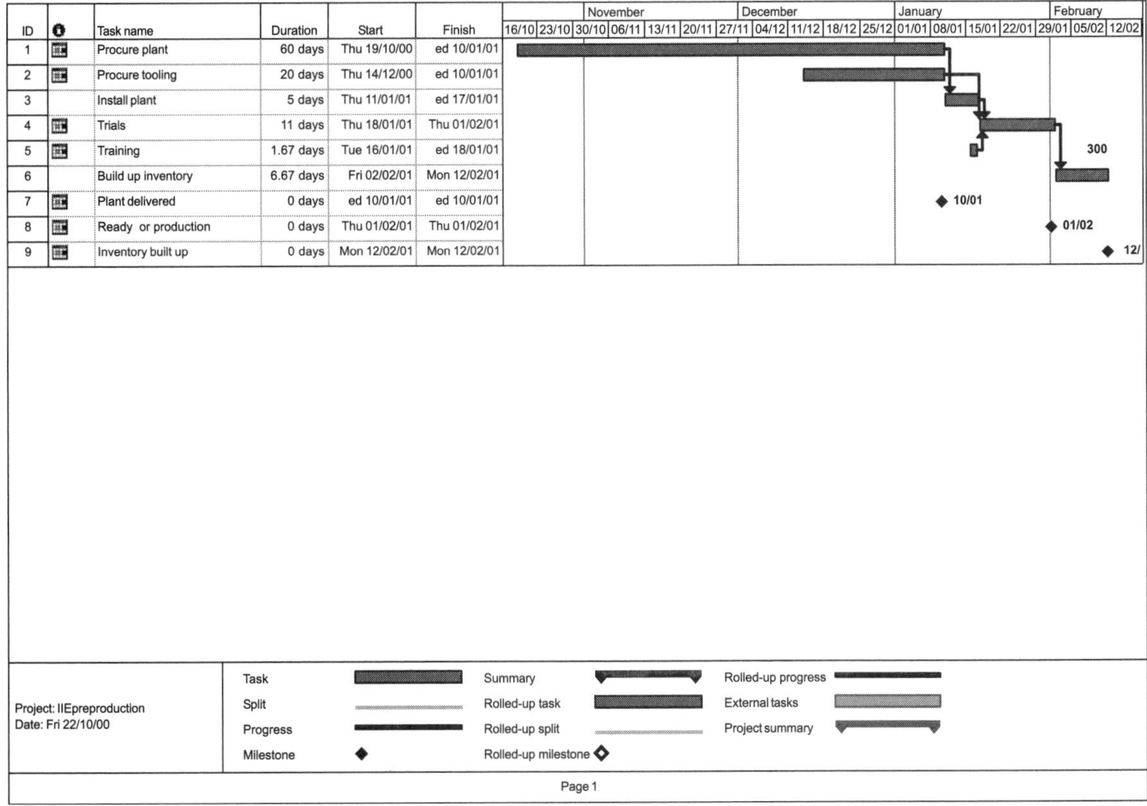

ID		Task name	Duration	Start	Finish
1		Procure plant	60 days	Thu 19/10/00	ed 10/01/01
2		Procure tooling	20 days	Thu 14/12/00	ed 10/01/01
3		Install plant	5 days	Thu 11/01/01	ed 17/01/01
4		Trials	11 days	Thu 18/01/01	Thu 01/02/01
5		Training	1.67 days	Tue 16/01/01	ed 18/01/01
6		Build up inventory	6.67 days	Fri 02/02/01	Mon 12/02/01
7		Plant delivered	0 days	ed 10/01/01	ed 10/01/01
8		Ready or production	0 days	Thu 01/02/01	Thu 01/02/01
9		Inventory built up	0 days	Mon 12/02/01	Mon 12/02/01

Figure 4.2.11 *Gantt chart for production project*

IIEpreproduction

ID		Resource name	Initials	roup	Max. units	Std rate	vt. rate	Cost/ se	Accrue at	Base calendar	Code
1		Plant operatives	P	peratives	300	£10.00/hr	£12.50/hr	£0.00	Prorated	Standard	

Page 1

Figure 4.2.12 *Resource information report*

- Cost/use – this is accumulated during project running based on hours and associated rate that have been used.
- Accrued – indicating basis of adding up the cost as the resource is actually used during the project.

When we tie the resource used to the activities, we can see on a resource usage calendar the total hours planned against each activity and the hours planned on any particular day (see Figure 4.2.13). We can also see on the overall project calendar which activities are planned for any day in any month (see Figure 4.2.14).

Changing the completion date

Having worked out a project's completion date, we may find that this date is not acceptable.

To move the completion to a later date is easily achieved by simply delaying the start of the project, or one of the activities along the critical path. But how do we proceed if we need an earlier completion date?

To shorten, or crash as it is often referred to, the overall duration of a project we have to shorten some of the critical path activities within the project. There are various ways of doing this, for example:

- Changing the specification so that less is achieved, e.g. in this project reducing the weeks scheduled to build up the inventory and ending up with a smaller inventory to support the product launch.
- Working extra hours on the project:
  - Using overtime, e.g. having the plant operatives working more than 40 hours per week.

Figure 4.2.13 *Resource usage calendar*

IIEpreproduction

ID	❶	Resource name	ork	Details	15 Jan 01 M	T		T	F	S	S
		nassigned	0 hrs	ork							
		Procure plant	*0 hrs*	ork							
		Procure tooling	*0 hrs*	ork							
		Install plant	*0 hrs*	ork							
		Plant delivered	*0 hrs*	ork							
		Ready for production	*0 hrs*	ork							
		Inventory built up	*0 hrs*	ork							
		Plant operatives	464 hrs	ork		16h	24h	24h	24h		
		Trials	*264 hrs*	ork				24h	24h		
		Training	*40 hrs*	ork		16h	24h				
		Build up inventory	*160 hrs*	ork							

Page 1

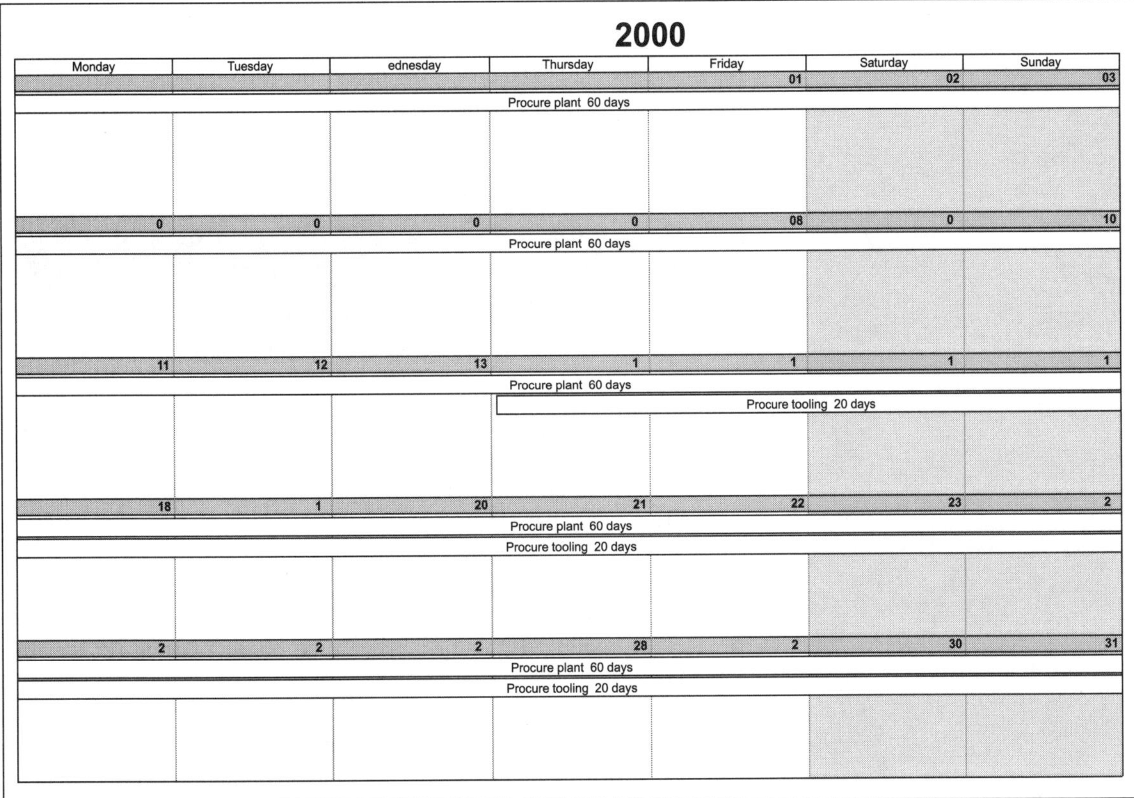

Figure 4.2.14 *Project calendar*

- Using more people. This may help some activities but we do not necessarily get a proportional reduction in time. Extra work time may be needed due to extra communication problems, using people without the required skill level or who are unfamiliar with the project who need to be brought up to speed.
- Working extra shifts, e.g. having the new machine put onto two, three or even four shifts.

- Reducing the lead time of externally supplied items:
  - Using special delivery channels, e.g. using air freight instead of road transport.
  - Having the supplier reschedule their production to release our items earlier.
  - Requesting the supplier to work extra hours to produce the item quicker.

Most of these will cost extra and the management skill is finding the balance between them that will deliver the needed completion date at the minimum cost. We have not applied any of the above to this project.

Combining projects

One of the main problems when organizations are undertaking many interconnected projects is the cross-scheduling of resources that are required on several projects. Microsoft™ Project allows this to be done.

Figure 4.2.15 shows the Gantt chart for three projects combined, but without any links established between them. This shows that the new plant is scheduled to be producing before the site preparation activity for the new factory will be started! If projects are connected we need the links established to determine when the combined project will be complete.

Activity 4.2.5

Can you think of any key links between the projects?

There are a few possible links that could be made, some of which may be based on using the same people resources at key times. If this happens we would need to reschedule to level usage or increase resources.

The two links that we will use are:

- We cannot install the new plant (P.3) until the new building is complete (B.11).
- Procure tooling (P.2) relies on the final product design (D.7) being done.

By establishing these links on the Gantt we arrive at a revised Gantt chart as shown in Figure 4.2.16. This took out some excess float by using the start delay function on the following activities within the new plant project to make use of the large amount of float they carried:

- Procure plant (P.1 and training (P.7) activities – start delay of 146 days from start of combined project, i.e. 20.8 weeks (7 day week).
- Procure tooling (P.2) – start delay of 75 days from end of the expected completion of final design activity (D.7), i.e. 10.7 weeks.

The design project has just been left in its original schedule which means there is a long float before the procure tooling needs to start. This gives a completion time for the combined project as 37.5 weeks. If this date was unsuitable, then we would need to adjust some of the activity durations on the critical path. Unfortunately most software packages do not show a PERT view of a combined project – only of the milestones on it.

More on network diagrams

Network diagrams are made up using branches and nodes. There are different conventions for placing and identifying activities using these branches and nodes.

Activity on arrow (AOA)

This method puts the description of the activity along the branch and the node represents an event as in Figure 4.2.17(a). This convention requires the use of dummy activities, e.g. activity D in Figure 4.2.17(a), as two activities cannot use the same two nodes.

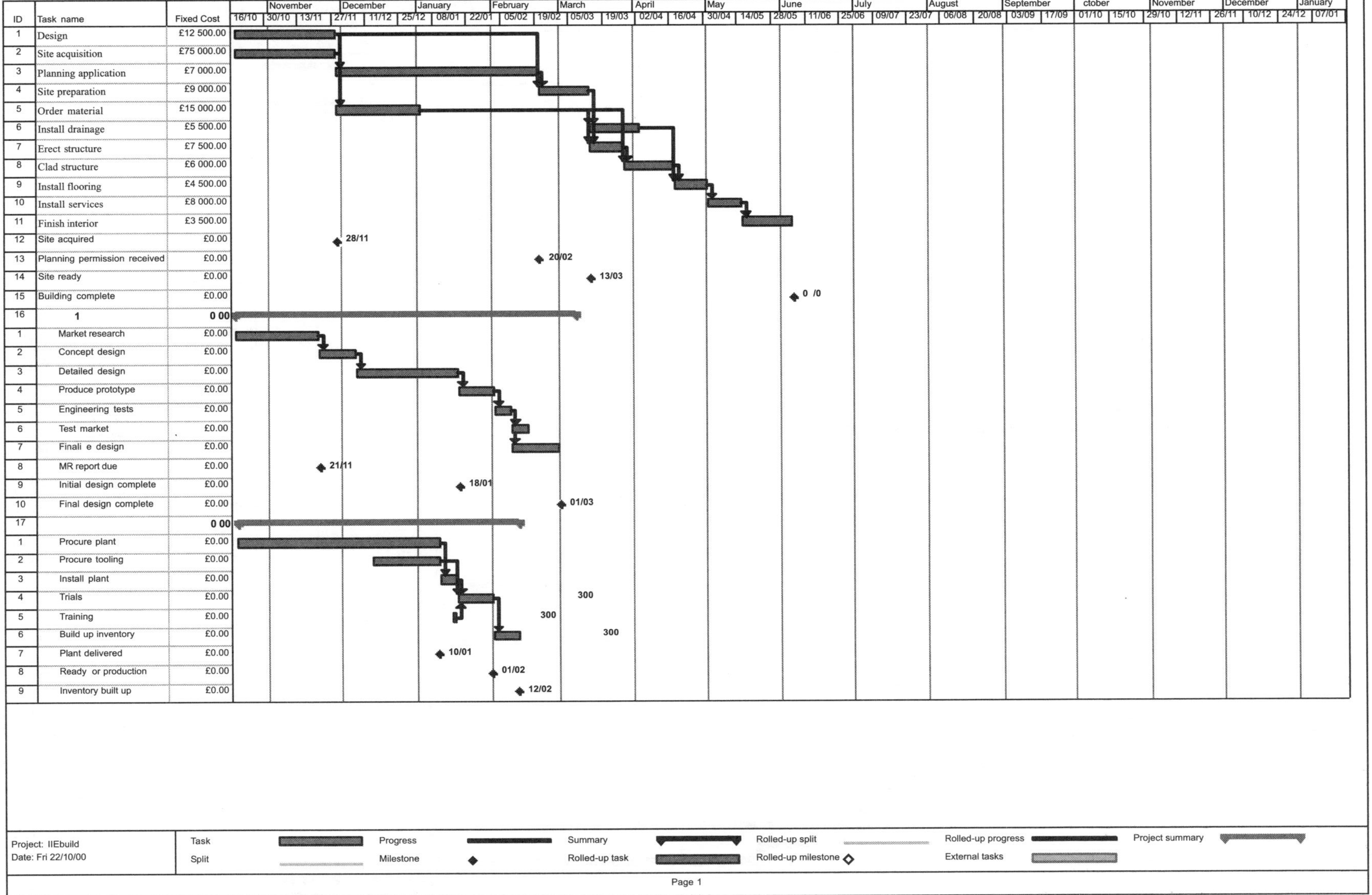

Figure 4.2.15 *Gantt chart for combined project – before linking*

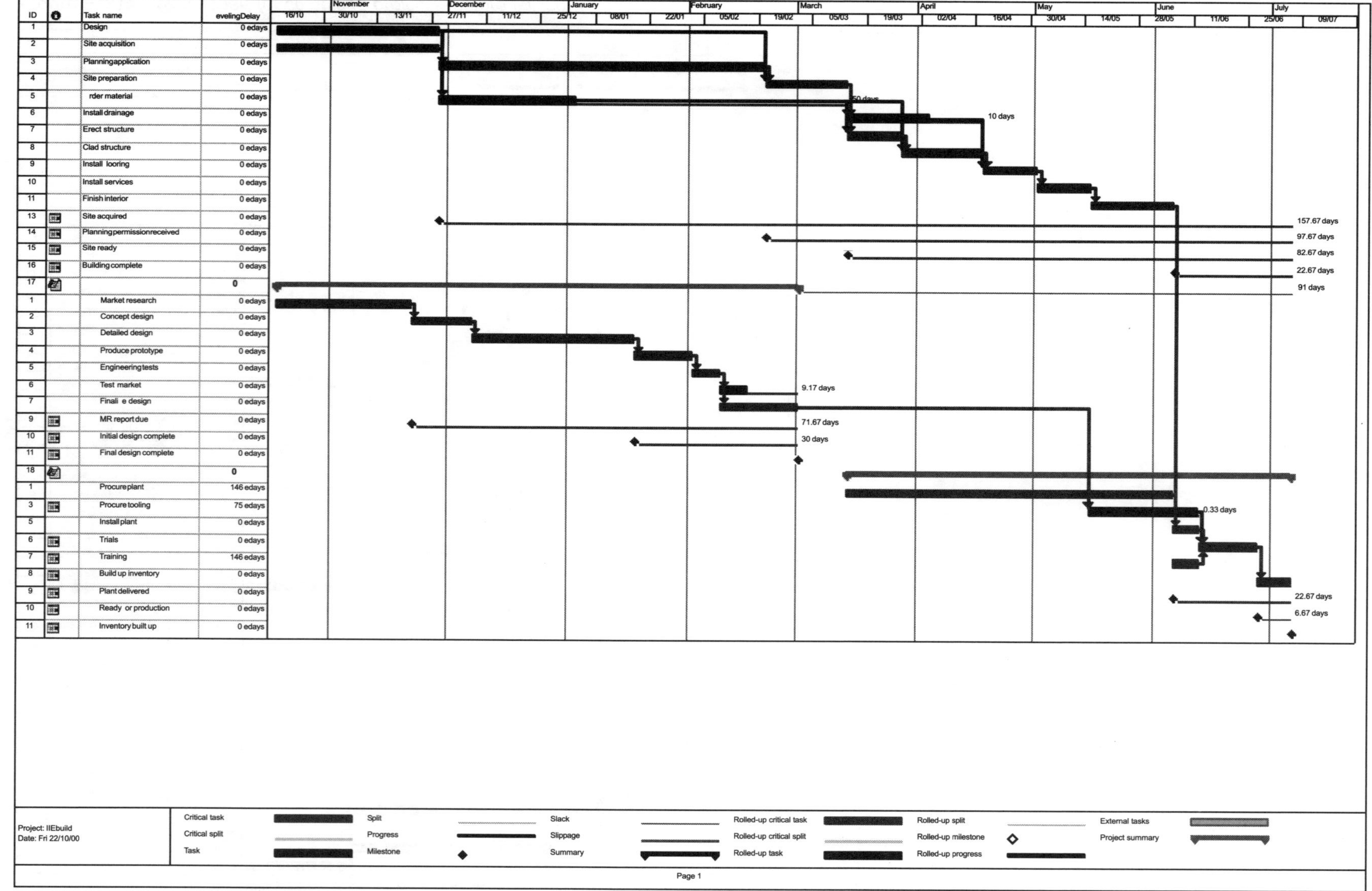

Figure 4.2.16 *Gantt chart for combined project – after linking*

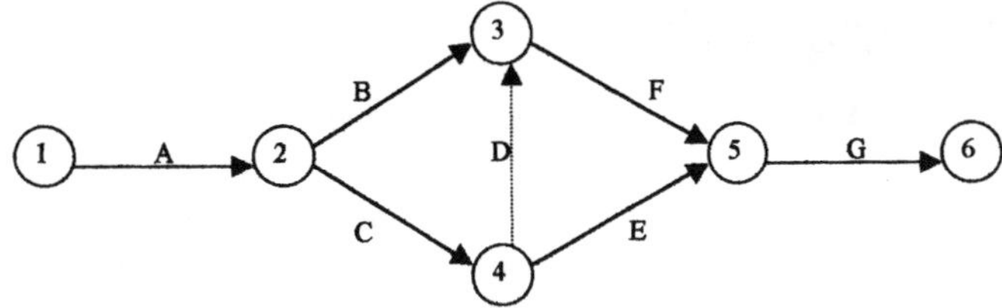

(a) Activity on Arrow (AOA) network diagram
(The numbers represent events and the letters represent activities. Activity D is a dummy, i.e. has zero time value)

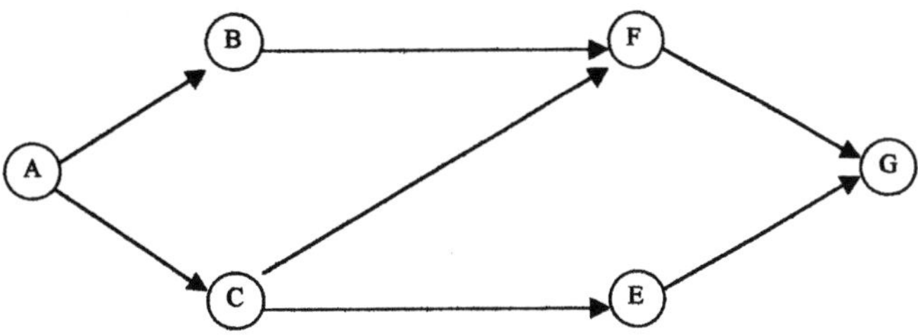

(b) Activity on Node (AON)
(The letters again represent activities, the arrows the logical relationship between them. In this case there is no need for the dummy activity, D, that appears in the AOA network)

Figure 4.2.17 *Two ways of drawing project network diagrams*

Activity on node (AON)

Here the activity information is placed at the node, and the branches represent the logical connection between the modes (activities) as in Figure 4.2.17(b). Many people prefer the use of AON because:

- It does not require the use of dummy activities.
- It is easier to follow the logic of dependencies.

AON has tended to become the norm in computer-based packages such as Microsoft™ Project, which uses a modified (i.e. reduced) AON format.

Some people find that the use of AON networks reduces the need to complete the matrices we have been using as in Table 4.2.2.

Problems 4.2.1

(1) If activity B.6 started two weeks later than its ES time, how much extra time could it then use before it affected the overall project time?

(2) What is the difference in completion time for the design project between the optimistic times and the pessimistic times?

(3) Under what circumstances would you accept a project based on the optimistic times?

(4) How difficult is it to estimate any time for a task you have done only once before?

(5) Draw your own network diagram to show all the activities on the critical path for the combined projects.

4.3 Dealing with uncertainty

This section looks in detail at why so many projects are partial, or complete, failures. The main problem appears to be uncertainties and how we cope with them. Eliyahu Goldratt's Critical Chain methodology is examined in detail as a possible solution.

As mentioned in Section 4.1, it is common for projects to end in partial, or complete, failure, i.e. having one, or more, of the following serious discrepancies:

- Taking longer than expected.
- Costing more than budgeted for.
- Failing to produce the full expected benefits.

This state of affairs has led participants to believe and even expect that large projects cannot succeed fully, mainly because of the many uncertainties involved in them. The American Production and Inventory Control Society has even postulated a set of 'laws' that reflect these expectations:

Laws of project management

(1) No major project is ever installed on time, within budget and with the same staff that started it.
(2) Projects progress quickly until they are 90% complete, then they remain at 90% complete forever.
(3) One advantage of fuzzy project objectives is that they let you avoid the embarrassment of estimating the corresponding costs.
(4) When things are going well, something will go wrong:
    - When things cannot get any worse, they will.
    - When things appear to be going better, you have overlooked something.
(5) If project content is allowed to change freely, the rate of change will exceed the rate of progress.
(6) No system is ever completely debugged. Attempts to debug a system inevitably introduce new bugs that are even harder to find.
(7) A carelessly planned project will take three times longer to complete than expected, a carefully planned project will only take twice as long.
(8) Project teams detest progress reporting because it vividly manifests their lack of progress.

It would appear that it is the uncertainties that lead to the disappointing outcomes. However, as mentioned in Section 4.1, there are many factors that can contribute to the failure. If the following factors are present, then failure is almost certainly guaranteed:

- Unobtainable objectives, usually changing as the project moves along.
- Lack of top management support.
- Lack of resources or skills, especially in project management.
- Poor selection of suppliers.
- Political resistance and sabotage.

Yet even with projects that do not suffer from these – failures are common. Does that mean that where there are uncertainties, failure is a natural consequence, as shown in the abbreviated cause–effect diagram in Figure 4.3.1?

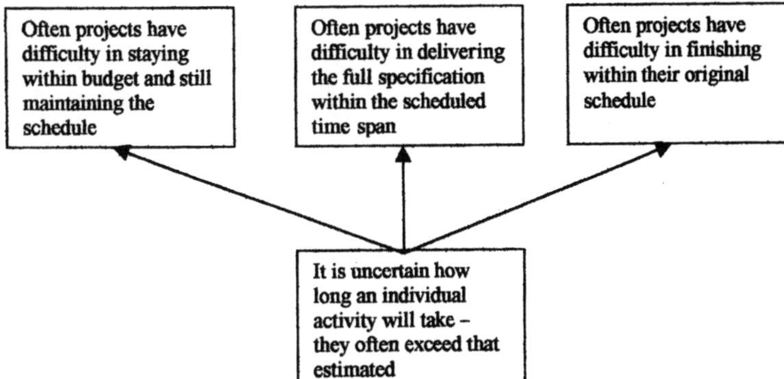

Figure 4.3.1 *Cause–effect diagram used to explain why projects often fail in meeting original time span specification and/or budget*

Eliyahu Goldratt, author of *The Critical Chain*, argues that although uncertainties can play a part – it is the way we manage these uncertainties that is the major reason for the failures. Goldratt states we should look at four points:

- How the estimate for the duration of an activity is derived.
- How activities are linked at the scheduling stage.
- How time is used before and during an activity.
- How the project is actually managed.

Why times extend

Estimating an activity's duration

You probably found it very difficult to answer a question on how long it would take you to build a house using your own labour, as there was no laid down specification (Law 3 on page 178) and you probably have never personally carried out all the tasks involved in such an undertaking. It is understandable that people find it difficult to estimate how long it will take them to complete an operation when what is involved in it is unknown to some degree.

In Section 4.2 (page 163), we made reference to the three-time-estimate method, but even this is mainly speculative guesswork, which may be better than a single time – or it may not. How then is an estimated time really arrived at?

Even with tasks that are fairly repetitive, there are variations in the time to complete them. This variation does not follow the normal distribution, but instead follows a pattern as shown in Figure 4.3.2(a). This is because there are two patterns superimposed on each other:

- One pattern is the normal distribution where the variations are limited and spread evenly around a mean, forming the bell-shaped curve as seen in Figure 4.3.2(b).
- The other pattern is a negative exponential one as shown in Figure 4.3.2(c). This distribution is made up of times which have a wide variance – often quite longer than the mean time under the normal distribution. This is due to spasmodic events occurring – for example, a crash will slow traffic flow which will then delay other drivers.

When we are asked to consider the probability of catastrophic events occurring, we tend to overestimate that probability because these events are easy to recall as having happened. Therefore we tend to not commit ourselves to a simple normal mean time, but want to make allowances for the possible happenings – even if it is highly improbable that they will.

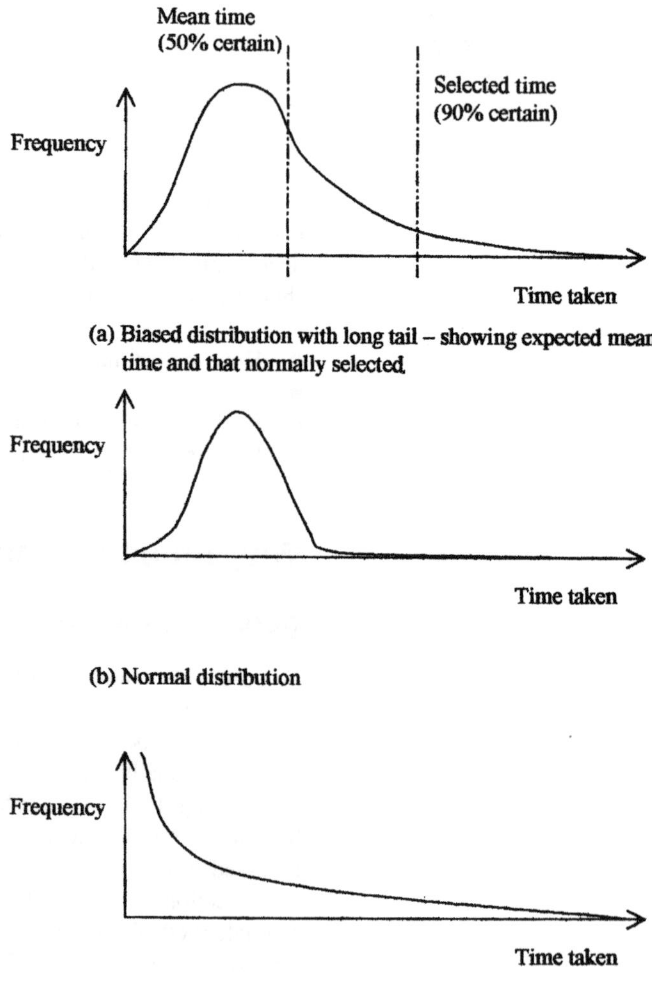

Figure 4.3.2 *Make up of times taken to carry out an activity*

In addition there is the natural tendency not to accept a target time when there is a 50% probability of it being exceeded. We would find a lower probability more acceptable – say only a 10% probability of being exceeded as shown in Figure 4.3.2(a). We prefer to have a good factor of safety, especially if we are being judged on the result.

If this is so, then the overall project timescale should be made up from a series of activities most of which have a high degree of being met. This should then, in turn, give the overall completion time a high degree of certainty in being met. Why then are so many projects late?

Preparing the schedule

The project schedule is prepared as in Section 4.2. As everyone has added safety time to the individual activities, the project completion date is extended – normally more than acceptable. It therefore must be reduced, but how?

Management knows about the tendency to add safety time to estimates of activity duration, therefore the first thing that happens is that everyone is *requested* to accept a reduced time target. As the estimate is already padded, the request is accepted if not too severe.

The net result is that the completion date for the project is reduced which earns kudos for the project manager. However, the reduction process lays the ground for an excuse if an activity is late – after all the revised date was not the original estimate!

Note that if you are a manager who has put in a tight estimate, you will find accepting a tighter target difficult. However, as other managers involved are likely to be accepting the new target, you will probably have to go along, or be seen as 'not being a team player'. Having learned this lesson, you won't make the same *mistake* again – especially if you fail to meet the revised target. The next time you *will* add a safety allowance. The system punishes you otherwise.

Operating to the schedule

When an activity is being completed in your area, there are three contact points with the schedule:

- Incoming, i.e. when the previous activity is completed.
- Operations, i.e. when the activity is being carried out.
- Outgoing, i.e. when the work is passed onto the next department.

At each of these points, the way we manage things can increase the probability of extending the time taken.

Incoming phase

There is uncertainty about when the previous activity will complete and the work will be passed on. This means that the department is probably not ready to start on the activity when it arrives. The people concerned will be working on other tasks and the incoming work will then queue until they become available. If they did stop what they were doing, they would lose time on that task and are therefore reluctant to do so.

A delay therefore enters the system, shortening the available time.

Operational phase

There will also be little urgency in starting the new activity as it is known that less time is normally needed than allowed for. Note that the estimation of time at the planning stage tended to be pessimistic, now the opposite occurs and an optimistic view is taken – often over-optimistic by making little, or no, allowance for things to go wrong. Yet, if we go by 'Law 3' – things will go wrong.

The student syndrome

The student syndrome as shown in Figure 4.3.3 is an example of this. The syndrome demonstrates the work pattern of a student when completing an assignment:

- The assignment brief is given out at time T_{out}.
- The student has a brief look at it and estimates the work involved.
- The student then uses that estimate to schedule back from the due date, T_{in}, to determine when to start working on the assignment, T_{start}.
- Sometimes the student forgets and starts later than planned.
- Once the student starts, it is found that there is more involved in completing the assignment than originally estimated.
- Net result is that as the due date approaches, the input has to steeply grow.
- Often time runs out and the report is late, or has to be handed in incomplete.

Figure 4.3.3 *Student centre syndrome. Demonstrates when students tend to put work into completing their project, i.e. at the last moment. They decide to start work using an optimistic time to schedule back from the deadline – the only thing is that it takes longer than they thought, therefore the work is hurried, incomplete and has errors in it*

This means that the activity is started late in the available time and the probability is that the remaining time will not be sufficient to complete it. The activity therefore will extend beyond the scheduled finish time, or the job will have to be passed on incomplete.

The latter just adds to the following department's tendency not to start work right away, because the job is not in a complete stage and may be altered during the completion phase. If altered this means that work done could be wasted – therefore rather than start an incomplete job, the follow-on department will await the full job. Therefore the next activity will not be started.

With pressure on to complete in time, extra resources are demanded, i.e. costs rise, or a request for a change in the specification is made.

Outgoing phase

The above shows that there is a high probability that this happens late – but what happens if the work is completed before the due date? Is it passed on? Two things conspire against an early pass-on:

- The next department is probably not ready to receive the work, and is reluctant to accept it early because this means a change in their internal schedule and probably a change to their scheduled finish date.
- If the work is passed on early, you may think praise is due. However, the reward is usually a forced reduction in your next estimate because 'last time you over estimated'.

Therefore even when completed early, the job will not be passed on. Instead it will be retained and polished or embellished. Therefore an opportunity to save costs will be lost – in fact the extra input may actually increase the costs which were budgeted for.

Activity 4.3.1

Try this little exercise to demonstrate the above effects.

Prepare a list of random whole numbers between 1 and 10 to represent a series of activity durations along a critical path. What is the sum of these and the average value (rounded up to a whole number)? This gives the project scheduled completion time.

Now we will run the project using the following rule:

> If a number in the list is less than the average reset it to the average value. Leave the other numbers as they are.

Now re-add up the list of numbers. How much of a percentage increase from the sum of the first list has resulted? This is the degree of extensions common in project completion times.

Net result

Any lateness in one, or more, activity is passed on throughout the project and does not get offset by occasions when an activity is completed early as early happenings do not get passed on.

It is therefore no wonder that major projects often have problems in holding their completion date, budgets or specification – even when the estimates of activity durations are quite slack. These symptoms are especially apparent in an organization with a blame culture where risks are normally avoided.

The problem has been analysed by applying Goldratt's theory of constraints which resulted in the identification of the core problem as depicted in Figure 4.3.4. Goldratt claims it is not the uncertainties in activity times that cause projects to fail, but how that uncertainty is managed.

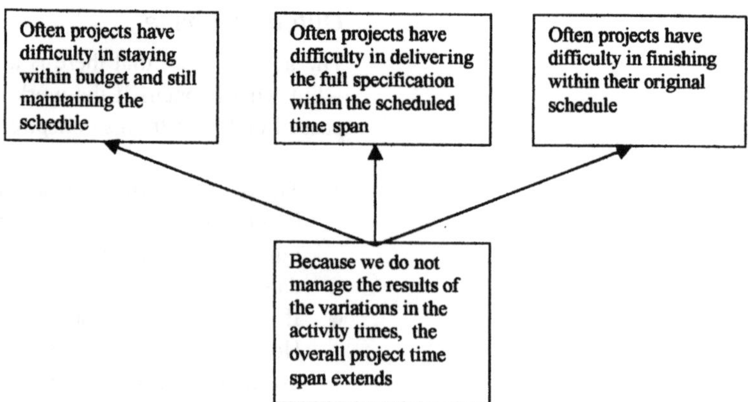

Figure 4.3.4 *Goldratt's core problem identification of why projects fail*

Goldratt's core problem resolution

In Goldratt's theory of constraints all the present system is expressed in a linked cause–effect diagram (not shown). This is then analysed to identify the underlying core problems.

A core problem is defined as one where there are two requirements towards a common objective but these rest on different assumptions which appear to be in complete conflict. The result is expressed in a *cloud* diagram (as in Figure 4.3.5) and the intention is to inject a solution that breaks both assumptions and hence enables both requirements and the common objective to be met.

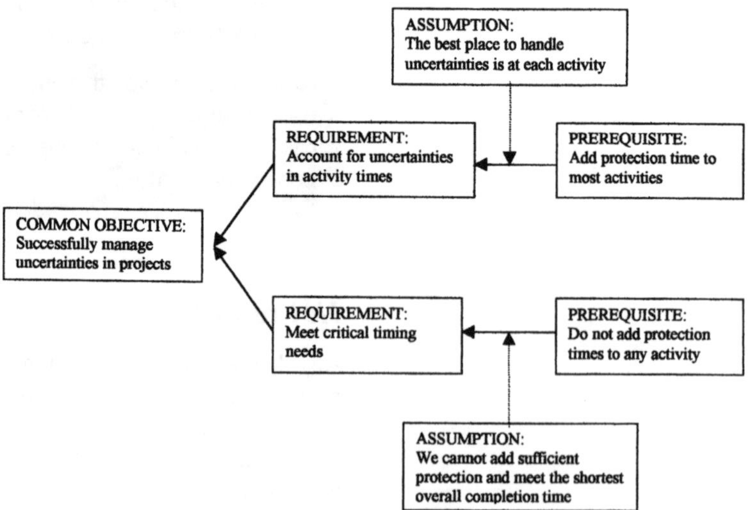

Figure 4.3.5 *Using Goldratt's 'cloud diagram' to express the assumptions behind the core problem where there appears to be two mutually exclusive requirements*

The solution

Now that we have identified many of the reasons for projects failing – is there anything we can do as managers to overcome them?

Goldratt states that there is – how we handle the fact that there will be variations in durations. This he recommends is looked at in two ways:

- The scheduling of activities, especially in allowing for the variability.
- How we actually manage the project thereafter.

Scheduling

The first aspect is that a time is selected that represents a fair, non-inflated duration, i.e. around the median value of that expected under normal circumstances. This may be based on the PERT three-time estimate initially. The schedule is made up from these values as in Section 4.2 (pages 163–169).

Next a value is calculated representing possible variations for the duration of each activity. Again the PERT three-time estimate method could be used to calculate individual variances and an overall variance along the critical path. The key comes in where these variances are applied. They are not applied to each activity, but are retained as a *project buffer* for the whole project. The difference can be seen in Figure 4.3.6.

Where we have activities not on the critical path, a *feeding buffer* can be calculated from the variances on their path and inserted into the schedule by working backwards from the point where they enter the critical path as in Figure 4.3.7.

We now have a schedule with some allowance for variations, but that by itself is little different from what we had after Section 4.2.

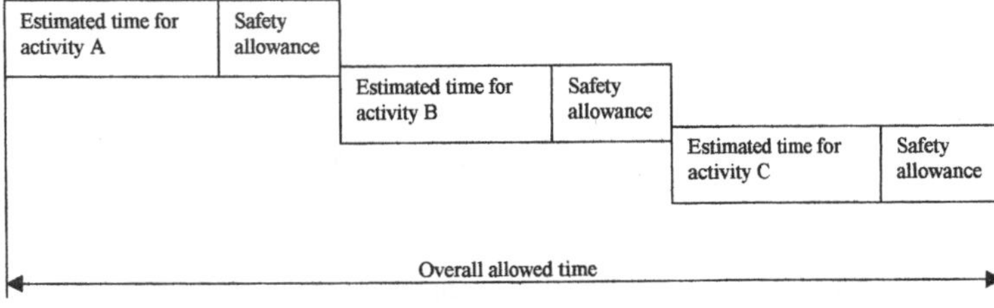

(a) Safety time allowed at each individual activity

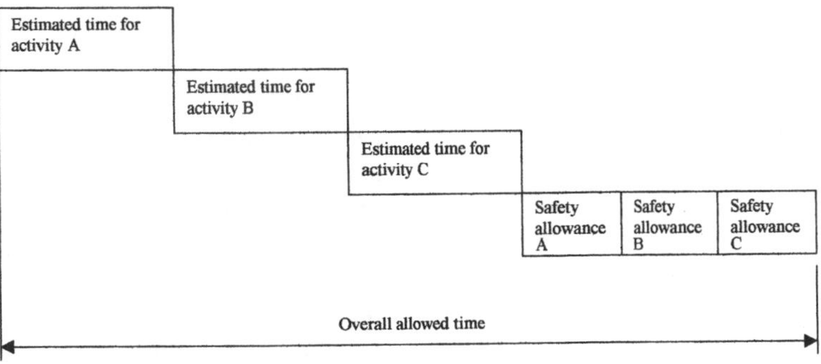

(b) Safety allowances for the individual activities collected to give one large buffer, i.e. safety allowance, for the overall project completion time

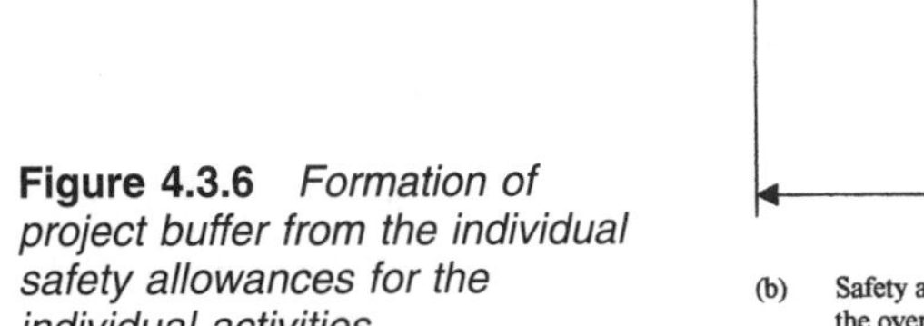

Figure 4.3.6 *Formation of project buffer from the individual safety allowances for the individual activities*

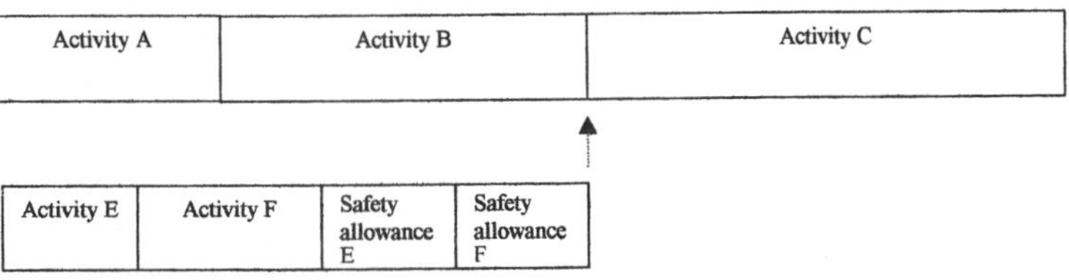

Figure 4.3.7 *Use of feeding buffers to give safety at merging activities*

Managing the project

Goldratt states we should not do this from an overview viewpoint, but in detail. This is not quite as difficult as it sounds because at any one time, only a small number of the activities will be operational, or about to be so. It is these we need to focus on.

- Operational activities: These need to be closely monitored to determine progress, especially how soon they will be completed.
- Follow-on activities: These need to know exactly when the preceding activity will complete so that they are ready to start immediately on their activity. In effect we use a *resource time buffer* (see Figure 4.3.8) before each activity so that as the preceding activity nears completion the buffer reduces through stages of:
  - Advance warning – but no action required as yet.
  - Start to get ready – begin to disengage from other tasks.
  - Get ready – disengage from other tasks.

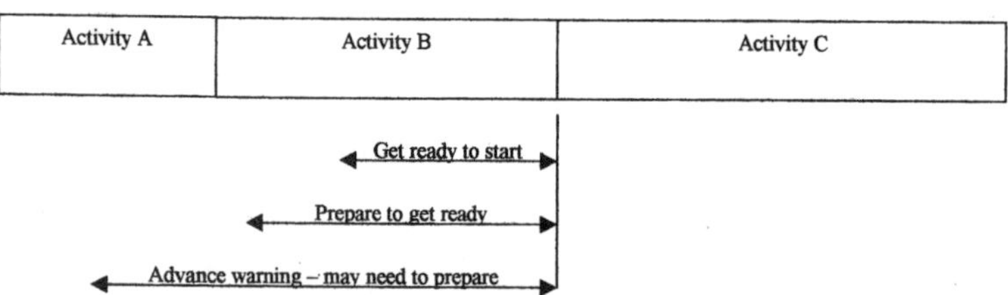

Figure 4.3.8 *Use of resource time buffers to give advance warning of when activity C should be prepared for based on safety allowance associated with previous activities*

The overall buffer is there to be used when difficulties arise, (Law 4 on page 178). The use of the resource time buffer means that jobs are handed over complete and can be worked on immediately with full commitment.

We now have a tight schedule and a method of managing it that will aid successful project completion. Have we covered everything – or can something still go wrong? Problems could still arise from multi-tasking and changes in the project specification.

Multi-tasking

Multi-tasking arises when a resource is committed at the same time to more than one task, or project. This normally arises because of the problems in scheduling and managing many projects at the same time.

One obvious solution is to reduce the number of projects alive at any one time. This may appear unacceptable at first glance, but one of the reasons that so many projects are alive at the same time is that they take a long time to go through the system and people could have a lot of waiting time on their hands if they are not kept fully loaded with work. Goldratt is very clear – the main results of having many things to do are:

- None gets the attention it really needs.
- Switching back and forward between tasks loses time and focus.
- Queues develop because only one task can be done at a time – the others wait.

The first two apply especially to managers – so having fewer projects can result in them getting completed more effectively and efficiently.

The other solution is to automate scheduling when there are many projects. Software such as ProChain™, from Creative Technology Labs, has been developed that makes use of Goldratt's methodology in using buffers (see screen in Figure 4.3.9) and this can identify clashes caused by different projects needing a resource at the same time. This software allows us to reschedule tasks to level demand on these resources and their interconnected activities across projects.

That leaves two problems – the alteration of the specification, and persuading managers to transfer ownership of their safety time to the project leader.

We know that the objective should be set before the project begins (see page 153), but even when the proper procedure is carried out there will always be a few occasions when specifications must be changed. The main advantage in this aspect that Goldratt's methodology gives is that if projects are completed quickly – there will be less opportunity to change the specification.

The problem of getting managers to let go of their safety factors is more problematic and needs the development of trust over time. This depends very much on the culture within the organization.

Goldratt's methodology can ensure that projects are:

- Completed on time
- Completed within budget
- Completed to specification.

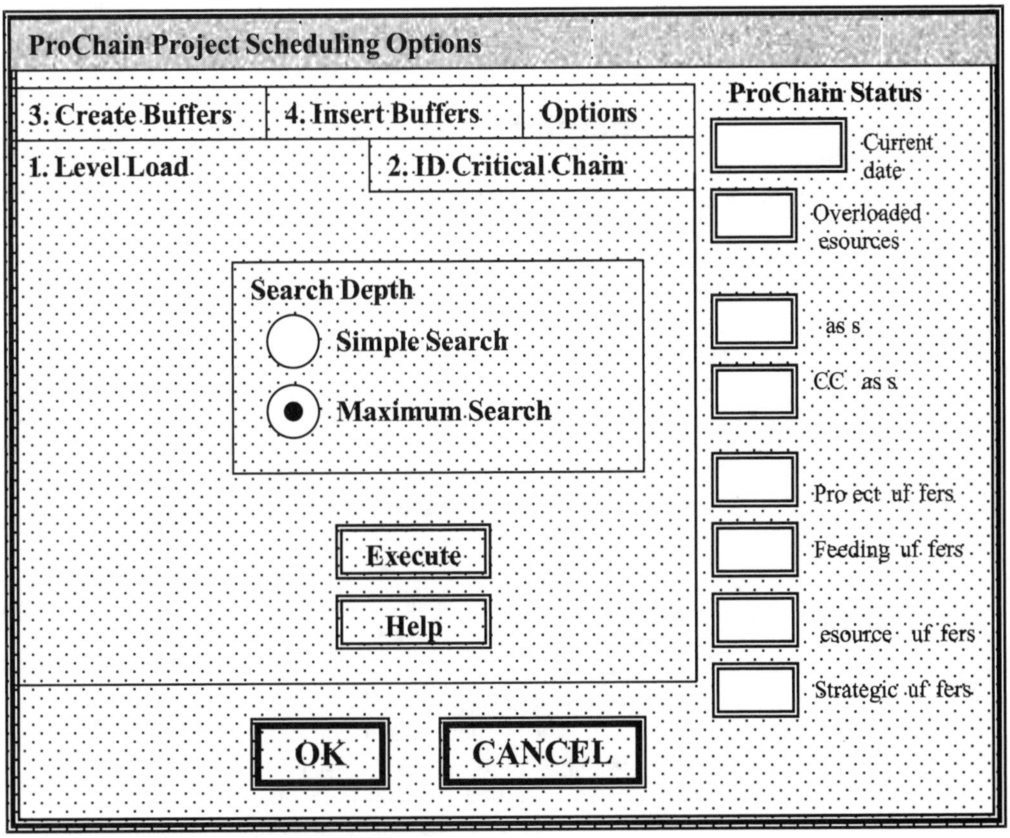

Figure 4.3.9 *Screen shot of ProdChain(TM) screen*

Problems 4.3.1

(1) How long would it take you to build a house using only your own labour?
(2) Draw a rough histogram of the times taken for your journeys from your home to your place of work, or study. What pattern does it show?
(3) Do you always start a task immediately it is given to you?
(4) When do you start a task when there is a set date for completion?
(5) If you were a manager, would you be happy about accepting an activity time without any safety factor built in?

4.4 Controlling projects

Once the scheduling has been done, the project gets underway. Even when the initial stages have been carefully carried out, the progress needs to be monitored and reported on. Corrective action may need to be taken if the uncertainties which were identified, along with those not foreseen, cause major disruptions beyond the contingency plans and buffers built in. Finally a review needs to be carried out to determine future plans and ensure that lessons learnt are retained.

Having set up the project schedule, it then requires careful management to turn the plan into reality, some of which we have dealt with in Section 4.3. There are two main aspects involved in managing projects:

- Introducing changes to the project aims and schedules.
- Managing the project to meet aims.

Managing change to project aims

If we have carried out well the initial stages of initiation and feasibility study, we should have an achievable project with aims that fit into the overall business plan for the organization. The detailed planning and scheduling that follows details the steps towards attaining the desired aims within the laid down constraints.

However, although it is an ideal that we should not change the aims of an ongoing project, on a few occasions this does become necessary. These exceptional occasions must be initiated and authorized by top management to match changes in the business plan. These should be mainly due to changes in the marketplace or the organization's circumstances, but may arise from the project itself as it progresses – or fails to.

There will always be occasions when the assumptions on which the project was authorized are found to be wrong, perhaps because of misjudgement or a mistake made. The important thing here is not to discover who to blame, but as quickly as possible reassess the situation and determine what needs to be done.

If a project is continually changing then perhaps the initial stages have been rushed, or incorrectly carried out. In this case, we need to stop and reassess the true situation. This may be difficult to do politically.

Alternatively it may be that the organization is going through a period of extreme turbulence. It may be better in the latter circumstance to suspend the project until that turbulence clears as otherwise there is a high probability of wasted effort and cost.

The change(s) that may have to be accommodated are:

- Change in requirements, perhaps because of new information, or that the original aims have been found to be unobtainable.
- Change in deadline, perhaps to meet competition's launch of a new product or to catch up lost time.
- Change in budget, perhaps because of significant changes in cash flow or spending in some areas exceeding that budgeted for.
- Change in priority, this may be on a time base of which phase to complete first, or even between projects.

In these cases the whole project plan has to be re-examined and will almost certainly have to be rescheduled and resourced to meet the new targets and constraints. In effect it means returning to the first stage of concept and going through the same processes as a new project albeit perhaps in less time. It may be, in extreme cases, that the project will need to be postponed or abandoned.

When top management decides upon changes to the project, it is important that these changes are controlled and the new aims and targets are conveyed to all concerned. This will mean a revision to the terms of reference and all the schedules, and should be handled in a similar manner to the organization's document control procedure:

Documentation control

Within organizations there are many documents for operating procedures, quality systems, design, etc. that lay down the present way of working. There tends to be many copies scattered around the organization.

Whenever one of these documents is changed it is very important that all previous copies are withdrawn and destroyed to prevent people working to outdated information.

A register therefore needs to be kept of all standard documents which lists all recipients of the document. When a change is introduced, details such as author, date, version and brief description of the revision are recorded. A copy of the new document is then sent to the authorized recipients and the previous copy recalled to be destroyed.

It is a constant battle to ensure that only the latest versions are in circulation because people have a tendency to take unofficial copies of documents which are not recorded. It is more effective that all possible users are included as recipients to ensure that only up-to-date documentation is in use.

Reporting on project progress

We need to monitor, and hence report on, the project's progress against the planned completion time, the cost budgeted for and the specification. We need to do this not only to ensure we are on target but that resources are available to maintain the momentum as we move towards completion.

To do this, we need to continually update the following for each ongoing activity:

- Has the specification been, or does it need to be, changed?
- What overall time has elapsed so far?
- How much time is required to complete the activity and the project?
- What portion of each activity has been completed and what remains to be completed?
- Any future problems apparent?
- What cost has been incurred so far?
- How much cost is required to complete the activity and the project?
- Has the overall cost of the project already changed or is it now anticipated to change?

Most project management software incorporates reports and some allow customized reports to be designed. For example, Microsoft™ Project has the following in-built reports, only two of which are illustrated herein:

Overview reports

- Project summary (see Figure 4.4.1): Shows overall start/finish information and variances of the overall project such as dates, duration, work, cost and activity status.
- Top level tasks: Shows all main activities, i.e. no sub-activities, and planned start/finish times and percentage complete.
- Critical tasks: As for top level task report plus preceding and follow-on activities.
- Milestones: Similar to top level task report, but for milestones only.
- Working days: Shows normal weekly pattern of available hours.

Current activities reports

- Unstarted tasks: Shows all activities which have not started.
- Tasks starting soon: Shows activities due to start between user-set dates.
- Tasks in progress: Shows all activities which have started, but not finished yet.
- Completed tasks: Shows all activities which have finished.
- Should have started tasks: Shows all activities which have not started, but that should have.
- Slipping tasks: Shows all activities which have not finished, but that should have.

Cost

- Cash flow: Shows total budget broken down into a weekly basis of the activity budget costs.
- Budget: Shows against each activity the budget cost, that spent and the amount still to spend (of the budget).
- Overbudget tasks: Shows how much should have been spent according to percentage of each activity completed against what was actually spent.
- Overbudget resources: Similar to overbudget task report, but based on actual cost of the resource as it is being used.

IIEbuild
University of Brighton

as of Sun 22/10/00

Dates

Start:	Wed 18/10/00	Finish:	Fri 06/07/01
Baseline start:	Wed 18/10/00	Baseline finish	Tue 05/06/01
Actual start:	NA	Actual finish	NA
Start variance:	0 days	Finish variance:	22.67 days

Duration

Scheduled:	187.67 days	Remaining:	187.67 days
Baseline:	165 days	Actual:	0 days
Variance:	22.67 days	Percent complete:	0%

Work

Scheduled:	544 hrs	Remaining:	544 hrs
Baseline:	0 hrs	Actual:	0 hrs
Variance:	544 hrs	Per cent complete:	0%

Costs

Scheduled:	£158 940.00	Remaining:	£158 940.00
Baseline:	£0.00	Actual:	£0.00
Variance:	£158 940.00		

Task status		Resource status	
Tasks not yet started:	17	Resources:	0
Tasks in progress:	0	Overallocated resources:	0
Tasks completed	0		
Total tasks:	17	Total resources:	0

Figure 4.4.1 *Project summary report*

- Earned value: Compares activity costs for variance on the following:
  - Budget cost by time of each activity.
  - Budget cost by percentage completed of each activity.
  - Actual cost of work done on each activity.
  - Estimated cost to complete each not finished activity.

Assignments reports

- Who does what: Shows by resource the number of hours budgeted against each activity.
- Who does what when: Shows on a calendar of the project the weekly resource budgeted for against each activity.
- To do list (see Figure 4.4.2): Produces a week-by-week list of any activities scheduled and the resource to be used.
- Overallocated resources: Shows any resource where the time for scheduled activities exceeds that available.

Workload

- Task usage: Shows on a week-by-week basis the scheduled activities planned and the resources required.
- Resource usage: Shows on a week-by-week basis each resources' scheduled activities planned.

To Do List as of Thu 02/11/00
IIEpreproduction

ID	☐	Task name	Duration	Start	Finish	Predecessors	Resource names
Week of 04 June							
7	☐	Training	5 days	Wed 06/06/01	Tue 12/06/01		Plant operatives [300%]
Week of 11 June							
7	☐	Training	5 days	Wed 06/06/01	Tue 12/06/01		Plant operatives [300%]
6	☐	Trials	11 days	Wed 13/06/01	Wed 27/06/01	5,3,7	Plant operatives [300%]
Week of 18 June							
6	☐	Trials	11 days	Wed 13/06/01	Wed 27/06/01	5,3,7	Plant operatives [300%]
Week of 25 June							
6	☐	Trials	11 days	Wed 13/06/01	Wed 27/06/01	5,3,7	Plant operatives [300%]
8	☐	Build up inventory	6.67 days	Thu 28/06/01	Fri 06/07/01	6	Plant operatives [300%]
Week of 02 July							
8	☐	Build up inventory	6.67 days	Thu 28/06/01	Fri 06/07/01	6	Plant operatives [300%]

Figure 4.4.2 *To do list for resources*

Custom reports

- These can be designed by the user, but this requires skill to know where to pick up information from, etc.

In addition the Gantt chart can be used to visually demonstrate project progress on an activity level (see Figure 4.4.3). On the chart for the build project, the current date is the dotted vertical line at the start of December.

Figure 4.4.3 *Progress Gantt chart for design project*

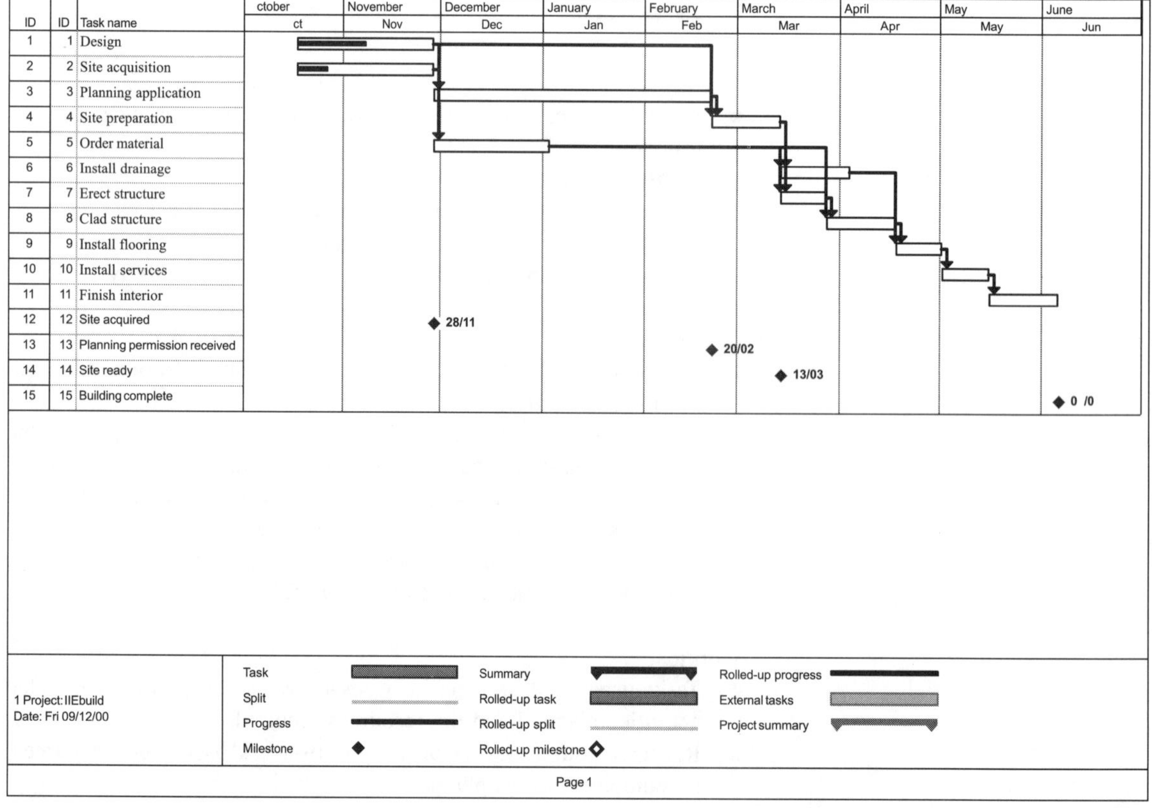

The planned activity duration is shown by a clear box and the actual completion by a horizontal line within the box. In this case, activities are behind schedule – the design activity is only 50% complete and the site acquisition activity is only 20% complete. The planning application and order material activities cannot start as they are tied to the others.

This report enables the managers of the activities to adjust ongoing input, including perhaps some rescheduling. For example:

- The site acquisition activity is critical for the follow-on planning permission activity and this needs checking to see why the finish date has slipped.
- If there is a chance of the finish date being delayed then the start dates of all other activities may have to be changed, or special changes will need to be made to some activities to catch up on the slippage to ensure that the overall completion date is held.
- As the design activity is behind schedule, it may be possible to switch staff to help some other activity, perhaps on a different project, within that section.

Some software packages have built-in graphic assistance to visually present progress information, but often the project manager designs his own. These can be on an activity level or for the project as a whole.

One main variance analysis is the earned value analysis. A chart of this using the data in Table 4.4.1 is shown in Figure 4.4.4:

Table 4.4.1 Cost data for earned value chart

Week number	*Budget cost work scheduled (£)*	*Actual cost work completed (£)*	*Budgeted cost work completed (£)*
	BCWP	ACWC	BCWC
1	0	0	0
2	5 000	2 500	4 000
3	15 000	12 500	12 000
4	25 000	25 000	20 000
5	40 000	45 000	30 000
6	55 000	60 000	45 000
7	65 000	70 000	55 000
8	75 000	80 000	60 000
9	90 000	100 000	80 000
10	100 000	115 000	90 000

From the table, we can see four variances:

- Time variance: One week, i.e. the value for that portion of the project completed by week ten, is that which should have been completed by week nine in the original cost estimate, i.e. £90 000.
- Schedule variance (SV) of £10 000 being the difference between what was budgeted to have been completed by week ten and what was actually completed, i.e. SV = BCWS – BCWC.
- Cost variance of £15 000 over what should have been spent by week ten, i.e. CV = BCWS – ACWC.

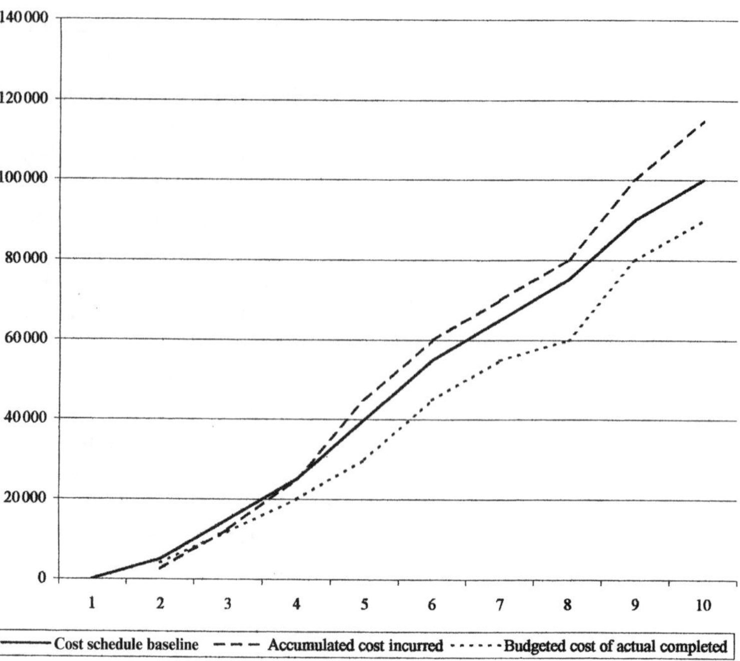

Figure 4.4.4 *Earned value chart*

- Total variance of £25 000 over the budgeted cost of what has actually been completed, i.e. TV = BCWC – ACWC.

These should all be analysed as to why they have occurred (see Chapter 5) and action taken to redress the variance, if possible. If the variance is not recoverable, the overall schedule and budget will need reforming to show new attainable targets. These must be communicated to senior management and other interested parties for agreement.

Progress meetings need to be held regularly with those involved in ongoing activities and advise those who are next to work on the project. These need not all be formal meetings, in fact most information should continually be gathered whilst visiting sections.

Taking action

- Assess the situation:
  - Note the effect on cost and time of present activity.
  - Work out what, if any, correction can be made to remainder of activity.
  - Find out the reason for the occurrence – not to pin the blame onto people, but to prevent recurrence.
- Impact analysis is required where the recovery cannot be achieved within the ongoing activity. This requires wide consultation with all concerned to reschedule if necessary the remainder of the project.
- Often the project manager will have to resolve issues between sections. This needs creativity in contingency usage, making available more resources, changing dates, or reducing scope.
- The main causes which normally need to be addressed are estimating skills, training, recruitment, and motivation, although often the problems are based in the organization's culture.

Completion

Almost as important as the initiation phases, the completion phases lay the ground for successful operation thereafter.

The final stage is the final commissioning of the new process, then the formal handover and acceptance of the same. This must be formally recorded to denote the full operational extent, and any problems remaining, of the new process.

Once the project has been handed over, then a careful review and measure of the project success can be made. This is not just to decide if all aims have been achieved but to:

- Determine further needs to bring the new process up to full operational requirements and to maintain it there.
- Identify lessons learnt from the project operation for future projects.

Although the project team will play a major part in the post-project review, it is best if this is conducted by a different person than the project manager.

Problems 4.4.1

(1) Find out what you can about any major project that failed and determine the root causes.
(2) Why is it difficult to change a project that has already been agreed?
(3) Can you think of several projects you have been involved with where the same mistake has been made?
(4) Why do you think many organizations fail to carry out a proper project audit?
(5) What problems could arise when the project review is carried out by someone other than the project manager?

5 Money in the organization

Summary

There is a song titled 'Money makes the world go round' and that is certainly true for organizations. Money is what enables organizations to survive now and into the future. As an engineer you will need to know how your organization structures its internal finances and use this in making decisions.

There is little real mystery surrounding financial aspects which you cannot grasp, if you understand the language used in the financial function. This chapter sets out to explain this key aspect of business.

Objectives

By the end of this chapter, the reader should:

- be able to carry out investment appraisal using the techniques of payback, rate of return and discounted cash flow whilst considering non-financial aspects (Section 5.1);
- understand how costs arise and vary, and be able to cost out the material, labour and overhead aspects of product costing including applying a depreciation allowance (Section 5.2);
- be able to produce a cash flow statement, a profit and loss statement and the balance sheet and appreciate how book-keeping is carried out (Section 5.3);
- be able to prepare and interpret budgets and carry out variance analysis including applying financial ratios (Section 5.4).

5.1 Investment appraisal

Any investment made by a business has one aim – to earn more than it costs in as short a time as possible. This is irrespective of the type of investment. Designing a new product, purchasing a new machine, employing another person or launching an advertising campaign should all be subjected to the same criteria.

We will cover investment appraisal as the first aspect of money in business. The questions to be asked include:

- How much will it cost?
- What are the benefits involved?
- How soon will we recover our spending?
- How much will we make in total?
- How risky is it?

Investment appraisal sets out to answer these and other questions.

Costs and benefits

As in any decision making, investment appraisal should be based on facts. Some of these will be hard facts, i.e. not in dispute, and some will be less certain. They all need to be identified, with any degree of uncertainty attached to them.

Costs are simply money spent and benefits are what is gained. Costs are normally tangible and usually can be measured in money. Benefits are sometimes less tangible, but should whenever possible also be expressed in money terms.

In a business, costs and benefits are classified as either capital or revenue.

- Capital items are those which are used within a business over a period longer than one year, i.e. the assets. They can be costs – such as purchasing and installation of new machinery, or development costs of a new product. They can also be benefits – such as the sale of plant or buildings.
- Revenue items are the normal day-to-day money flowing in and out of the business.

A complete checklist of possible costs and benefits is worth compiling when making an appraisal. There are several particular aspects which are worthwhile considering here and now.

Indirect costs

Many projects make use of functions such as design and development sections, production engineering and facility management. They are normally treated by accountants as overhead departments. If any such functions are used in a project, their actual costs should be included.

Overheads

The normal accountancy practice is to apportion onto direct costs an extra cost to cover overheads such as rent, rates, general lighting and heating, supervision, administration and sales which cannot be directly attributed to any product or process. When carrying out investment appraisal, only where a real change will occur in any of these items should the value of that change be included.

Financing costs

Almost all investment will involve an initial amount of money to be spent. As discussed in Chapter 1, this money may come from:

- Borrowing from a bank or investment house – costs involved are setting up the loan and repaying the capital and interest.
- Money already in the business – costs involved are the return on investment that could be made elsewhere in the business. At the least the money could be invested in an interest earning bank account.
- Extra issue of shares (borrowing from shareholders). Costs here are those involved in the share issue, and the future dividend expected by the investors. In addition there may be discounts or premiums involved in the share price which may have to be taken into account.

No matter which method is used, the cost of financing the project should be used to compare the possible returns.

Grants and subsidies

There are many grants and subsidies available for projects. These can be from EC, national or local sources. Some will be offset against capital purchases and others against revenue items such as recruitment and training.

Depreciation

This is covered in detail in Section 5.2 (page 223). It is sufficient now to say that when using the payback and average rates of returns methods, depreciation is not used as full repayment of the initial investment is made. In the discounted cash flow methods, however, depreciation charges are considered as they are an identifiable revenue cost affecting tax payable.

Investment appraisal techniques

All appraisals start with the preparation of a cash flow statement which lists all the incomings and outgoing money and when they occur. Figure 5.1.1 shows a detailed cash flow for the replacement of a machine showing the full range of actual spending incurred.

Figures 5.1.2 and 5.1.3 show simplified cash flows for the development of two new products – A and B. We shall use these simpler examples to compare the various techniques for a company which can only afford to invest £150 000 at this point in time.

Payback period

This is the simplest and most commonly used technique. It basically calculates how long it takes to recover the initial investment. Taking the values in Figure 5.1.1, they come down to:

OUTGOINGS

DESCRIPTION		VALUE	TIMING
			(Week number)
MACHINE SELECTION			
Visits to exhibitions		£900	1–4
Early consideration of possible machines		£250	5
Visits to suppliers		£1200	6–8
Desk evaluation of possible machines		£250	9
Tests on products		£1500	10–12
Evaluation of possible machines		£500	13–14
Contract awarded		£350	18
Machine purchase (total £85000)			
	with order	£17000	18
	on delivery	£59500	26
	after one year	£8500	78
Transportation		£3500	26
First year's spares		£4250	26
	Subtotal	£97700	
MACHINE INSTALLATION			
Site clearance		£1250	22
Rerouting of power supplies		£450	24
Foundations		£1500	24
Installation		£2500	26–28
Proving trials		£3500	29–32
Removal of old machine		£500	33
Remaking good floor		£500	34
	Subtotal	£10200	
OTHER COSTS			
Operator training		£600	28
Maintenance training		£350	28
Extra insurance – first year		£500	26
	Subtotal	£1450	
	TOTAL	£109350	

INCOMINGS AND SAVINGS

REDUCTION IN ANNUAL COSTS			
Power		£2500	
Reduced scrap		£12500	
Reduced direct labour		£15000	
Maintenance		£3500	
	Subtotal	£33500	Ongoing from week 32
SALES OF ASSETS			
Sale of old machine		£3500	34

Figure 5.1.1 *Cash flow statement for replacement of a machine*

DEVELOPMENT COSTS	£150 000
EXTRA REVENUES	
Year 1	£25 000
Year 2	£50 000
Year 3	£100 000
Year 4	£90 000
Year 5	£30 000

Figure 5.1.2 *Cash flow for product A*

DEVELOPMENT COSTS	£150 000
EXTRA REVENUES	
Year 1	£70 000
Year 2	£80 000
Year 3	£40 000
Year 4	£20 000
Year 5	nil

Figure 5.1.3 *Cash flow for product B*

Investment

Total outgoing	=	£109 350
Less sale of old machine	=	£3500
Net investment	=	£105 850
Savings	=	£33 500

$$\text{Payback period} = \frac{\text{net investment}}{\text{savings}}$$

$$= \frac{£105\,850}{£33\,500}$$

$$= 3.16 \text{ years}$$

This means that the investment will be paid back in three years and two months after the money starts to come in.

Activity 5.1.1

What is the payback period for the new product developments shown in Figures 5.1.2 and 5.1.3?

(Product A = 2.75 years, product B = 2 years)

This method is often used as a hurdle with organizations, with management setting a minimum payback period – depending on the economic circumstances. In hard times, projects with more than a one year payback period could be rejected in many organizations.

Although this method is simple, it does suffer from several drawbacks:

- It ignores investment after the payback period.
- It ignores the time value of money.

Annual rate of return

As can be seen in the payback examples, there can still be considerable benefits continuing after the payback period has been reached. To overcome the first of the drawbacks in the payback method, we have various methods. One way of comparing the total income over a project's life against the investment is to calculate an annual rate of return. There are two ways to do this.

Average gross annual rate of return (AGARR)

The easiest way to demonstrate this is by working through the data for product A from Figure 5.1.2:

Initial investment	=	£150 000
Total income over the product's life	=	£295 000
Life of product	=	5 years

$$\text{Average annual income} = \frac{£295\,000}{5 \text{ years}}$$

$$= £59\,000$$

$$\text{Therefore, AGARR} = \frac{£59\,000 \times 100}{£150\,000}$$

$$= 39.3\%$$

Activity 5.1.2

Carry out an AGARR calculation on product B in Figure 5.1.3.

(35% over the four years only)

As can be seen the two products have different AGARR figures. Product A has a higher overall return, gives a higher AGARR, and is maintained for five years rather than the four years for product B.

It should be noted that product B has a higher initial recovery rate than product A. As we can surmise, high early returns are probably better than later returns, as later income can be reduced by unforeseen events, e.g. high inflation, new competition, drying up of supplies, etc. If we calculate a factor to take into account the time value of money, then the comparison between the investments may show up in different preferences. We will look at this after considering the other method of calculating annual rate of return:

Activity 5.1.3

What would be the difference in AGARR if we considered product B over the same five years as product A. Should we use the same time period?

(AGARR over five years is 28%)

As you can see the time period being considered makes a significant difference. The AGARR for product B now shows an even lower figure than that for product A as there is no income in the fifth year. The decision of what time to consider when comparing the two projects would depend on what happens at the end of product B's life.

Average net annual rate of return (ANARR)

This method is similar to the AGARR method but with several key differences in calculating income and considering the investment. Again we will use the data for product A (from Figure 5.1.2) to show the process involved:

Total income over the product's life = £295 000 (as before)

Initial investment = £150 000 (as before)

Net proceeds = £295 000 – £150 000

= £145 000

Life of product = 5 years (as before)

$$\text{Average annual net proceeds} = \frac{\text{net proceeds}}{\text{product life}} = \frac{£145\,000}{5 \text{ years}} = £29\,000$$

End value of investment = zero (no plant or equipment for resale)

$$\text{Average Investment} = \frac{(\text{initial investment} - \text{end value})}{2} = \frac{(£150\,000–0)}{2} = £75\,000$$

$$\text{Therefore ANARR} = \frac{\text{average annual net proceeds} \times 100}{\text{average investment}} = \frac{£29\,000 \times 100}{£75\,000} = 38.7\%$$

The reason for taking an average for the investment is that the initial investment is being recovered, or paid back, by equal amounts each year. Therefore at the end of the product life the initial investment is fully paid back using this procedure.

Note that the ANARR is only slightly different than the AGARR value of 39.3% we calculated above in this example.

Activity 5.1.4

Now calculate the ANARR for product B in Figure 5.1.3 for both four and five years.

(20% over four years; 16% over five years)

For product B, these ANARRs are significantly different from its AGARRs of 35% and 28% respectively. In this case, product A has a much better ANARR value of 38.7%. The ANARR method is a more severe method, especially when the net proceeds are small in relation to the initial investment.

However, both ANARR and AGARR ignore the time value of money.

Discounted cash flow (DCF) methods

When considering the value now of money earned or spent in the future, we can apply commonsense to see it will not be the same as the face value. What would you rather have – £1000 today, or £1000 next year?

How much extra would you wish to be paid one year from now in place of the £1000 today?

£1010, £1050, £1100, or more?

Obviously your answer would depend on a few factors such as how much you needed the £1000 today, how risky would it be to wait the year before payment, etc.

Businesses have the same problem – how much are future earnings worth today? The method they use to calculate the value of future earnings is called discounting. The discounting calculation is the inverse of the compounding calculation used for calculating the interest earned, or paid, on bank accounts, loans and mortgages.

Maths in action

Compound interest

If we deposited a sum of £1000 in a bank account that paid 10% per annum interest each year and left it there for a number of years, we would expect it to grow:

Deposit in year 0 = £1000 (year 0 is taken to mean time now)

Balance at end of year 1 = £1000 + (10% of £1000) = £1100

Balance at end of year 2 = £1100 + (10% of £1100) = £1210

Balance at end of year 3 = £1210 + (10% of £1210) = £1331

Balance at end of year 4 = £1331 + (10% of £1331) = £1464.10

If we express compounding as:

$$\text{Balance after one year} = \text{old balance} \times \frac{(100 + \%\text{ interest rate})}{100}$$

Then discounting is the inverse of compounding:

$$\text{Value now} = \frac{\text{value one year from now} \times 100}{(100 + \%\text{ discount rate})}$$

Table 5.1.1. Table of discount factors

	Discount factor																							
Years	1%	2%	3%	4%	5%	6%	7%	8%	9%	10%	11%	12%	13%	14%	15%	16%	17%	18%	19%	20%	21%	22%	23%	24%	25%
0	1	1	1	1	1	1	1	1	1	1	1	1	1	1	1	1	1	1	1	1	1	1	1	1	1
1	0.9901	0.9804	0.9709	0.9615	0.9524	0.9434	0.9346	0.9259	0.9174	0.9091	0.9009	0.8929	0.8850	0.8772	0.8696	0.8621	0.8547	0.8475	0.8403	0.8333	0.8264	0.8197	0.8130	0.8065	0.8000
2	0.9803	0.9612	0.9426	0.9246	0.9070	0.8900	0.8734	0.8573	0.8417	0.8264	0.8116	0.7972	0.7831	0.7695	0.7561	0.7432	0.7305	0.7182	0.7062	0.6944	0.6830	0.6719	0.6610	0.6504	0.6400
3	0.9706	0.9423	0.9151	0.8890	0.8638	0.8396	0.8163	0.7938	0.7722	0.7513	0.7312	0.7118	0.6931	0.6750	0.6575	0.6407	0.6244	0.6086	0.5934	0.5787	0.5645	0.5507	0.5374	0.5245	0.5120
4	0.9610	0.9238	0.8885	0.8548	0.8227	0.7921	0.7629	0.7350	0.7084	0.6830	0.6587	0.6355	0.6133	0.5921	0.5718	0.5523	0.5337	0.5158	0.4987	0.4823	0.4665	0.4514	0.4369	0.4230	0.4096
5	0.9515	0.9057	0.8626	0.8219	0.7835	0.7473	0.7130	0.6806	0.6499	0.6209	0.5935	0.5674	0.5428	0.5194	0.4972	0.4761	0.4561	0.4371	0.4190	0.4019	0.3855	0.3700	0.3552	0.3411	0.3277
6	0.9420	0.8880	0.8375	0.7903	0.7462	0.7050	0.6663	0.6302	0.5963	0.5645	0.5346	0.5066	0.4803	0.4556	0.4323	0.4104	0.3898	0.3704	0.3521	0.3349	0.3186	0.3033	0.2888	0.2751	0.2621
7	0.9327	0.8706	0.8131	0.7599	0.7107	0.6651	0.6227	0.5835	0.5470	0.5132	0.4817	0.4523	0.4251	0.3996	0.3759	0.3538	0.3332	0.3139	0.2959	0.2791	0.2633	0.2486	0.2348	0.2218	0.2097
8	0.9235	0.8535	0.7894	0.7307	0.6768	0.6274	0.5820	0.5403	0.5019	0.4665	0.4339	0.4039	0.3762	0.3506	0.3269	0.3050	0.2848	0.2660	0.2487	0.2326	0.2176	0.2038	0.1909	0.1789	0.1678
9	0.9143	0.8368	0.7664	0.7026	0.6446	0.5919	0.5439	0.5002	0.4604	0.4241	0.3909	0.3606	0.3329	0.3075	0.2843	0.2630	0.2434	0.2255	0.2090	0.1938	0.1799	0.1670	0.1552	0.1443	0.1342
10	0.9053	0.8203	0.7441	0.6756	0.6139	0.5584	0.5083	0.4632	0.4224	0.3855	0.3522	0.3220	0.2946	0.2697	0.2472	0.2267	0.2080	0.1911	0.1756	0.1615	0.1486	0.1369	0.1262	0.1164	0.1074
11	0.8963	0.8043	0.7224	0.6496	0.5847	0.5268	0.4751	0.4289	0.3875	0.3505	0.3173	0.2875	0.2607	0.2366	0.2149	0.1954	0.1778	0.1619	0.1476	0.1346	0.1228	0.1122	0.1026	0.0938	0.0859
12	0.8874	0.7885	0.7014	0.6246	0.5568	0.4970	0.4440	0.3971	0.3555	0.3186	0.2858	0.2567	0.2307	0.2076	0.1869	0.1685	0.1520	0.1372	0.1240	0.1122	0.1015	0.0920	0.0834	0.0757	0.0687
13	0.8787	0.7730	0.6810	0.6006	0.5303	0.4688	0.4150	0.3677	0.3262	0.2897	0.2575	0.2292	0.2042	0.1821	0.1625	0.1452	0.1299	0.1163	0.1042	0.0935	0.0839	0.0754	0.0678	0.0610	0.0550
14	0.8700	0.7579	0.6611	0.5775	0.5051	0.4423	0.3878	0.3405	0.2992	0.2633	0.2320	0.2046	0.1807	0.1597	0.1413	0.1252	0.1110	0.0985	0.0876	0.0779	0.0693	0.0618	0.0551	0.0492	0.0440
15	0.8613	0.7430	0.6419	0.5553	0.4810	0.4173	0.3624	0.3152	0.2745	0.2394	0.2090	0.1827	0.1599	0.1401	0.1229	0.1079	0.0949	0.0835	0.0736	0.0649	0.0573	0.0507	0.0448	0.0397	0.0352

Therefore if we wish to know what £1000 of next year's money would be worth now if we discount at 10%, then:

$$\text{Value now} = \frac{£1000 \times 100}{100 + 10}$$

$$= £1000 \times \frac{100}{110}$$

$$= £909.09$$

Rather than have us calculate a discounting ratio each time, tables are produced to show a range of present values for different discount rates over a number of years. They are similar to that in Table 5.1.1 and are simple to use:

To find the present value of £1000 due in three years at a 10% discount rate, we look at the column headed 10% and move down it until we are on the three year row. We take that value – in this case 0.7513 – and multiply the future monetary value by it to find the present value:

Present value = £1000 × 0.7513 = £751.30

Activity 5.1.5

Try out a few for yourself. First find the discount ratio then the present value of the following:

Future value	*Year*	*Discount rate*
£1000	5	20%
£3000	2	5%
£400	4	15%

(0.4019 and £401.90, 0.9070 and £2721, 0.5718 and £228.72)

We can see now how we discount future money to find present values. The only problem is 'What do we select as the discount rate?' The same hurdles are applied as with payback periods – high rates in hard times, easing when things look better.

To show the full process, let us again use the data for product A in Figure 5.1.2 with a discount factor or rate (DCF) of 20%. Where the flow is outward, i.e the initial investment, we use a negative value. Using a spreadsheet program such as Excel© to enter the data means it can easily be changed (see Figure 5.1.4).

Figure 5.1.4 shows a net present value (NPV) of £18 885.03, i.e. a positive value demonstrating that at a 20% discount rate the project would show a net profit of £18 885.03 at present values.

YEAR	CASHFLOW	DISCOUNT RATIO 20%	PRESENT VALUE
0	–£150 000.00	1	–£150 000.00
1	£25 000.00	0.8333	£20 833.33
2	£50 000.00	0.6944	£34 722.22
3	£100 000.00	0.5787	£57 870.37
4	£90 000.00	0.4823	£43 402.78
5	£30 000.00	0.4019	£12 056.33
TOTAL NET PRESENT VALUE			£18 885.03

Figure 5.1.4 *Net present value calculations for product A*

Activity 5.1.6

Now try the example in Figure 5.1.3 using the same discount factor of 20%.

(– £3317.90)

The calculation for product B shows a negative value which means it fails to show a profit in present value terms at the 20% discount rate. It would not proceed if 20% was the desired rate to be used for discounting.

Internal rate of return (IRR)

Using set rates of discount means that some projects can be rejected for failing to meet that criteria. It is probable that an organization could use different discount factors to take into consideration any risks involved. These discounted hurdle rates are often set arbitrarily.

There is, however, another method using discounting which can be employed. If we find the discount rate that would give exactly zero profit, we can compare different projects by using their equivalent discount rate. This rate where the discounted project returns zero is called the internal rate of return (IRR). The higher the IRR the better.

This rate can be found by a process of trial and error to find a rate that returns a net present value of zero.

A second way is to find an approximate value for the IRR by ratios. If we recalculate the net present value for product A (Figure 5.1.2) using a new discounted factor of 30%, we would arrive at a net present value of – £16 075.44. Therefore the IRR must lie between 20% and 30%.

To find the approximate IRR:

At 20% DCF, NPV = £18 885

At 30% DCF, NPV = –£16 075

Difference of +10% DCF gives a drop of £34 960 in NPV

If the relationship followed a straight line, then a graph showing the NPV between 20% and 30% DCF would cross the zero value in proportion to the differences:

Where ZV = point at which NPV = zero

$$\frac{ZV}{10\%} = \frac{£18\,885}{£34\,960}$$

$$ZV = \frac{£18\,885 \times 10\%}{£34\,960}$$

$$= 5.41\%$$

Therefore IRR = 20% + 5.41 = 25.41%

The true line joining the NPV values is actually a concave curve – so the straight line taken is only approximate. By substituting the 25.41% into the CDF spreadsheet, we find a value of –£1519.76 showing that this approximate value is high.

The correct value is 24.97% found using the goal-seeking function in Excel© to discover the IRR (see Chapter 7 for this use of Excel©).

Activity 5.1.7

Try the same approximation for the data for product B (Figure 5.1.3). First, find the net present value using a 15% discount factor and then the approximate IRR and compare it to the actual IRR by using the goal-seeking function in Excel©.

(£9096.77, 18.4%, 18.59%)

Both DCF and IRR are often a keystone in project appraisal.

Cash flow diagram

A good way to visualize the cash flow in a project is to draw a simple graph. Taking product A this would give a graph as shown in Figure 5.1.5.

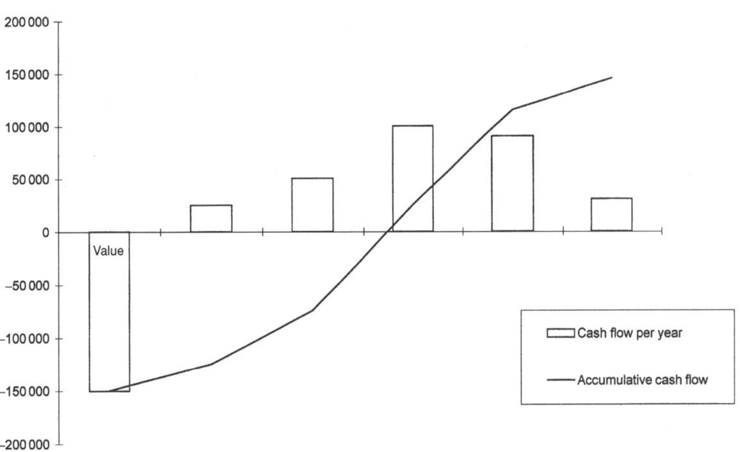

Figure 5.1.5 *Cash flow diagram for product A (Figure 5.1.2)*

Looking at the graph it can be seen that the cash flow line crosses the x-axis at 2.75 years. This is the payback period or the breakeven point. The slope of the line shows how quickly the cash is flowing. Note, if you use the automatic Excel© graph function, it draws the line from the centre of each division and is therefore more difficult to read correctly.

Activity 5.1.8

Draw a similar graph for product B (Figure 5.1.3) and check that the cash flow crosses the x-axis at the payback period.

Comparison between the appraisal techniques

As can be seen in Table 5.1.2, each technique gives a different judgement on the projects. Product B appears better in the payback period method, but product A gives better values for the remaining techniques, but not all to the same relative degree.

Table 5.1.2 Comparison between products using different appraisal techniques

	Product A	*Product B*
Payback period	2.75 years	2 years
AGARR	39.3%	35% over four years 28% over five years
ANARR	38.7%	20% over four years 16% over five years
DCF @ 20% rate	£18 885	–£3317
IRR	24.97%	18.59%

It is good practice to subject any appraisal figures to more than one process.

Integrating non-economic considerations into the selection process

Each of the projects can be compared under as many aspects as deemed desirable, perhaps by ranking them using the paired comparison technique (see Chapter 2 – page 93) to reduce them to the few main contenders.

The remaining contenders then need to be put through the organization's economic appraisal system to determine their economic standing. There may be projects which require combining to determine their cross-effect on benefit and shared costs.

Once these stages have been completed, any remaining projects may need further ranking in priority. This can be carried out by producing a further matrix scoring each alternative on a scale from 1 (little impact) up to 10 (critical impact) under four or five significance criteria:

- Technical importance: Some projects must be completed before others can be advanced. This may be due to the lack of technical ability of the organization or a change in resources such as a new building or CAD systems.
- Financial investment: The full cost benefit analysis will show up the total amount of investment required. Here very high investment scores 1.
- Financial return: The time to pay back and total return can be judged here to determine future financial benefit.
- Competitive advantage: The impact on ability to deliver the order winning criteria demanded by our customer. These should tie in with the business plan.
- Intangible benefits: These could be factors such as staff turnover, public image, safety, environmental, etc. where the cash equivalent is difficult to determine. Scoring here could be on a perception basis.

All the criteria need not carry the same weighting. A survey of top management may decide at any one time that the criteria may be weighted as follows:

- Technical importance: 1
- Financial investment: 1
- Financial return: 2
- Competitive advantage: 3
- Intangible benefits: 1

At another time they may increase the value of say, the financial investment to 2 because of cash flow problems.

From the scoring we can produce a priory grid as in Figure 5.1.6. The total in the second end column shows the total score against three projects under both weighted and non-weighted conditions. From these, the project with the highest weighted score would be selected for first priority – in this case alternative Y.

Criteria	Weighting	Alternative X		Alternative Y		Alternative Z	
		Raw score	Weighted score	Raw score	Weighted score	Raw score	Weighted score
Technical importance	1	5	5	6	6	8	8
Financial investment	1	2	2	3	3	6	6
Financial return	2	6	12	9	18	7	14
Competitive advantage	3	5	15	9	27	5	15
Intangible benefits	1	8	8	3	3	6	6
TOTALS		26	42	30	57	32	49

Figure 5.1.6 *Decision matrix comparing alternative solutions to select priority*

Time and other risks

It is important that a suitable time frame is selected to judge projects which do not have a clear life span.

If a short time frame is used then figures can be fairly accurately predicted. However, too short a frame can mean longer-term benefits could be missed. If too long a frame is used this increases the inaccuracy through uncertainty and risk due to unforeseen events, changing markets or changes in technology. Long term would also see plant deterioration, higher maintenance and even renewal.

There are other risks involved in any attempt to forecast markets (see Chapter 6, Section 6.2). These vary from changes in economic conditions, market behaviour, poor designs, competitors' actions, war and natural disasters which may affect customers or suppliers.

Some allowance can be made for possible risks by scenario analysis (see Chapter 6, Section 6.4). It is also worthwhile to carry out a sensitivity analysis by varying some of the input information – such as plus or minus 10% on the sales figure, or the product life.

Non-financial factors

Although it is normally best to use monetary terms, these are sometimes difficult to determine, and resist attempts to express them in pure money terms. The benefits still need to be compared to the costs involved even if they cannot be quantified.

An example of this would be where new leading edge technology is being developed, but the actual improvements can be difficult to quantify initially. In Chapter 4, page 154, we examine one way of examining the overall effect of a project by considering the competitive advantage gained, the technical importance and other intangible benefits in addition to the financial value.

Perhaps the best question to ask in these circumstances is a simple one: 'What happens if we do not make the investment?'

Problems 5.1.1

(1) Think about the last project you have been involved in. What were the benefits gained? Can you express all these benefits in money terms?
(2) Why do you think so many projects fail to make the returns envisaged in the case made for the investment?
(3) Why do you think that many companies request a payback period of less than one year for small capital projects?
(4) Do you think that all extra revenue spending, e.g. new product development, should be subjected to the same appraisal process as capital?
(5) Make out a case to justify spending your own money on a top-of-the-range personal computer for home use.

5.2 Cost determination

Management accounting is the name given to the allocation and control of internal costs within an organization. We need to understand how costs arise and vary to determine how to set them against products and services. For simplicity, we shall define services as being an intangible product.

As we mentioned in the previous section on investment appraisal, a business has one main aim – to earn more than it pays out over as short a time as possible. This applies especially to normal day-to-day running to accumulate profits which can then be used to fund investment and payback borrowings. No organization can survive unless it does make a profit. Even not-for-profit organizations such as charities cannot spend more than they receive in income, whatever the source.

The incoming revenue is based on the price we receive for products, or services. Profit is what is left after deducting all money spent.

Therefore to ensure that we do receive more money than we spend we have to ensure that prices more than cover all the costs involved. Prices themselves are difficult to control as they are often determined by what people will pay – not what a product costs. This leaves costs which are to be controlled. Unfortunately all the costs are not easy to set against particular products, or services.

Classifying costs

Type of costs

Costs can be described in two main ways.

Direct costs. This is a cost for any material or activity that is used directly in making the product. These costs include:

- Raw materials which go into the product.
- The labour operating machines, or assembling a product.
- Power and other consumables used up when making the product.

The easiest way to decide if a cost is direct is to ask the question – 'If we make one more of this product, what actual money will need to be spent?'

There are other costs which are related to making a product such as machinery deterioration, maintenance, setting up the machine and inspection which we could perhaps try to classify as direct. This is often difficult and we shall return to how to set these against products later – for now, we shall treat these as indirect costs.

Indirect costs. These are all the costs which do not directly go into the making of the product. They include items such as supervision, general heating and lighting, administration, sales, etc. Note although these do not go directly into the making of a product, they are only there because the products are being made.

Behaviour of costs

Fixed costs. These do not vary as the amount of product made varies – at least in the short term as in Figure 5.2.1. In the long term they actually do vary – e.g. business rates change from year to year. Examples of these are rent and business rates, supervision, administration, etc. It should be obvious comparing this with indirect costs that fixed costs are in fact mainly indirect.

Variable costs. These should vary directly in relation to the products being made as in Figure 5.2.2. These can be seen as similar to direct costs.

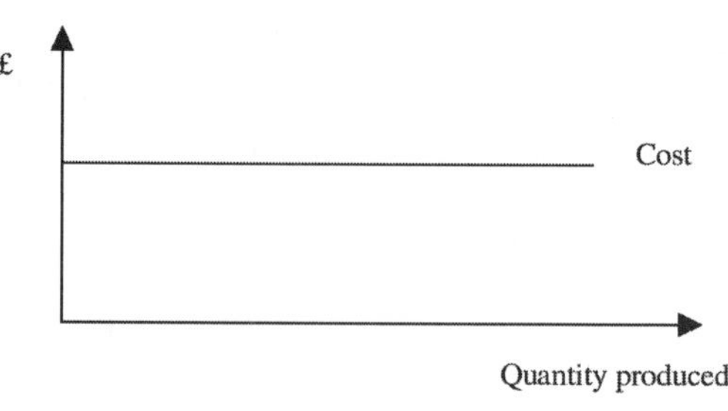

Figure 5.2.1 *Behaviour of fixed cost in relation to quantity produced*

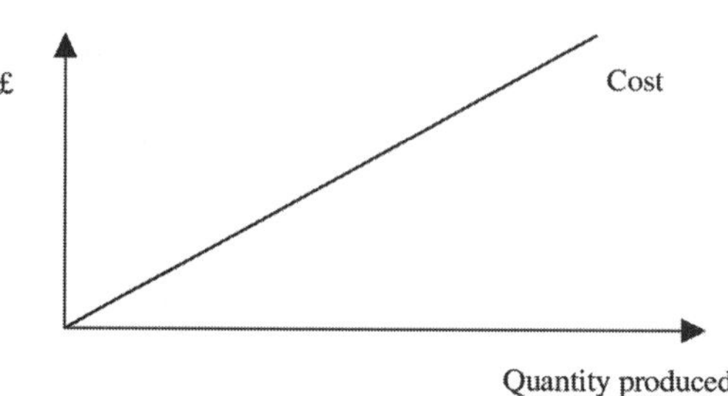

Figure 5.2.2 *Behaviour of variable cost in relation to quantity produced*

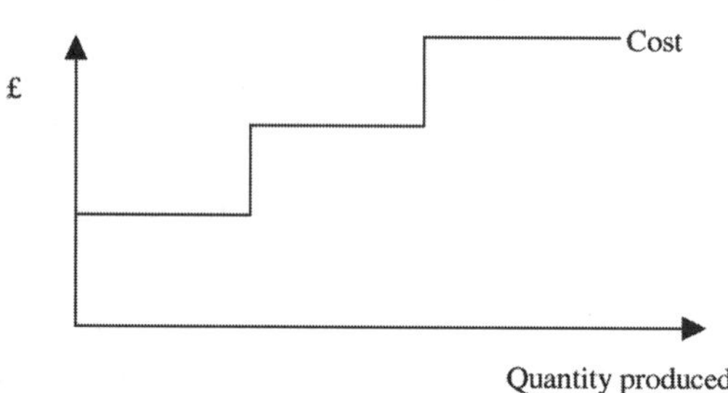

Figure 5.2.3 *Behaviour of semi-variable (stepped) cost in relation to quantity produced*

Semi-variable costs. These are costs which do not vary in relation to slight variations in activity, but do in relation to significant changes. An example of this would be supervisors:

One supervisor may suffice to oversee 10 operatives, but if the number of operatives rose to 20, then an extra supervisor may have to be appointed. This behaviour is often stepped as in Figure 5.2.3.

What is slightly confusing is that many costs which are classified as variable can behave at times like a fixed cost, or behave in a less than directly proportional way. All supposedly fixed costs also vary – at least from year to year. We shall discuss this more later when looking at labour costs.

Costing

We shall now have a look at the way we set costs against products and services.

Labour

Labour costs are often thought of as easy. Just set the wages paid against product costs. In reality, it is more complex than that.

For example, it is difficult to employ part of an employee. We normally have to employ people for a set time period and may not have sufficient orders to keep them working for all the time we pay for. Their wage cost is then behaving as a fixed cost, or at least as a stepped cost.

Similarly an organization may have a human resource policy of not hiring and firing as trade fluctuates, making labour costs act very much like a fixed cost. Some organizations attempt to get round this by employing flexible part-time workers, but these are not always easy to find with the required skills.

Similarly the cost per hour of operatives can vary according to premiums for overtime, shift working and degree of skill. It is common within some industries to pay people a premium for having the capability of performing more than one skill e.g. being capable of operating more than one type of machine.

Labour cost elements

Labour costs must include more than just the basic pay that an employee receives. The extra costs involved are:

- National Insurance
- Contribution to pension funds
- Holidays
- Sick pay

In addition, every minute of an employee's time is not spent directly working on a product. There will be:

- Tea breaks
- Meetings
- Training
- Receiving instruction
- Idle time due to:
  - Machine breakdowns
  - Shortage of work or material
  - Awaiting decisions

Everything done may not result in good products – some defective items will be produced involving either scrapping of work done, or reworking – which may be done by the same employee or a different one.

Finally there are overheads which are determined by the number of employees such as supervision, wages calculations, welfare and perhaps even recruitment if there is turnover of staff. However, let us leave these, to the discussion on overheads.

An example calculation is shown in Example 5.2.1.

Example 5.2.1

The build-up of an employee's cost rate per hour.

COSTS	
Hourly paid rate	£6.00
Basic working week	37.5 hours
Weekly wage	£225.00
Annual wage	£11 700.00
Employer pension contribution @ 8%	£936.00
Employer National Insurance contribution @ 6% average	£702.00
Yearly cost	£13 334.00
TIME	
Paid time – 52 weeks @ 37.5 per week	1950 hours
Less	
Leave	
Annual – 5 weeks	190 hours
Public holidays – 10 days	75 hours
Sickness – 5 days per year	37.5 hours
Meetings – 2 hours per month	24 hours
Training – 2 weeks per year	75 hours
Subtotal non-producing	401.50 hours
Available productive time	1548.50 hours

$$\text{Therefore cost per hour} = \frac{£13\,334.00}{1548.50 \text{ hours}} = £8.61$$

As we can see from Example 5.2.1, the actual cost rates per hour for employees are more than they are paid per hour. In this case, almost 44% more.

Note that this rate further assumes that all available time is actually spent on producing work. When allowing for non-productive time, cost per hour can readily approach double what an operative is paid for basic work time. If the employee in Example 5.2.1 was only 80% producing then the applied cost per hour would rise to £10.76–80% above his paid rate.

Controlling labour costs

From the above, it should be obvious that determining and controlling labour costs necessitates looking at a combination of factors:

- Paid rate of employees.
- Manning levels.
- Productivity.
- Mode of operation.
- Lost time – by cause.
- Premiums paid – for overtime, shifts, skills.
- Turnover (hidden costs of recruitment and training).
- Additional on-costs – pension, welfare, holidays, etc.

Being able to correctly identify the true cost rate means that product costs are accurate and decisions such as what product mix should be made or make-or-buy are based on reality.

Material costs

Similar to labour, at first sight determining material costs should be simple. Again, however, there are complications when we get down to details.

All material purchased does not leave the premises in the form of product:

- All processes have waste associated with them and this has to be included in any cost set against the product.
- There are losses due to scrap or reworking.
- Some materials are used in a process, but do not end up in the product – examples include emery powder (used for polishing), gas consumed in heat treatment and flux used in soldering.

In addition:

- Material which enters the premises causes other costs to arise such as receiving, handling, storage, stock-keeping, stock checking, deterioration, obsolescence, theft, etc.
- Money tied up in stock has a cost associated with it, e.g. interest.
- Even the purchase operation itself costs money to source the item, prepare an order, perhaps then to chase the order and, of course, eventually to pay the supplier.

Where possible therefore, we should determine all the associated costs and include them in the material cost when they happen.

Where material is ordered specifically for a job, then the cost of that material should be put against the job. This should happen even if there is material left over – that should only be excluded when it definitely will be used for another job or sold. In the latter instance the selling price of the material left over can be excluded. If reuse is only probable, then it cannot be excluded. If the material is thought not to be of use but is later sold, or used then it is a bonus.

An example of material bought for a particular job could be a 2 m by 1 m sheet of 15 mm thick mild steel chequered plate (i.e. 2 m^2). A rectangular piece measuring 1.23 m by 1 m is used (1.125 m^2) leaving a piece measuring 0.25 m by 1 m. If there is no definite use for the off-cut then it must either be sold (and the price gained deducted from the initial material cost) or, if it cannot readily be sold, then the full amount of the material must be charged to the job.

Valuation of stock material

In times when the prices of raw material fluctuate even determining the cost of standard stock material becomes more complex. This is because there are several methods of valuing stock material as it is withdrawn.

Even where materials are being used on a first-in-first-out basis, they can be valued in different ways. To demonstrate the differences, we shall look at the following material stock movements under four different valuation methods:

1 Jan	Zero stock
2 Jan	Delivery: 1000 units @ £0.50 each
9 Jan	Delivery: 1000 units @ £0.55 each
11 Jan	Issue 800 units
14 Jan	Issue 800 units
16 Jan	Delivery: 1000 units @ £0.60 each
18 Jan	Issue 800 units

FIFO (first-in-first-out)

As the name implies, material usage is based on the earliest prices, until all material bought at that price is used up. It then moves onto the next price. The example is shown in Excel© format in Table 5.2.1.

As can be seen, the first received was 1000 units @ £0.50 each. These were the first to be issued for the 800 on 11 Jan – leaving 200 of the first receipt still in stock. When the next 800 were issued on 14 Jan only 200 units @ £0.50 each remained to be issued, the remaining 600 units had to be issued from the second receipt, i.e. @ £0.55 each.

Another receipt was on 16 Jan of 1000 units @ £0.60 each. For the issue on 18 Jan only 400 remained of that valued at £0.55. This left 400 to be issued at a value of £0.60. The stock remaining is all valued at £0.60 each.

LIFO (last-in-first-out)

This time value for the latest receipt is used for issues. The example is worked out in table 5.2.2.

As can be seen, at the time of the first issue on 11 Jan, the last received were 1000 units @ £0.55 each. These were the first to be issued for the 800 on 11 Jan – leaving 200 of the next latest receipt still in stock. When the next 800 were issued on 14 Jan only 200 units @ £0.55 each remained to be issued, the remaining 600 units had to be issued from the first receipt, i.e. @ £0.50 each.

Another receipt was on 16 Jan of 1000 units @ £0.60 each. This value was used against the 800 issued on 18 Jan. The remaining stock consisted of 200 @ £0.60 and 400 @ £0.50.

Average cost

In this method, the average price of the stock is recalculated after each receipt and this new average value is then used for any issues made before the next receipt (when average values need again be calculated). The example is shown in Table 5.2.3.

The two receipts on 2 and 9 Jan average out at a cost of £0.525. Note the Excel© sheet shown in Figure 5.2.3 has been set to show currency values to two places, but uses the true value to calculate values charged out.

This average figure is then used for issues until the receipt of 1000 units @ £0.60 each on 16 Jan when a recalculation of the average stock price has to be made. This new average price is then used for the material issued on 18 Jan.

Standard cost
In order to prevent recalculating an issue value each time an expected average, or standard, cost is often set at the beginning of a period for all issues. The example is shown in Table 5.2.4.

This time a standard issue cost of £0.55 was expected. Note that the actual cost of the material only agreed with the standard cost on one occasion. On the other two occasions, the actual price showed a variation from the standard set.

Table 5.2.1 Material charged out and remaining stock value for FIFO method

Date	*Received*	*Value/ unit*	*Value in*	*Issue out*	*Value/unit*	*Value out*	*Stock quantity*	*Stock value*
01-Jan							0	0
02-Jan	1000	£0.50	£500.00				1000	£500.00
09-Jan	1000	£0.55	£550.00				2000	£1050.00
11-Jan				800	£0.50	£400.00	1200	£650.00
14-Jan				200	£0.50	£100.00	1000	£550.00
				600	£0.55	£330.00	400	£220.00
16-Jan	1000	£0.60	£600.00				1400	£820.00
18-Jan				400	£0.55	£220.00	1000	£600.00
				400	£0.60	£240.00	600	£360.00
Totals	2000		£1650.00	2400		£1290.00		

Table 5.2.2 Material charged out and remaining stock value for LIFO method

Date	*Received*	*Value/ unit*	*Value in*	*Issue out*	*Value/unit*	*Value out*	*Stock quantity*	*Stock value*
01-Jan							0	0
02-Jan	1000	£0.50	£500.00				1000	£500.00
09-Jan	1000	£0.55	£550.00				2000	£1050.00
11-Jan				800	£0.55	£440.00	1200	£610.00
14-Jan				200	£0.55	£110.00	1000	£500.00
				600	£0.50	£300.00	400	£200.00
16-Jan	1000	£0.60	£600.00				1400	£800.00
18-Jan				800	£0.60	£480.00	600	£320.00
Totals	3000		£1650.00	2400		£1330.00		

Table 5.2.3 Material charged out and remaining stock value for average value method

Date	*Received*	*Value/ unit*	*Value in*	*Issue out*	*Value/unit*	*Value out*	*Stock quantity*	*Stock value*
01-Jan							0	0
02-Jan	1000	£0.50	£500.00				1000	£500.00
09-Jan	1000	£0.55	£550.00				2000	£1050.00
11-Jan				800	£0.53	£420.00	1200	£630.00
14-Jan				200	£0.53	£105.00	1000	£525.00
				600	£0.53	£315.00	400	£210.00
16-Jan	1000	£0.60	£600.00				1400	£810.00
18-Jan				400	£0.58	£231.43	1000	£578.57
				400	£0.58	£231.43	600	£347.14
Totals	2000		£1650.00	2400		£1302.86		

Table 5.2.4 Material charged out and remaining stock value for standard cost method

Date	*Received*	*Value/ unit*	*Value in*	*Issue out*	*Value/unit*	*Value out*	*Stock quantity*	*Stock value*
Standard cost of material per unit charged out set at £0.55								
01-Jan							0	0
02-Jan	1000	£0.50	£500.00				1000	£500.00
09-Jan	1000	£0.55	£550.00				2000	£1050.00
11-Jan				800	£0.55	£440.00	1200	£610.00
14-Jan				200	£0.55	£110.00	1000	£500.00
				600	£0.55	£330.00	400	£170.00
16-Jan	1000	£0.60	£600.00				1400	£770.00
18-Jan				800	£0.55	£440.00	600	£330.00
Totals	3000		£1650.00	2400		£1320.00		

Comparison of methods

Table 5.2.5 shows the values of the issues out and of the remaining stock derived from the four methods. They are listed in order of issue value. No two have the same values. These differences will show up in the product cost calculation and the main profit and loss statement and the balance sheet.

Any method can be decided on – as long as the organization always presents its information in the same way from one year to the next. It depends on whether the organization decides to make more emphasis on the items going as costs (value out) against profits or the valuation of its

Table 5.2.5 Comparison of material valuation methods

Method	*Value out*	*Stock value*
FIFO	£1299.00	£360.00
Average value	£1302.86	£347.14
Standard cost	£1320.00	£330.00
LIFO	£1330.00	£320.00

assets (stock value). In the long run, it does not matter too much as long as the method is clearly stated and consistently applied.

Controlling material costs

- Ensure all material, including waste and scrap is charged to the job.
- Determine true holding costs.
- Identify slow moving or obsolete stock.
- Determine true procurement costs.
- Minimize stocks held to match production requirements.

The latter gives rise to systems such as material requirement planning (MRP) and just-in-time (JIT).

Process costs

We should build up a true picture of all the costs involved in operating an item of plant in a similar manner to that used to build up the cost of an operative-hour. First we have to determine what items of cost are caused by our operating that particular plant item. We then have to determine the monetary value to be set against the plant (see Example 5.2.2) for the budgeted number of running hours.

Example 5.2.2

Build-up of a machine hour rate

OVERHEAD ITEM	ANNUAL COST
Depreciation	£2500.00
Insurance	£1200.00
Maintenance (labour and spares)	£1400.00
Power	£1150.00
Consumables	£600.00
Tools (non-product specific)	£1500.00
Total annual running costs	£8350.00

Annual hours running = 30 hours/week × 50 weeks

= 1500 hours

Overhead cost $= \dfrac{£8350.00}{1500 \text{ hours}}$

= £5.67/hour

Operative cost/hour = £8.92

Setter's cost (10% of hours cost) = £1.12

Total machine rate = £15.71/hour

The overhead items of depreciation and insurance are independent of the running hours, but the remainder depend on the usage of the machine.

Note that the figure used for the operative rate was that calculated for the worker being available 100% of the time. However, only an 80% utilization has been used to calculate the overhead rate per hour. The remaining 20% could be set against a different cost area for control purposes. It should also be included in the machine hour rate, in which case this would rise to £18.93/hour.

Only 10% of a machine setter's hourly cost rate has been set against the machine as that was calculated to be the amount of time the setter spent on the machine.

Where tooling is required specifically for a product then that cost should be set against the quantity to be produced by each tool.

Overheads

There are a multitude of activities which have to be carried out in a business which are difficult to directly attribute to a product, or process. These are the overheads in the business.

Allowing for ways to recover the cost of overheads is a major problem area in accounting. They must be covered for when deciding if a product is profitable or not at the price received.

If they are wrongly allowed for it could lead to incorrect decisions being made. Therefore it is always best to approach decision making on the same basis as investment appraisal, i.e. only consider overheads which will actually change when making decisions such as make-or-buy.

There are several ways to consider treating overheads.

Absorption costing

Under this method, we attempt to allocate all overheads to specific products, or processes. These overheads are then, in theory, *recovered* by producing products. This is sometimes referred to as full costing. The stages are:

(1) Select identifiable cost centres to collect and allocate costs from. This could be a particular group of machines or an indirect function such as maintenance.
(2) Estimate all overheads.
(3) Allocate these overheads to cost centres where possible to identify their usage therein.
(4) Apportion unallocated costs using estimation of usage. This is often done arbitrarily.
(5) Apportion service departments into direct departments.
(6) Calculate absorption rates.
(7) Charge overheads to products.

The absorption rates used vary considerably.

If a company had an overhead cost of £25 000 associated with a machine shop to be allocated it can choose from a variety of bases:

- Per unit produced: Cost centre overheads divided by number of units produced:

  Example £25 000 over 100 000 units = £0.25 per unit

- Per labour hour: Cost centre overheads divided by direct labour hours.

  Example £25 000 over 12 500 direct hours = £2.00 per hour

- Per machine hour: Cost centre overheads divided by direct machine hours.

  Example £25 000 over 1000 machine hours = £25.00 per machine hour

- Per material usage: Cost centre overheads divided by direct material costs.

  Example £25 000 over £50 000 material = £0.50 per £ material

- Floor space: Cost centre overheads divided by total floor area.

  Example £25 000 over 10 000 m^2 = £2.50 per m^2

Only usage over time will dictate if the method selected is reasonable. Problems that arise are:

- An inappropriate unit for recovery is selected.
- An assumption that arbitrary allocation is accurate.
- Changes in product mix leading to under- or over-recovery of overheads.

Marginal costing

Because of the problems of inaccuracy in overhead allocation, marginal costing is often used in decision making and determining the profitability of a product. In this method the variable costs, including accurately allocated overheads, are compared to the selling price. The difference is called the contribution.

Contribution = selling price – variable costs

The contribution has both to cover the unallocated overheads and add to the profit. This allows companies to gauge if there is a sufficient gap between selling price and variable costs.

Profit = contribution – overheads

As an example, if we sold 30 000 units of a product @ £3.50, which has £2.25 variable costs then:

Unit contribution = selling price – variable cost
= £3.50 – £2.25
= £1.25

Total contribution = sales × unit contribution
= 30 000 × £1.25
= £37 500

If we have unallocated overheads of £25 000 then

Profit = total contribution – overheads
= £37 500 – £25 000
= £12 500

Marginal costing can be used to determine breakeven points. This is achieved by combining two costs (variable and overheads) into one graph and comparing it to the income against quantity produced as in Figure 5.2.4. Quite often this graph is used with fixed costs being used instead of overheads and as such it is similar to the cash flow diagram shown in Section 5.1.

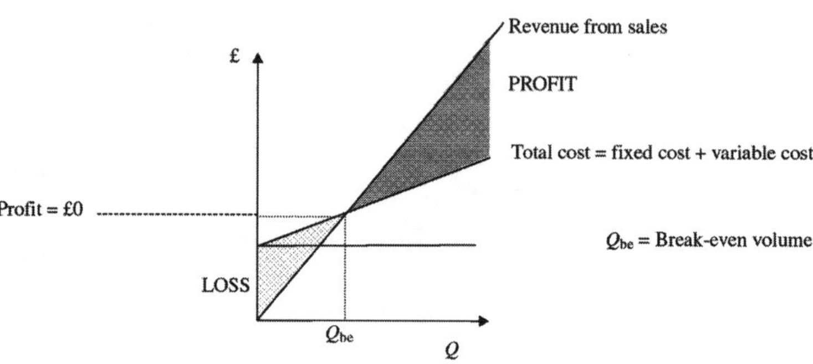

Figure 5.2.4 *Breakeven analysis*

However, as in absorption costing, marginal costing also has inherent complications:

- Variable and overheads costs do not always follow simple patterns.
- Where a variety of products are being made, effort may not be made to allocate overheads towards products.
- Efficiency and usage levels may not be taken fully into account.

Control of overhead costs

Overhead functions' costs are not directly variable with units produced. They must still be controlled as they determine the profit left from the total contributions.

There are a few methods which are used to set initial budgets.

- Zero budgeting: This method is very similar to investment appraisal as in Section 5.1. Each overhead section has to justify its contribution towards profitability. In effect they have to prove that profits would be lower if they did not exist at certain levels.
- Activity based costing (ABC): It is recognized that different levels of activity result in cost level variations. ABC attempts to derive these causes, labelled cost drivers, so that the variations can be foreseen. An example of an activity would be salary and wage administration. Activity varies according to the number of people being paid and the wage systems employed.

The main activity after setting budgets is budgetary control.

Depreciation

When an organization purchases for use over several pieces of plant or equipment it does not put that cost in as a direct cost in the year of purchase. What it does is to assume that the plant's value gets used up, or loses value, over a period of time. It therefore slowly feeds a part of that original cost into the direct cost over the plant life. This is termed depreciation.

Depreciation is a method used by accountants to transfer the loss in value of assets into an expense that is then transferred into the trading/ profit and loss account. A similar approach can be taken to transfer development costs into the operating costs.

It is not like the other overhead costs as no actual money flows out of the company – that took place when the asset was purchased, or the development work took place.

There are several different methods for calculating depreciation within an organization, but only the two most commonly used methods are covered here – the straight line method and the reducing balance method.

The straight line method

This allows the same monetary value to be used each year.

$$\text{Annual depreciation} = \frac{\text{original cost} - \text{residual value}}{\text{estimated useful life}}$$

For example, take an organization purchasing a new car for £20 000 and it plans to sell it in four years. It anticipates it will receive £10 000 for the car then:

$$\text{Annual depreciation} = \frac{£20\,000 - 10\,000}{4} = \frac{£10\,000}{4} = £2500$$

If the same organization buys a new PC for £2400 that it estimates it will replace in 3 years with no resale potential then:

$$\text{Annual depreciation} = \frac{£2400}{3} = £800$$

These examples are shown graphically in Figures 5.2.5 and 5.2.6 respectively.

The reducing balance method

This allows a different value, i.e. a reducing amount, to be used each year.

Annual depreciation = latest book value multiplied by a depreciation factor

If the same organization used the reducing balance method then if they decide that the car loses 20% per year, then they would have depreciation values as in Table 5.2.6.

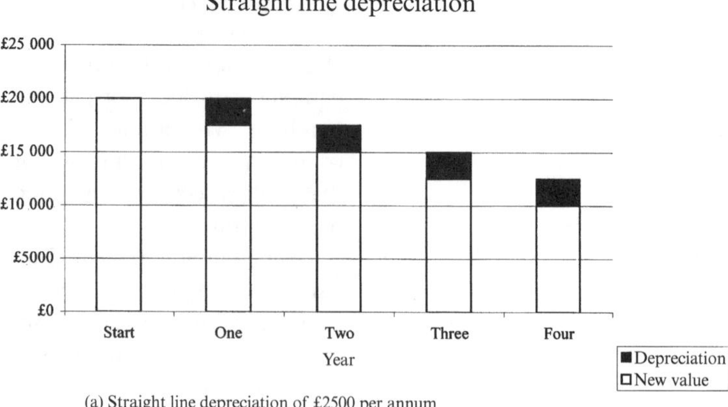

(a) Straight line depreciation of £2500 per annum

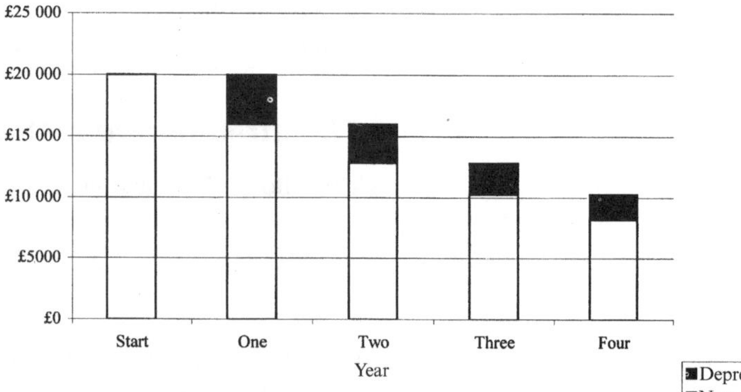

(b) Reducing balance depreciation using 20% per annum

Figure 5.2.5 *Comparison of depreciation techniques used for a car*

Table 5.2.6 Depreciating of car using reducing balance factor of 20%.

Year	*Values £*		
	Start	*Depreciation*	*End*
1	20 000	4000	16 000
2	16 000	3200	12 800
3	12 800	2560	10 240
4	10 240	2048	8192

In the case of the PC, if the company decides to use a 50% depreciation per annum factor then the depreciation in each year is as shown in Table 5.2.7.

As can be seen the closing values using the reducing value method are highly dependent on the factor used. In fact the residual value will never reach zero under this method. Graphs of these are also shown in Figures 5.2.5 and 5.2.6 which demonstrate the change in the depreciation from year to year.

The decision on which method to use is made by the organization itself. However, for tax purposes, a company is limited to the method,

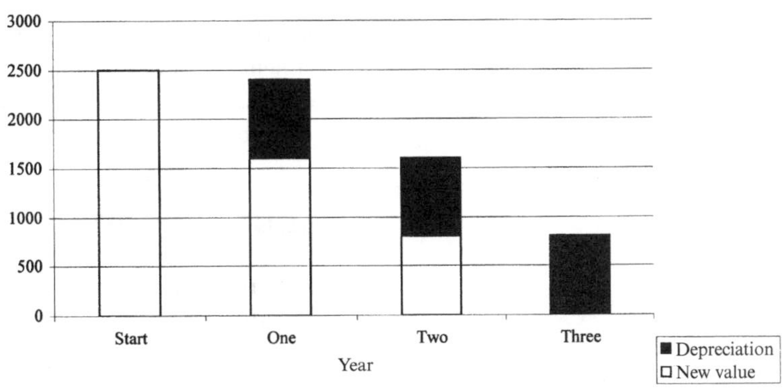

(a) Straight line depreciation of £800 per annum

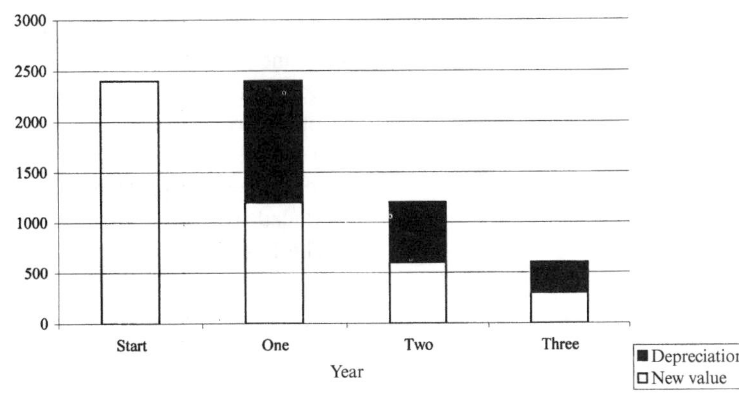

(b) Reducing balance depreciation using 50% per annum

Figure 5.2.6 *Comparison of depreciation techniques used for a computer*

Table 5.2.7 Depreciation of computer using reducing balance factor of 50%

Year	*Values £*		
	Start	*Depreciation*	*End*
1	2400	1200	1200
2	1200	600	600
3	600	300	300

and values, laid down by government in what is termed capital allowances.

Typical UK government capital allowances are:

- Industrial buildings – 4% per annum straight line
- Plant and machinery – 25% per annum reducing balance
- Fittings and fixtures (if permanent part of building) – 4% per annum straight line
- Motor vehicles – 25% per annum reducing balance

These are periodically subject to change and must be checked with the tax authorities each year. It is not uncommon for the values to be temporarily changed when a government wishes to encourage capital investment – for example, the UK government for the period 1999–2002 is allowing 100% capital allowances for IT investment, including developing E-commerce facilities.

Product costing

It is important when working out a product cost to agree beforehand the way overheads will be treated. This is very much down to company policy.

Probably it is best to use marginal costing with as many directly allocated overheads as possible in the calculation. Non-allocated overheads can be shown as a distinctly separate item if required.

The next is to ensure *all* costs are identified and how they will be applied to the costing. An important factor is the quantity to be produced as that will determine how items such as tooling, setting up and scrap allowance will be applied.

Detailed costings are required especially when considering different manufacturing processes.

If we take a one-off component as per Figure 5.2.7. There are many different ways to produce this. For this example let us assume it will be produced by machining from a 125 mm square bar.

The first stage is to produce a process plan showing:

- What operations are required.
- What plant will be used to meet tolerance requirements.
- What time is involved for producing each one and for any set-up involved.
- Any special tooling required.
- Inspection required.
- Starting size.
- Protection and/or packaging required.
- Delivery cost.

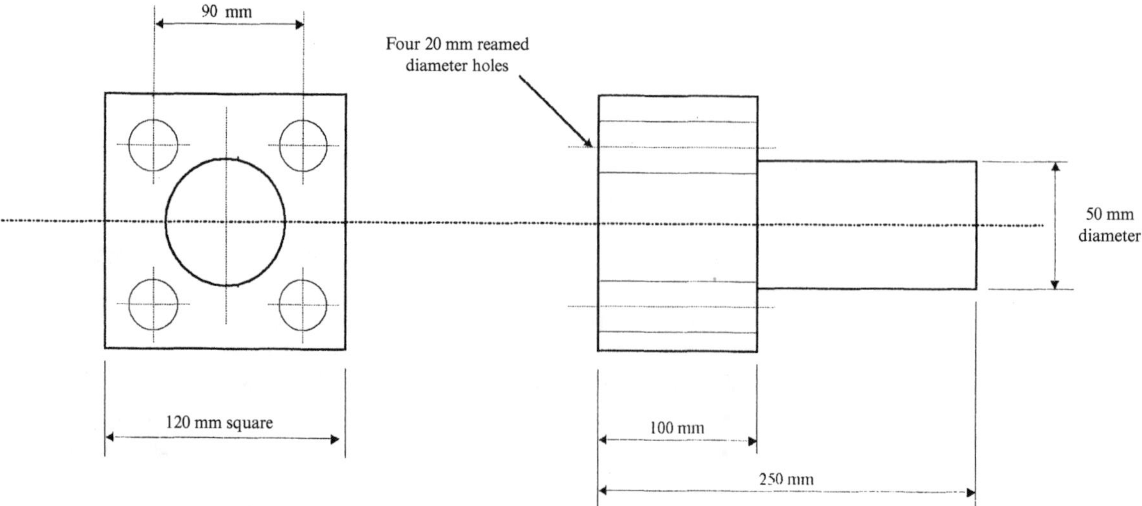

Figure 5.2.7 *Component – material mild steel*

Table 5.2.8 One-off operation cost sheet

Operation number	*Operation*	*Plant*	*Set-up time*	*Process time/item (hr)*	*Cost/hour (£)*	*Cost per run (£)*
1	Cut to length	Saw 105	N/A	0.15	12.00	1.80
2	Mill 120 mm square	Mill 311	N/A	0.33	18.00	6.00
3	Turn and face	Lathe 503	N/A	0.45	17.00	7.65
4	Mark off, drill/ream	J/B 721	N/A	0.33	20.00	6.67
5	Inspect	CC 25	N/A	0.10	15.00	1.50
TOTAL						23.62

From these the costing can be done (see Table 5.2.8).

To this should be added the following costs:

Raw material – bought cut to length @ £9.50
Protection – zero
Delivery @ £3.50

As it is a one-off, then we should also consider extra costs such as raising the production order and the purchasing costs – say £30.

This means a direct cost of £23.62 + £9.50 + £3.50 + £30.00 = £66.62 to which should be added a mark-up percentage to contribute to non-allocated overheads and profits. This would probably mean a selling price of about £100 to cover all costs and give a reasonable profit.

It is worthwhile comparing what this cost would calculate as if we only used an average operator wage rate of £7.00 per hour. The total time involved is 1.03 hours. Multiplying by £7 gives £7.21. If we add material @ £9.50 a total of only £16.71 is arrived at. This is much less than the total arrived at when a detailed costing is carried out hence stressing the importance of carrying out a full costing.

Now what sort of change is involved when considering a large order quantity?

The first is that different processes would probably be selected and special tooling can be invested in.

This time we have a repeating order scheduled at 1000 off per month of the same component as per Figure 5.2.7. Let us assume this time it will be produced by machining from a special ordered 120 mm pre-finished square bar.

The first stage is again to produce a process plan from which the new costing can be done (see Table 5.2.9).

To this should be added the following costs:

- Raw material – bought in bars at £47.50 per metre. Required 250 mm finished length. Required length would be at least 250 mm plus an allowance for parting-off cutter and scrap at bar ends – say allow 280 mm per item. This gives a cost per item of £13.30, which is more than the one-off. This is because we have called for a special starting size, but it saves the operation of machining the 120 mm square.
- There is no allowance for inspection as this is considered to be an overhead in this case because 100% inspection is not necessary.
- Because it has moved operations (to higher producing machines which cost more per hour) we have to invest in special tooling.

Table 5.2.9 1000 off operation cost sheet

Operation number	*Operation*	*Plant*	*Set-up time*	*Process time/item (hr)*	*Cost/hour (£)*	*Cost per batch (£)*
1	Turn and face and part-off	NC lathe 503	0.50	0.06	35.00	2117.50
2	Drill	M/D 333	1.00	0.03	25.00	775
TOTAL BATCH COST						2892.50
Cost per item						2.90

Special chuck jaws for the numerical controlled lathe and a drill jig for the multi-head drill machine. If these cost £2000 and £750 respectively there is a total tooling cost of £2750. We then have to decide how many units each tool will be written off against – in this case we shall use a year's requirements of 12 000 units. This gives a spread cost of £2750 divided by the 12 000, i.e. 0.23 per unit.

- Delivery @ £0.50 each – because we have higher load usage, meaning lower cost.
- As it is a repeat order, we could also consider extra costs such as raising production order and purchasing costs as recoverable from the contribution. This means a direct cost of £2.90 + £13.30 + £0.50 + £0.23 = £16.93 to which should be added a slightly larger mark-up than in the case of the one-off mark-up percentage to contribute to the greater non-allocated overheads and profits. This would probably mean a selling price of around £30 to cover all costs and give a reasonable profit.

Problems 5.2.1

(1) What are the cost items which would be classified as direct costs when considering a car service at a garage?
(2) What is the cost rate per hour of overtime for the employee in Example 5.2.1, if a payment of 25% is made for overtime working? Hint: Only pension and National Insurance is added and there are no deductions for leave, etc. (£8.55)
(3) Compare the Automobile Association's published figure of how much a small car costs per mile against a simple calculation of the cost of petrol used.
(4) How would you allow for recovery (payback) of the costs of developing a new product in the overhead costs?
(5) If you buy a new car, how much value does it lose in the first three years?

5.3 Producing the accounts

Financial accounting is the name given to the determination of the flow of money into, out of, and tracking what happens to this money inside, the organization. It measures the health of the organization.

There are three important documents which describe the financial health of the organization:

- The balance sheet – showing on one particular date where the money is within the organization and also what it is owed from and owed to the outside world. This is known as a snapshot of the financial position.
- The profit and loss account – showing over a set period the sales income, the cost to produce this, other operating expenses and the net profit. This shows the financial performance of the organization.
- Cash flow statement – showing the movement of money into and out of the organization.

Accounting standards

In order to ensure that potential suppliers and investors can examine all the company's financial health in the same way, published accounts must conform to similar standards in presentation and content.

Legal requirements

The Companies Acts 1985 and 1989 prescribe that annually private companies must:

(a) Publish their profit and loss account.
(b) Publish a balance sheet.
(c) Adopt historical accounting rules (or permitted alternative).
(d) Give information on:
  (i) trading of the company
  (ii) important events
  (iii) likely future developments.
(e) Publish Research and development activities.
(f) Publish a Source of increase or decrease in fixed assets, including revaluations.
(g) Publish accounting policies.
(h) Publish an Auditor's report.

Note: small and medium private companies need publish only limited information.

Accounts tend to follow recognized guidelines such as SSAP (Statement of Standard Accounting Practice) but this is not a legal requirement.

Concepts involved in financial accounting

- Money measurement: Only include items which are measured in monetary terms.
- Realization: When items are invoiced is normally the date used for determining when transactions have taken place. This means a cut-off date is selected to determine which account is credited/debited.
- Conservation or prudence – 'Do not anticipate profits' and 'Provide for all possible losses' are the guidelines. A good example is to allow for probable bad debts which will not be recovered.

- Matching: Accounts should show all money spent but only on activities carried out within the period.
  - Accruals cover activities carried out, but not yet paid for, e.g. in arrears such as electricity.
  - Prepayments are money paid in advance of future activity, e.g. rent.
- Consistency: Accounts must be calculated in same way from period to period.
- Disclosure: Any changes which have a significant effect on the accounts must be stated, e.g. a change in accounting practices.
- Objectivity: Endeavour to limit personal bias.
- Duality: The double entry principle. Every transaction is entered in two accounts – once as a debit and once as a credit.
- Verifiability: Accounts should be capable of being verified by independent auditors.

Cash flow statement

A very important day-to-day financial control that an organization carries out is over cash flow. More companies go out of business by running out of cash to pay their employees and suppliers than for being unprofitable.

Table 5.3.1 demonstrates a simple cash flow for a company in its second year of trading. Points to note are:

- All values are shown in £s, unless otherwise stated.
- Sales – shown in units (in italics) and value. Charged out at £20 each but paid for one month later (in arrears). The latter figure is shown in the Revenue row. This is when the cash is actually received and is the figure used when working out the cash flow.
- Materials – shown in units (in italics) and value. Purchased in quantities to make 500 units. Paid for @ £4.80 per unit on delivery.
- Premises are paid for six months in advance.
- Carriage is @ £1.50 per unit, paid in advance.
- The company borrowed £12 500 from the bank in the first year. Interest only is paid on this @ 16% per annum with bank charges of £50. Bank payments are made quarterly in arrears.
- Direct wages are paid to a part-time operator and include National Insurance.
- Indirect wages are the owner's withdrawings as a salary and also include National Insurance.
- The following amounts are part of this year's cash flow although they are items in the previous year's profit and loss account and balance sheet (realization/matching concepts):
  - Revenue of £4000 received in January for sales in December of previous year.
  - Expense of £550 paid to bank in January to cover period Oct. – Dec. of previous year.
  - Expense of £450 paid to accountant in March for preparing previous year's accounts.

For simplicity, we are ignoring Value Added Tax (VAT), National Insurance Contribution (NIC) and pay-as-you-earn (PAYE) payments to

Table 5.3.1 Cash flow for year two of company. All values are shown in £s, unless otherwise stated. (Note figures in shaded background are paid in this year, but were incurred in the previous year and show up in the previous year's profit and loss statement and balance sheet)

Month:	*Previous year figures*	*1*	*2*	*3*	*4*	*5*	*6*	*7*	*8*	*9*	*10*	*11*	*12*	*Total for year*	*Outstanding at year end*
Sales – units		200	170	180	150	200	150	150	200	170	250	200	150	2 170	
Sales – value		4 000	3 400	3 600	3 000	4 000	3 000	3 000	4 000	3 400	5 000	4 000	3 000	43 400	
CASH FLOW															
Revenue		4 000	4 000	2 500	3 400	3 100	900	3 100	2 000	4 600	3 600	4 900	3 700	39 800	7 600
Outgoings															
Material–units		200	500		500			500			500		500	2 700	
Cost		960	2 280		2 280			2 280			2 280		2 280	12 360	
Direct wages		500	500	500	500	500	640	640	680	740	950	760	640	7 550	
Indirect wages		500	500	500	500	500	1 000	1 100	1 200	1 200	1 200	1 200	1 200	10 600	
Rent		2 500							2500					5 000	
Carriage		300	255	270	225	300	225	225	300	255	375	300	225	3 255	
Misc.		500	150	150	150	100	100	125	175	180	60	45	95	1 830	
Car		200	120	130	150	100	120	750	130	150	100	70	70	2 090	
Accountant					450				350					800	750
Bank		550			550			550			550			2 200	550
Tax paid														0	
CASH FLOW CALCULATION (Note that payments against the first year accounts are shown in shaded boxes)														Check	
Start		8 673	6 663	6 858	7 808	6 403	8 003	6 818	4 248	913	2 988	1 073	3 598	8 673	
In		4 000	4 000	2 500	3 400	3 100	900	3 100	2 000	4 600	3 600	4 900	3 700	39 800	
Out		6 010	3 805	1 550	4 805	1 500	2 085	5 670	5 335	2 525	5 515	2 375	4 510	45 685	
End	8 673	6 663	6 858	7 808	6 403	8 003	6 818	4 248	913	2 988	1 073	3 598	2 788	2 788	

government agencies. These were covered in Chapter 1 when dealing with external money transfers.

We can see from the cash flow figures that at all times a positive balance was maintained in cash.

Should this have become negative, then action would have to have been made. This could be arranging an overdraft or delaying some payment, but may be difficult to achieve in this example as most of the expenses had to be paid in advance due to little history of past successful trading by this organization. Companies with more history can normally arrange credit with their suppliers.

It is vital to make a budgeted cash flow for the future to determine when spending will occur and ensure that sufficient funds are available to cover it. Otherwise you would have to arrange an overdraft rather urgently.

The profit and loss account and the balance sheet

Using the data in Table 5.3.1, we will prepare these two interconnected accounting documents. The format we will use is called the vertical as all figures are complete in the one set of columns.

We require some further information on present material and product stocks, depreciation, and some information from last year's balance sheet.

Stock and material usage

Material usage = purchases + opening stock (closing material stock in previous year) – closing stock

= £1104 + £12 360 – £1710

= £11 754

In this case we have no product stock. A similar calculation would have to be made if there was.

Note that the closing stock figure is entered into the balance sheet as a current asset.

Depreciation

This is calculated as shown in Figure 5.3.1 as part of the balance sheet and then transferred to the profit and loss account. In this case, the calculations have all been made on the basis of 25% of the old value. The new value of the asset is shown in the balance sheet.

Preparation of the profit and loss account

This proceeds using the above figures and the remaining information is transferred from the cash flow statement and shown in Figure 5.3.2.

This gives two important values:

- Gross profit after taking off the cost of manufacturing of £19 079, and
- Net profit after tax of –£5429, i.e. a loss of £5429.

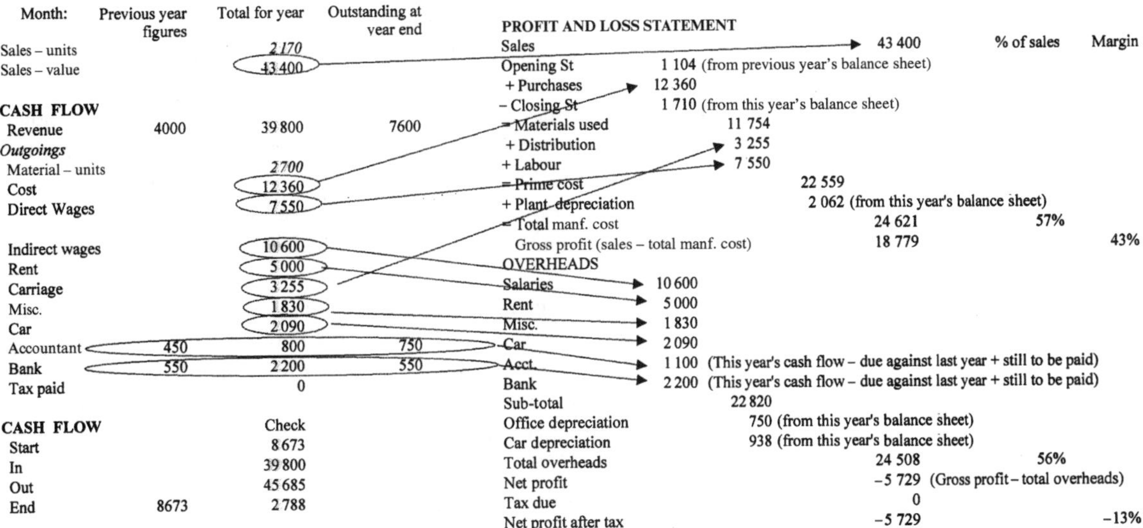

Month:	Previous year figures	Total for year	Outstanding at year end
Sales – units		*2 170*	
Sales – value		43 400	
CASH FLOW			
Revenue	4000	39 800	7600
Outgoings			
Material – units		*2700*	
Cost		12 360	
Direct Wages		7 550	
Indirect wages		10 600	
Rent		5 000	
Carriage		3 255	
Misc.		1 830	
Car		2 090	
Accountant	450	800	750
Bank	550	2 200	550
Tax paid		0	
CASH FLOW		Check	
Start		8 673	
In		39 800	
Out		45 685	
End	8673	2 788	

PROFIT AND LOSS STATEMENT				% of sales	Margin
Sales			43 400		
Opening St	1 104 (from previous year's balance sheet)				
+ Purchases	12 360				
– Closing St	1 710 (from this year's balance sheet)				
= Materials used		11 754			
+ Distribution		3 255			
+ Labour		7 550			
= Prime cost			22 559		
+ Plant depreciation			2 062 (from this year's balance sheet)		
= Total manf. cost			24 621	57%	
Gross profit (sales – total manf. cost)			18 779		43%
OVERHEADS					
Salaries	10 600				
Rent	5 000				
Misc.	1 830				
Car	2 090				
Acct.	1 100 (This year's cash flow – due against last year + still to be paid)				
Bank	2 200 (This year's cash flow – due against last year + still to be paid)				
Sub-total		22 820			
Office depreciation		750 (from this year's balance sheet)			
Car depreciation		938 (from this year's balance sheet)			
Total overheads			24 508	56%	
Net profit			–5 729 (Gross profit – total overheads)		
Tax due			0		
Net profit after tax			–5 729		–13%

Figure 5.3.1 *Preparation of profit and loss account from cash flow statement. All values are shown in £s, unless otherwise stated*

Month:	Previous year figures	Total for year	Outstanding at year end
Sales – units		*2 170*	
Sales – value		43 400	
CASH FLOW			
Revenue	4 000	39 800	7600
Outgoings			
Material – units		*2 700*	
Cost		12 360	
Direct wages		7 550	
Indirect wages		10 600	
Rent		5 000	
Carriage		3 255	
Misc.		1 830	
Car		2 090	
Accountant	450	800	750
Bank	550	2 200	550
Tax paid		0	
CASH FLOW		Check	
Start		8 673	
In		39 800	
Out		45 685	
End	8 673	2 788	

BALANCE SHEET				
FIXED ASSETS	Prev	Deprec @ 25%	New	
Office equipment	3 000	750	2250	
Process equipment	8 250	2063	6188	
Car	3 750	938	2813	
Sub-total	15 000			
Depreciation		3 750		
Net fixed assets			11 250	
Debtors	7 600			
Stock	1 710			
Cash	2 788			
Net current assets		12 098		
Creditors	1 300			
Net current liabilities		1 300		
Net working capital			10798 (Current assets – current liabilities)	
NET ASSETS (Fixed assets + working capital)				22 048
Bank loan			12 500	
Owner's invest	17 500			
Reserves	–7 952			
Net owner's capital			9 548	
NET CAPITAL EMPLOYED				22 048
CHECK BALANCE (Net assests – net capital employed)				0

Figure 5.3.2 *Preparation of balance sheet from cash flow statement. All values are shown in £s, unless otherwise stated*

Note: no tax is due as no profit was made – the loss can be carried forward to set against any future profits.

We shall examine how well the company is doing by evaluating some key ratios in Section 5.4 later.

Completion of the balance sheet

This is completed by transferring data from the cash flow statement and the profit and loss accounts and is shown in Figure 5.3.1.

Net working capital shows the difference between short-term liabilities and short-term assets. The short-term liabilities include what is owed, but not paid. Net working capital should be positive to show that, in the short term, the business can repay its short-term liabilities through its short-term assets.

The reason for the balance sheet being so named is simple – it shows that the organization has a monetary balance between what it owns (the net short-term and long-term, or fixed, assets) and what it owes (the long-term liabilities, i.e. the capital employed made up of the owner's investment and long-term loans). There should be no difference, i.e. they are in balance.

Any profit produced belongs to the owner – so any loss must also be borne by the owner, up to the limit of his liability. The reserve heading shown under owner investment denotes profit which has been retained in the business (i.e. not paid out as dividends, but still owed to the owners). It is produced by adding this year's profit (or loss) retained to the previous year's reserves.

In this case both the previous year's and the current values have negative values. This means the net owner's investment is decreasing – this is fairly common in a new company but requires careful monitoring so that a profitable stage is reached soon.

On the good side, there has been a positive cash flow over the last four months.

Some common items which have not occurred in this particular profit and loss account and balance sheet, but may in other companies, include:

- Allowance for bad debt. Some of the money owed may not be recovered because of customers going out of business, or fraud. This must come off the sales figure in the profit and loss account.
- Intangible assets:
  - Goodwill – an amount paid for another business over and above the physical value of the assets gained – this is depreciated just as a fixed asset.
  - Intellectual property such as patents – sometimes difficult to value, but may be a source of earnings, or may be sold off to another business.
  - Prepayment and accruals: If any made, or due to be paid then this should be entered. Prepayments would appear under current assets and accruals under current liabilities. Tax due could come under these headings, as could rent paid in advance.
  - Revaluations: Where carried out, it must be described in accompanying notes with the basis of that revaluation.

Keeping the books

The process of book-keeping is one which often gives people problems in understanding. The basic process itself is simple – anyone who has a bank account should be able to understand entries. The difficulty comes in the multitude of different books and ledgers used, identifying which money flows where and making the many entries. Thankfully, modern accounting packages mean that the computer does most of the detailed entries.

Using the same data as in the cash flow shown in Table 5.3.1, the process is as follows.

The books

There are a multitude of separate accounts called books and ledgers (as that is what they once physically were). All transactions are recorded in them:

- Cash books: A book of first entry. Records cash and bank transactions.
- Day books: Also a book of first entry. Records cash and credit sales and purchases.
- Ledgers: Used to store and accumulate accounting information:
  - General/nominal ledger – stores different accounts of the items in profit and loss account and balance sheet.
  - Sales ledger – what customers owe (debtors).
  - Purchase ledger – what is owed to suppliers (creditors).

Meaningful classification of financial information is fundamental to its eventual usefulness. There are four basic types of accounts:

- Expenses and sales: These are normally used during the accounting period. They will be closed off and transferred to the trading, profit and loss account at the end of a trading period. They will restart in the next period with no entries.
- Assets and liabilities: These remain open at the end of a period as they normally have a residual value that will continue to be used by the organization in the future. They will be balanced and then their value will be copied to the balance sheet. This balance remains in these accounts to start off the next period.

The double entry system

Every financial transaction is listed in these accounts and involves an exchange of equal value shown as two associated entries – in different accounts. This is the double entry system.

Key point

Every transaction will appear as two entries – once as a debit entry and the second as an equal credit entry. Normally in a different account.

- Whatever comes into the business is debited to an appropriate account.
- Whatever is going out of the business is credited to an appropriate account.
- Transferred amounts between accounts are shown as a debit entry and a corresponding credit entry.

You may think that accountants do the books in a mirror image to what you are used to seeing in your bank statements. This is because we are used to interpreting our bank balance from our viewpoint. The statement is correctly prepared looking at matters from the bank's viewpoint. It is our definition of credit and debit that is different to those used in accounting. The problem is that our definition is becoming the norm outside accountancy – which causes problems when we first examine book-keeping.

Your deposit coming in is a debit on the bank's cash book and this must have a corresponding credit entry in another account – yours. It is a liability on the bank to pay you. It sounds confusing; you may want to read these paragraphs again after we work through some examples.

It is important that you think of the terms debit and credit in the opposite way to that when you read your bank statement. This is the key to understanding book-keeping. Any money going in is a debit and money coming out is a credit.

Book and ledger accounts

If we take the data shown in the balance sheet (Figure 5.3.1), then we would have the following accounts. We will identify them, give them a code and classify them under the four headings.

ACCOUNT	CODE	DESCRIPTION	CLASSIFICATION
Shares capital	(CL.1)	Money invested by owners	Liability
Reserves	(CL.2)	Profits retained in business	Liability
Loan	(CL.3))	Long-term loans	Liability
Cash book	(CB.1)	For money in and out of the business	Asset
Sales	(SB.1)	For sales made (all credit)	Sales
Purchases	(PL.1)	For raw material purchases	Expense
Stock	(SL.1)	For movements in and out of stock	Expense and asset
Equipment	(CA.2)	Value of existing equipment	Asset
Direct wages	(NL.1)	Payment to direct staff	Expense
Indirect wages	(NL.2)	Payment to indirect staff	Expense
Rent	(NL.3)	Payment for rent of building	Expense
Carriage	(NL.4)	Payment for carriage of product	Expense
Miscellaneous	(NL.5)	Payment for small items, e.g. stationery, postage, etc.	Expense
Travelling	(NL.6)	Payments such as fuel, insurance, road tax, etc. for car	Expense
Accountant	(NL.7)	Payment to accountant	Expense
Bank charges	(NL.8)	Interest and bank charges	Expense

The codes are to ease identification and reduce writing within the accounts when completing them. The terms used are historic and do not signify any real difference today. The codes used are:

CL = capital ledger – where long-term liabilities are recorded.
CA = capital asset – where long-term, i.e. fixed, assets are recorded.
CB = cash book – money in the bank.
SB = sales book – sales made.
PL = purchase ledger – purchases made.
SL = stock ledger – records of the value of stock (inventory).
NL = nominal ledger – other expenses.

There will also be other accounts such as tax due and dividend paid which are not activated yet. Later in the company's life there could also be further accounts, or some accounts such as equipment could be split into transport, office and process plant.

It will take some time to prepare full detailed accounts, so let's start right at the beginning to show what happens with the initial money in the bank, i.e. the cash book. We shall assume that money flows take place on the last working day of the month:

CASH BOOK (CB.1)
All figure are in £s

DEBIT				CREDIT		
Date	Transaction	Amount		Date	Transaction	Amount
1.1	b/f (CB.1)	8 672		31.1	Purchases (PL.1)	960
31.1	Sales (PL.1)	4 000		31.1	Wages (WL.1)	500
				31.1	Wages (WL.2)	500
				31.1	Rent (NL.3)	2 500
				31.1	Carriage	300
				31.1	Misc. (NL.5)	500
				31.1	Car	200
				31.1	Bank	550
				31.1	c/f to Feb. (CB.1)	6 662
	Sub-total	12 672			Sub-total	12 672
31.1	b/f from Jan. (CB.1)		6 662			

These entries should balance against each other at the end of January. The last entry on the credit side of £6662 is required to give this balance. Where there is a credit entry there must be an equal debit entry. That is made to reopen the cash book at the beginning of February. The terms c/f and b/f are abbreviations for carry forward and brought forward respectively.

All the credit entries above would have a corresponding entry in the noted account, for example:

PURCHASE LEDGER (PL.1)
All figure are in £s

DEBIT			CREDIT		
Date	Transaction	Amount	Date	Transaction	Amount
31.1	Purchase (CB.1)	960	31.1	To stock (SB.1)	960

The value of the purchases are immediately transferred over to the stock account, which is a special account supplying information for both the balance sheet and the profit and loss account. It appears as a debit entry.

At the end of the accounting year, the b/f figure in the asset and liability accounts is copied into the balance sheet, but remains in the account as that b/f value:

CASH BOOK (CB.1)
All figure are in £s

DEBIT				CREDIT		
Date	Transaction	Amount		Date	Transaction	Amount
1.12	b/f (CB.1)	3598		29.12	Purchase (PL.1)	2280
29.12	Sales (PL.1)	3700		29.12	Wages (WL.1)	640
				29.12	Wages (WL.2)	1200
				29.12	Carriage	225
				29.12	Misc. (NL.5)	95
				29.12	Car	70
				29.12	c/f next year CB.1	2788
	Sub-total	7298			Sub-total	7298
29.12	b/f (previous year CB.1)		7 298			

This value opens the next year's cash book account and is also copied (note – not transferred) to the balance sheet where it appears as a debit.

The stock account is actually an asset account and is the source for the stock figure copied to the balance sheet. However, it must contain a transaction to show the material usage.

If we look at this account, we find that up until the year end only the value of the opening stock and the purchases appear (in the debit column) to give a sub-total of £13 464. This is not the value that we have as physical stock as much of it has been used in manufacturing products.

At the year end, a physical stock check finds the actual closing stock to be £1710. The difference of £11 754 is attributed to the material used up in the manufacturing process.

Therefore on the credit side we show the closing stock value and an entry which is transferred to the profit and loss account, where it appears as a debit entry – remember double entry principle.

STOCK (SB.1)
All figure are in £s

DEBIT			CREDIT		
Date	Transaction	Amount	Date	Transaction	Amount
1.1	b/f (SL.1)	1 104	29.12	To P&L	11 754
31.1	Purchases (PL.1)	960	29.12	c/f Closing stock	1 710
29.2	Purchases (PL.1)	2 280			
28.4	Purchases (PL.1)	2 280			
31.7	Purchases (PL.1)	2 280			
31.10	Purchases (PL.1)	2 280			
29.12	Purchases (PL.1)	2 280			
	Sub-total	13 464		Sub-total	13 464
29.12	b/f (opening stock – SB.1)	1 710			

The opening/closing stock value is copied to the balance sheet.

This time we will use the credit/debit format to display the results. All asset accounts will appear on the debit side under long- or short-term headings. Liabilities appear as credit entries:

BALANCE SHEET
All figure are in £s

	DEBIT			CREDIT	
	Account	Amount		Transaction	Amount
	CURRENT ASSETS			CURRENT LIABILITIES	
29.12	Bank (CB.1)	2 788	29.12	Accountant (NL.7)	750
29.12	Stocks (SB.1)	1 710	29.12	Bank charges (NL.8)	550
29.12	Debtors (SL.1)	7 600	29.12	Working capital	10 798
	Sub-total	12 098		Sub-total	12 098
	LONG-TERM ASSETS			LONG-TERM LIABILITIES	
29.12	Working capital	10 798	29.12	Bank loan (CL.3)	12 500
29.12	Equipment (CA.2)	11 250	29.12	Owner's investment	17 500
			29.12	Reserves (CL.2)	–7 952
	Sub-total	22 048		Sub-total	22 048

Both sides are in balance.

The working capital is the balancing value in the current assets and its double entry appears on the debit side in the long-term capital and liability section.

Similarly, we produce the profit and loss account by closing off all sales and expense accounts and transferring their balances to the profit and loss account, leaving them empty ready for the start of the new year's trading.

Taking the direct wages account, we have:

DIRECT WAGES (NL.1)
All figure are in £s

DEBIT			CREDIT		
Date	Transaction	Amount	Date	Transaction	Amount
31.1	Wages (CB.1)	500	29.12	Closing (P&L)	7550
29.2	Wages (CB.1)	500			
31.3	Wages (CB.1)	500			
28.4	Wages (CB.1)	500			
31.5	Wages (CB.1)	500			
30.6	Wages (CB.1)	640			
31.7	Wages (CB.1)	640			
31.8	Wages (CB.1)	680			
29.9	Wages (CB.1)	740			
31.10	Wages (CB.1)	950			
30.11	Wages (CB.1)	760			
29.12	Wages (CB.1)	640			
	Sub-total	7550		Sub-total	7550

Note: the sub-total for the debit entries has been transferred to the profit and loss account leaving a debit balance of zero to start off the next year.

The completed profit and loss account looks like:

PROFIT AND LOSS ACCOUNT
All figure are in £s

DEBIT			CREDIT		
	Account	Amount		Transaction	Amount
	TRADING ACCOUNT				
29.12	Materials (SB.1)	11 754	29.12	Sales (SL.1)	43 400
29.12	Distribution (NL.4)	3 255			
29.12	Direct wages (NL.1)	7 550			
29.12	Plant deprec. (DL.1)	2 060			
29.12	Gross profit	18 779			
29.12	Indirect salaries (NL.2)	10 600	29.12	Gross profit	18 779
29.12	Rent (NL.3)	5 000			
29.12	Misc. (NL.5)	1 830			
29.12	Car (NL.6) (CL.2)	2 090			
29.12	Account (NL.7)	1 100			
29.12	Bank charges (NL.8)	2 200			
29.12	Office deprec. (DL.1)	750			
29.12	Car deprec. (DL.1)	938			
29.12	Loss (CL.2)	–5 729			

Again the totals balance.

The flow of the written transactions is shown in Figure 5.3.3.

In order to check during the financial year that all entries are being made correctly, many organizations carry out a trial balance (sheet) at other times than the year end. This is just a check process, it does not go into an end of year statement. Again with the increase in computer-based accounts this can be done speedily.

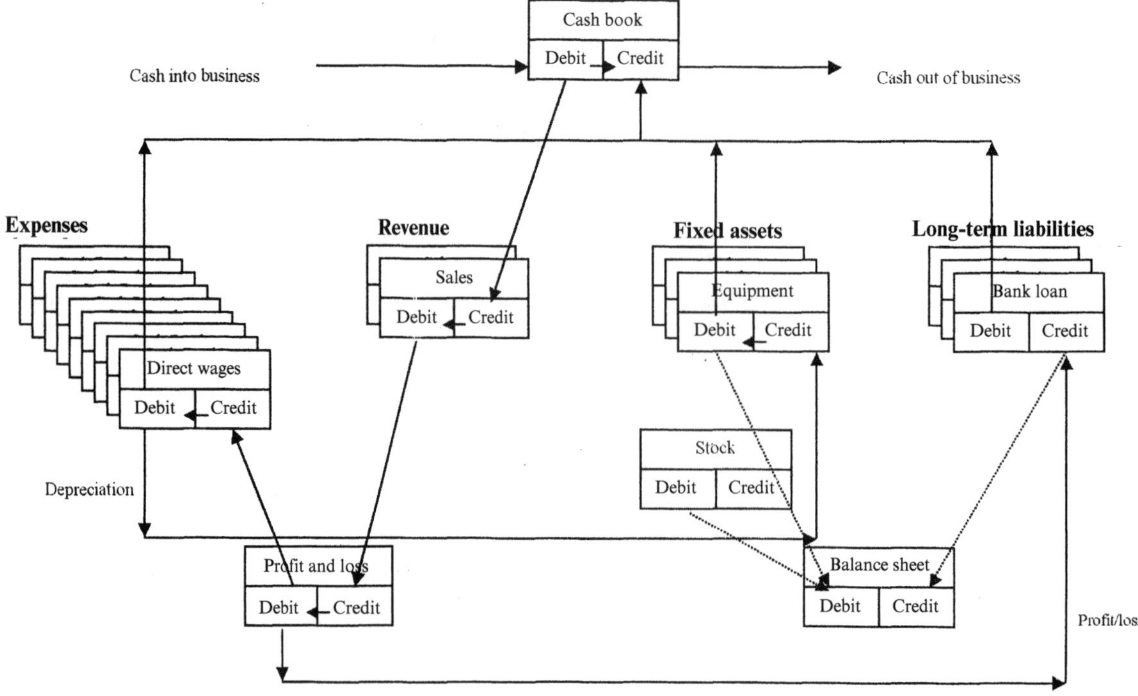

Figure 5.3.3 *Cash flow through the accounts*

Problems 5.3.1

(1) What do you think will be a bank manager's attitude if you suddenly had to seek an overdraft in the middle of your trading year?

(2) Can you think of other separate accounts that a company may wish to keep to add to the list on page 236?

(3) Try to complete the equipment account yourself. (Hint: use values already shown in Figure 5.3.1 for start values and depreciation.)

(4) Complete entries for another expense account – such as the indirect salary. (Hint: you can check your answer by looking at the balance sheet in Figure 5.3.1.)

(5) Get a public company's published account and see if you can follow the cash flow over the year that produced the new balance sheet.

5.4 Budgeting and the interpretation of results

Once operations have been completed, it is possible to analyse how effective they have been in financial terms. This involves first developing a standard, or budget, breakdown of the expected costs and analysing why differences have occurred.

Everyone has personal experience of having to spend within a set budget. We have a finite income and many things we would like to buy, but we cannot buy all of them, as our money does not go far enough.

Every organization has the exact same problem as individuals do. Pressure to spend on items which will make life easier, more prestigious or more enjoyable, but a limited income. Every section has something they would like to spend money on – such as more people or a new computer.

We discussed in Section 5.1 techniques for justifying expenditure. These are difficult and time consuming to apply to every small decision that is made. We need another method that will enable managers, and increasingly other employees, to know how much they can spend and what they should be spending that money on. This is what budgeting is all about.

As when we produce a manpower plan (see Chapter 2, Section 2.1), we have to look at the future business plan for the organization. We then have to look at each department and section to see what resources are considered necessary to meet that plan. We cost these resources to produce a budgeted cash flow of the outgoing expenses and capital spending for each section (see Figure 5.4.1).

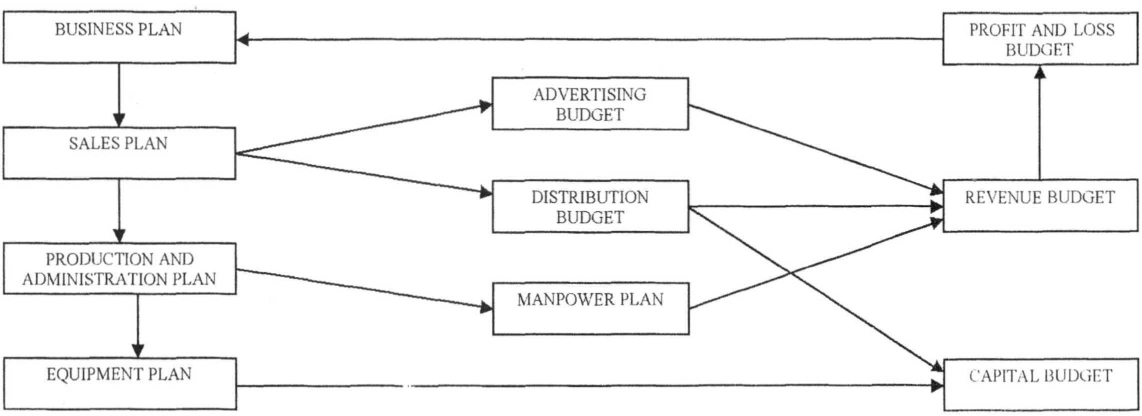

Figure 5.4.1 *Preparing the budget*

From these sections' cash flows and the incoming money, we calculate what the overall resultant profit and loss will be. This is a long, and fairly tedious but necessary, task, which requires the involvement of all managers as this review sets what resources will be available during the budget period.

Only if the anticipated end result meets the business plan, can we say that we are relatively happy with the budget. Should the anticipated cash flow not meet the business plan then the budgets need to be revised until it does.

However, that is not the end of the story, we need to be continually comparing the actual cash flow with the budgeted cash flow to ensure that the targeted performance is met.

Preparing the budget

The best way to demonstrate this is to build a complete budget. We will take as our example the business that we showed the actual results for in Section 5.3. We will go back to the year before and build up a budget for the business.

Sales budget

The owner anticipated the annual sales would be 2400 @ £20 each. These he broke down into the following monthly figures:

150, 150, 160, 170, 180, 180, 200, 220, 220, 240, 250, 280

Associated with these were materials, carriage and labour:

- Material is £4.80 per product unit.
- Carriage is £1.50 per product unit.

Labour was a little more problematic. The owner himself had been producing the units, but he was finding it difficult to find time to make them, sell them and spend time on developing a new product. He considered himself capable of making four units per hour. He planned to take on a part-time worker to train to manufacture the product. This would release the owner for selling and development.

This part-time operative was planned to be taken on in the fourth month. An initial allowance of £500 per month was made to cover 50 hours per month. This would start as a training budget and would reduce as the operative gradually came up to speed over a period of four months.

It was decided to make the standard labour cost of producing the sales figure £2.50 per unit. This was based on an expected cost of £10 per hour for the operative during which four units would be produced. The owner's salary initially would be split using this figure with the remainder going into indirect wage cost. As the operative came up to speed, the portion set against the owner's salary would reduce until by the eighth month it reduced to zero.

The owner then prepared a cash flow budget as in Table 5.4.1. The figures for miscellaneous and car expenses were estimated and spread over the year. The bank charges were considered to be the same as the first year and the accountant's cost expected to be a little more.

PROFIT & LOSS STATEMENT

Sales				48 000	% of sales	Margin
Open St	1 104					
Purchase	11 520					
Close St	1 104					
Materials used		11 520				
Distribution		3 675				
Labour		6 850				
Prime cost			22 045			
Plant depreciation			2 062			
Total manf. cost				24 107	50%	
Gross profit				23 893		50%
Salaries	9 856					
Rent	5 000					
Misc.	1 200					
Car	1 850					
Acct.	700					
Bank	2 200					
Sub-total		20 806				
Office depreciation		750				
Car depreciation		938				
Total overheads				22 494	47%	
Net profit				1 399		3%

Figure 5.4.2 *Budgeted profit and loss account*

Table 5.4.1 Budgeted cash flow

Month:	*Previous year figures*	*1*	*2*	*3*	*4*	*5*	*6*	*7*	*8*	*9*	*10*	*11*	*12*	*Total for year*	*Outstanding at year end*
Sales – units		150	150	160	170	180	180	200	220	220	240	250	280	2 400	
Sales – value		3 000	3 000	3 200	3 400	3 600	3 600	4 000	4 400	4 400	4 800	5 000	5 600	48 000	
CASH FLOW															
Revenue		4 000	3 000	3 000	3 200	3 400	3 600	3 600	4 000	4 400	4 400	4 800	5 000	46 400	5 600
Outgoings															
Material – units		150	150	160	170	180	180	200	220	220	240	250	280	2 400	
Material – cost		720	720	768	816	864	864	960	1 056	1 056	1 152	1 200	1 344	11 520	
Training wages					500	250	50	50						850	
Std labour		375	375	400	425	450	450	500	550	550	600	625	700	6 000	
Indirect wages		675	675	600	184	386	286	1 050	1 200	1 200	1 200	1 200	1 200	9 856	
Rent		2 500							2 500					5 000	
Carriage		300	225	240	255	270	270	300	330	330	360	375	420	3 675	
Misc.		100	100	100	100	100	100	100	100	100	100	100	100	1 200	
Car		100	100	100	100	100	100	750	100	100	100	100	100	1 850	
Accountant					450				200					650	500
Bank		550			550			550			550			2 200	550
Tax														0	
CASH FLOW CALCULATION (Note that payments against the first year accounts are shown in shaded boxes)														Check	
Start		8 673	7 353	8 158	8 950	8 770	9 750	11 230	10 570	8 534	9 598	9 936	11 136	8 673	
In		4 000	3 000	3 000	3 200	3 400	3 600	3 600	4 000	4 400	4 400	4 800	5 000	46 400	
Out		5 320	2 195	2 208	3 380	2 420	2 120	4 260	6 036	3 336	4 062	3 600	3 864	42 801	
End	8 673	7 353	8 158	8 950	8 770	9 750	11 230	10 570	8 534	9 598	9 936	11 136	12 272	12 272	

This produced a budgeted profit of £1399 (see Figure 5.4.2) which, although low, was reasonable as it was the business's second year and the owner anticipated drawing £13 200 as a salary in addition. Although a profit was expected, no allowance was made for tax as the first year's losses (£2223) could be offset against the anticipated year's profits.

Analysing the results

Unfortunately, as we saw in Figure 5.3.2, the anticipated profits turned out to be a loss of £5729 – a difference of £7128 less than that budgeted for.

Accountants use the terms favourable and adverse instead of good and bad.

- Favourable: Expenses less (negative) than anticipated or income greater (positive) than anticipated.
- Adverse: Expenses more (positive) than anticipated or income less (negative) than anticipated.

The anticipated figures are termed as the standard values. The differences we call variances and it is important that we analyse these variances and take action to rectify them, or make allowance for them if they are not changeable. To do this we should break the variance down into constituent causes:

Variance analysis

There are different control formulae which can be used to analyse variances, i.e. the difference from that budgeted for:

Overall control ratios

These are the key ratios which we should keep our eye on to ensure that we are working efficiently and making effective use of our facilities on a time basis:

- Productivity performance: A measure of how effective our people are actually working. See appendix to Chapter 3 for further discussion on how to determine this. A suitable measure is:

$$\frac{\text{Standard hours for items produced} \times 100\%}{\text{Actual hours worked}}$$

- Capacity usage: This looks at how much of our people's time we actually used in relation to what we set out to do in the budget. A suitable measure is:

$$\frac{\text{Actual hours worked} \times 100\%}{\text{Budget hours}}$$

- Volume index: Although this looks like the same as the productivity measure, it is using the hours that we originally set out to work as the base. It indicates how successful we were at bringing in work for them to do. A suitable measure is:

$$\frac{\text{Standard hours for items produced} \times 100\%}{\text{Budget hours}}$$

Note that these measures connect as follows:

$$\text{Productivity performance} = \frac{\text{volume index}}{\text{capacity usage}}$$

Sales

- Sales revenue variance = (actual price/unit × actual sold) – (standard price/unit × budgeted sales)
- Selling price variance = (actual price – standard price)/unit × actual sold
- Sales effectiveness = (actual sold – budgeted sales) × standard price/unit

Note: Sales revenue variance = selling price variance + sales effectiveness.

These indicate the reason for having a revenue different to that anticipated. Was it the quantity sold, i.e. the sales effectiveness, or the price that was charged, i.e. the selling price variance.

Materials

- Material cost variance = (actual price/unit × actual usage) – (standard price/unit × standard for number produced)
- Purchase price variance = (actual price paid – standard price)/unit × actual usage
- Material usage variance = (actual usage – standard usage for number produced) × standard price/unit

Note: Material cost variance = purchase price variance + material usage variance.

These indicate the reason for having a material cost different to that anticipated. Was it the price paid for the material, i.e. the material price variance, or the amount of material used, i.e. the material usage variance?

Labour

- Labour cost variance = (actual hourly rate × actual hours) – (standard hourly rate × standard hours for units made)
- Labour rate variance = (actual hourly rate – standard hourly rate) × actual hours
- Labour efficiency = (actual hours – standard hours for number produced) × standard hourly rate

Note: Labour cost variance = labour rate variance + labour efficiency.

These indicate the reason for having a labour expense different to that anticipated. Was it how well we used the labour, i.e. the labour efficiency, or the hourly rate they were paid, i.e. the labour rate variance.

These formulae allow us to look beyond the straight variance and find underlying reasons. It enables us separate numerical and cost/price causes.

Let us go back to the example and compare the items in the actual profit and loss account against those budgeted for (see Figure 5.4.3).

When we see so many figures it can easily become a little confusing. What are the reasons for all these variations? For example, this analysis shows a favourable variance for distribution – is this good? Initially the answer may be yes, but as less was sold than budgeted for, we would expect a drop in spending on distribution.

To adjust those items which should vary with output, we can produce what is termed a flexed budget. This is simply adjusting the variable costs in line with sales by multiplying by the ratio of actual unit sales to

	Budget	*Actual*	*Variance*
Sales units	2 400	2 170	–230
Sales value	48 000	43 400	–4 600
Materials used	11 520	11 754	234
Distribution	3 675	3 255	–420
Labour	6 850	7 550	700
Prime cost	22 045	22 559	514
Plant depreciation	2 062	2 062	0
Total manf. cost	24 107	24 621	514
Gross profit	23 893	18 779	–5 114
Salaries	9 856	10 600	744
Rent	5 000	5 000	0
Misc.	1 200	1 830	630
Car	1 850	2 090	240
Acct.	700	1 100	400
Bank	2 200	2 200	0
Sub-total	20 806	22 820	2 014
Office depreciation	750	800	50
Car depreciation	938	938	0
Total overheads	22 494	24 508	2014
Net profit	1 399	–5 729	–7 128

Figure 5.4.3 *Profit and loss variance*

	Flexed budget	*Actual*	*Variance*
Sales units	2 170	2 170	0
Sales value	43 400	43 400	0
Materials used	10 416	11 754	1 338
Distribution	3 255	3 255	0
Labour	6 194	7 550	1 356
Prime cost	19 932	22 559	2 627
Plant depreciation	2 062	2 062	0
Total manf. cost	21 797	24 621	2 824
Gross profit	21 603	18 779	–2 824

Figure 5.4.4 *Flexed budget variance analysis*

the budgeted unit sales. We take units rather than monetary value, as we may have sold items at less than the standard price:

$$\text{Flexed budget} = \text{original budget} \times \frac{\text{actual sales in units}}{\text{budget sales in units}}$$

Most of the items going into the manufacturing costs can be taken as variable. If we apply the above calculation to them, we can then produce a variance analysis of flexed budget against actual costs as in Figure 5.4.4. The plant depreciation is left the same as budgeted for as this will not change.

Looking at this we can see that the distribution costs are in line with the actual sales. The revenue is also in line, hence price gained per item was as budgeted. However, materials and direct labour variances are even worse than they appeared before flexing. This needs further analysis, which will need more information than the cash flow statement and the profit and loss account figures show.

There were three reasons for the direct labour adverse variance of £1356 in the flexed budget comparison:

- The part-time operative was taken on in January rather than in April. However, the hourly cost for this was as expected – therefore no rate variance needs to be calculated.
- The training took longer than expected. This gave an adverse variance of £250.
- The expected output of four units per hour was not achieved – therefore we had an adverse efficiency variance of £1106 (based on the flexed budget figures). Probably the estimate of four units per hour made no allowance for break times, reworking and other lost time – it requires checking.

This may be slightly confusing as the unflexed direct labour variance is only an adverse one of £700. The difference in the figures is due to a favourable variance from the original budget of £656 due to lower actual sales. The latter only appears favourable because it is less than that original budgeted for.

When we look into the material costs, we find that two changes from the budget basis have taken place. More materials were used – but some were bought at a cheaper price. This calls for some examination of units used and prices paid per unit before we do the detailed variance calculations.

- Products produced were 2170.
- Budgeted material cost per unit was £4.80.
- After January's purchases of 200 units, the owner started buying in bulk which gained him a discount of 5%, making the new cost of £4.56.
- Opening stock was £1104 valued at £4.80 each giving a unit stock of 230 (£1104 ÷ £4.80).
- Closing stock was £1710 valued at £4.56 each giving a unit stock of 375 (£1710 ÷ £4.56).
- Total units purchased was 2700 (200 at £4.80 and 2500 at £4.56).
- Usage was 2555 (230 + 2700–375).

Therefore we have a gain through buying cheaper (after the initial 200) but this has been offset by using more material. If we exclude the 200 bought at the budgeted cost, we can now see the monetary effect of these changes, using the following basis:

- Number produced = 1970 (2170–200)
- Material units used = 2355 (2555–200)
- Standard cost per unit = £4.80
- Actual cost per unit = £4.56

Cost variance = (actual price/unit × actual usage) – (standard price/unit × standard for number produced)
= (£4.56 × 2355) – (£4.80 × 1970)
= (£10738.80) – (£9456)
= £1282.80

Price variance = (actual price – standard price)/unit × actual usage
= (£4.56 – £4.80) × 2355
= –£565.20

Usage variance = (actual usage – standard usage for number produced) × standard price/unit
= (2355 – 1970) × £4.80
= £1848.00

Check: does Price variance + usage variance = cost variance?
–£565.20 + £1848.00 = £1282.80

We can see that there was a favourable variance of £565.20 due to a lower purchasing price. This was not sufficient to offset the adverse variance of £1848.00 due to excess usage. Again this needs checking as it is about 20% extra usage which is a high figure. Perhaps some was consumed during early training which had not been allowed for.

Looking at the remaining expenses, the amount budgeted looks as if it may have been optimistic – but again this needs to be examined and explained.

Interpreting company accounts

In accountancy there are a number of ratios that are used to judge the effectiveness of an organization's operations. These are worked out from the figures contained in the profit and loss account and the balance sheet.

The ratios normally require two consecutive years' figures to work out an average over a company's financial year. However, we shall use the one year's result contained in Figures 5.3.2 and 5.3.2 for simplicity.

Return on investment

In any investment, there is an associated risk. Normally the possibility of a high return is linked to a high risk of getting considerably less – perhaps even losing the initial investment.

In the accounts the return is the net profit after all deductions for items such as depreciation and tax but before any dividend is paid. We need to divide this by the capital employed – however, there are several figures we can use as the divisor. Remember always check what figures are being used.

One way is to use the shareholder's funds as the bottom line. Using the figures from our data, we have:

$$\text{ROCE (Return on capital employed)} = \frac{\text{Net profit}}{\text{shareholders' funds (at year beginning)}}$$

$$= \frac{-5729 \text{ (a loss)}}{15\,277 \text{ (initial investment + reserves)}}$$

$$= -37.5\%$$

This means that the shareholders' funds would be reduced to zero in just under three years if losses continued at this rate.

We could also use net assets, but we have to ensure that if a long-term loan is included in the funding then we add back onto the net profit any interest charges incurred by that loan.

The bank, however, will not let matters approach a point where their investment is too much at risk. They are liable to call in their loan before a further year has passed unless returns substantially increased.

Liquidity

If we examine the working capital of a company, we can see how quickly it can convert the current assets into cash to pay off the current liabilities. If we have a minus figure as working capital it demonstrates that we cannot meet our immediate needs and we are in trouble unless we can quickly find more cash.

Working capital ratios

The ratios used here indicate how quickly we should be able to pay off our short-term creditors. The second is a more severe test as it looks at a smaller base, i.e. as if the sales suddenly stopped. The ratios are:

$$\text{Current ratio} = \frac{\text{current assets}}{\text{current liabilities}}$$ Should always be > 1, preferably > 2

$$\text{Quick ratio} = \frac{\text{debtors + cash}}{\text{current liabilities}}$$ Should normally be > 1

In our example, we can see

- Net current assets are £12 098.
- Net current assets less stock are £10 388.
- Net current liabilities are £1300.

Therefore,

$$\text{Current ratio} = \frac{12\,0984}{1300} = 9.3, \text{ which is high}$$

$$\text{Quick ratio} = \frac{10\,388}{819} = 7.9, \text{ which is also high}$$

Although both ratios are high, this is not necessarily a good indication. For example, if we could immediately convert our stocks and debtors into cash we could pay off 75% of our bank loan and save £1500 in interest charges.

We need to know the reason behind the figures and see if they can be changed to advantage:

- Being a new company, most of the purchasing is paid for in cash. We do not have credit with our supplier, as yet. If supplier credit could be obtained this would immediately reduce the need for cash, which could help to reduce the bank loan. Better to use our supplier's money than our own.
- Customers should be paying for the goods after one month. The figure shown as owing indicates that they are taking longer and this puts pressure on cash flow. If they could be persuaded to pay early, even on delivery, again this would reduce the need for borrowing. Excess pressure to pay may lose us business and is a negative aspect we need to watch for.
- Raw material stocks represent almost two months' supply – reducing these and only having enough to meet current needs would reduce this figure. It may be more costly buying more frequently, however.

Margins and mark-ups

In commercial terms margins can be expressed in two ways. It is important that we know exactly which figures are being used:

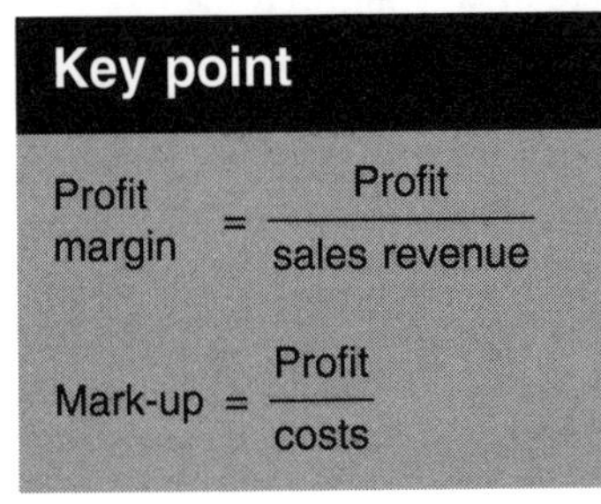

Key point
Profit margin = Profit / sales revenue
Mark-up = Profit / costs

- Margins are based on the selling price – can look at gross and/or net figures.
- Mark-ups are based on costs (buying price or manufacturing) – again can look at gross or net figures.

Both are usually expressed as percentages.

Value added is another way of expressing mark-ups, and could be expressed as a monetary value.

In our examples, we have the following figures.

- Sales = £43 400 i.e. 100% of sales revenue figure.
- Manufacturing costs = £24 621 = 57% of sales revenue.
- Gross profit = £18 779 = 43% of sales revenue (i.e. margin on manufacturing cost).
- Overheads = £24 508 = 56% of sales revenue.
- Total costs = £49 129 = 113% of sales revenue.
- Net profit = –£5 729 = –13% of sales revenue (i.e. margin on total cost).

We could alternatively express the third of these items as gross profit = 76% mark-up on manufacturing costs. Similarly we could express the fourth item, Overheads, as being 99.5% of manufacturing costs.

Therefore to find the sales breakeven point, we need to apply the concept we used in Section 5.2 (page 222).

These figures are useful in letting us know by how much, if any, we can adjust prices when under customer pressure to do so to increase sales.

Turnover ratios

Turnover simply means the number of times in a period that an average item is used up and replaced. There are three important ratios relating to debtors, creditors and stock:

Debtor ratios. Only includes sales made on credit. In our example, no sales were for cash.

$$\text{Debtors' turnover} = \frac{\text{sales on credit}}{\text{end of year creditors}} = \frac{43\,400}{7600} = 5.71.$$

$$\text{Debtor days} = \frac{\text{days in period}}{\text{debtors' ratio}} = \frac{365}{5.71} = 64 \text{ calendar days.}$$

Therefore the customers are taking over twice as long to pay as the terms allow. Extra effort is required to shorten this time, but this may make our customers less eager to deal with us.

Creditor ratios. In our data, suppliers were paid on delivery. If, however, we assume that one month's credit was allowed then:

$$\text{Creditor turnover} = \frac{\text{purchases}}{\text{last material purchased (creditors)}} = \frac{12\,360}{2700} = 4.58$$

$$\text{Creditor days} = \frac{\text{days in period}}{\text{creditor turnover}} = \frac{365}{4.58} = 80 \text{ days}$$

This figure is a little misleading as the business does not purchase every month, but gives an idea of what levels can easily be attained to counteract debtors taking a long time to pay. Organizations should use creditors to balance the effect on cash flow of debtors.

Stock ratios. We can base this particular example on the cost of raw material only as no product stock is held. Where work-in-progress or product stock is held, we need to include it at the cost of manufacture – note not anticipated selling price.

$$\text{Stock turnover} = \frac{\text{cost of sales}}{\text{closing stock}} = \frac{12\,360}{1704} = 7.25$$

$$\text{Stock days} = \frac{\text{days in period}}{\text{stock turnover}} = \frac{365}{7.25} = 50 \text{ days}$$

As we mentioned when looking at the current ratio, this is high and should be reduced to release cash.

Gearing ratios

These indicate how the long-term funding of the company is supplied – loans or owners' capital (shares and reserves). In the example at the end of year 2:

$$\text{Gearing ratio (A)} = \frac{\text{long-term loans}}{\text{net assets}} = \frac{12\,500}{22\,048} = 57\%$$

$$\text{Gearing ratio (B)} = \frac{\text{long-term loans}}{\text{owners' capital}} = \frac{12\,500}{9548} = 131\%$$

This indicates that currently the bank has a higher investment than the owner. This will make it difficult for the owner to ignore any suggestion made by the bank.

If you are looking for a guide then a simple one could be:

- Where risks are heavy or rewards are low – risk other people's money rather than your own.
- However, if rewards are high then it may be less expensive to borrow and pay interest than share out the profits.

Share price ratios

The last set of ratios we are going to look at are related to share prices. Up until now we have not considered shares as our example has been based on a single owner.

In Chapter 1, we looked at the various patterns of ownership, and concluded that most of the large companies are funded by the issue of ordinary shares. These shares, once issued, are bought and sold in stock markets by their individual owners.

The shareholders are the owners of the company and as such can, and often do, have considerable influence on the management. The shareholders' primal concern is not so much the present net value of the company assets but more the value of their shares. These two values can be considerably different.

Share prices can change considerably over a short period of time. The price is based on immediate factors and a mixture of other aspects regarding their future potential. We will not attempt here to value a business, just to describe how various ratios connected with shares are worked out.

To do this, we will need a new set of data, for a different organization. Again this has been simplified for ease of understanding:

- Issued shares: 5 million @ £1, equal to £5 million.
- Net profit after tax, but before payment of dividend: £1.2 million.
- Declared dividend for year: £0.12 per £1 share, giving a payout of £0.6 million.
- Shares trading at a mid-market price of £2.25 each.

The various concerns are:

Share price growth and stability

$$\text{Earnings per share} = \frac{\text{net profit}}{\text{number of shares}} = \frac{£1\,200\,000}{5\,000\,000} = £0.24 \text{ per share}$$

$$\text{Yield per share} = \frac{\text{dividend}}{\text{price of share}} = \frac{£0.12}{£2.25} = 4.8\%$$

$$\text{Price/earnings (P/E)} = \frac{\text{price of share}}{\text{earnings per share}} = \frac{£2.25}{£0.24} = 9.37$$

The P/E ratio of 9.37 represents the number of years to recoup the present share price based on present profits.

The actual share price will be affected by results from the organization and this must be taken into account when setting a strategy.

Problems 5.4.1

(1) Set down what your weekly income is and what your outgoings are. Are they the same? What can you do if your spending is more than your income?

(2) Compare the individual items in the actual profit and loss against that budgeted for in Figure 5.4.3. Why are some of the items up and some down. Is every figure bad?
(3) Is the drop in distribution costs in Figure 5.4.3 in line with the drop in sales from that budgeted for?
(4) Work out the percentage sales increase required to achieve breakeven, assuming no extra overhead spending is incurred. (31%)
(5) Get the published account of two public companies and see if you can explain the reason why the share prices are different.

6 Meeting customer needs

Summary

It is by satisfying customers in the marketplace that an organization grows and prospers. To satisfy the customer the organization needs to offer more than just physical products or bare services, the customer is looking for a number of benefits.

This chapter will help you think about all that the customer needs and the best way to meet them by aligning your internal operations to support the marketing effort. As an engineer you will need to understand how the market works and what you can do to ensure that you gain and retain customers. This chapter introduces you to marketing techniques.

Objectives

By the end of this chapter, the reader should:

- understand the role of marketing and the structure of the marketplace in which a product exists and the factors affecting demand (Section 6.1);
- understand the need for and be able to apply forecasting techniques, allowing for their degrees of uncertainty (Section 6.2);
- appreciate the sources of information for and the techniques used in market research exercises (Section 6.3);
- be able to analyse the factors demanded by customers and ensure that the organization retains a competitive position (Section 6.4).

6.1 The marketplace

This section takes a general look at the concept of marketing. It starts with explaining the role of marketing and aspects of supply and demand. The importance of a product's life cycle is introduced. The section completes with examining the distribution routes available and aspects of promotion and selling.

Historically everyone lived in small communities where all demand was catered for by local personnel, who were often multi-skilled. As communities grew in size, some specialization occurred forming specific trades, such as cobblers and saddlers. Transport facilities severely limited the area able to be serviced by a single tradesman, except in the larger towns. Some small businesses started up employing a few people, but opportunities were limited as all work was manual.

As transport developed, so did trade but initially only in luxury goods which could not be easily supplied by local craftsmen. This luxury trade was mainly limited to the richer portion of society.

The industrial revolution made the mass production of certain goods an economic proposition, as long as the new transport system could economically and swiftly move the product from the factory to the customers. From this scenario, came the need for marketing to ensure that the customer and the goods being produced matched.

It still tended to be a seller's market, but the spread of the industrial revolution soon introduced competition. Today that competition is truly worldwide with local products often competing against ones from half way round the world.

Marketing is no longer an optional choice – it has become a necessity for survival.

The role of marketing

The basic role of marketing is to match the resources of an organization to the needs and wants of its environment. To do this it must:

- Determine the internal factors which govern the total output of the organization, i.e. its ability to supply goods or services.
- Determine the external factors which influence the selection of goods, i.e. the demand.

Where there is a considerable mismatch between the internal resources, then management must bridge that gap. They do this either by changing the internal resources, or finding a market where the internal resources are useful.

The functions covered by marketing are:

- Forecasting future trends in market demand.
- Market research on their own present product and the competitions'.
- New product needs: To find needs which are not being met.
- Pricing strategy: To determine what price their products should be set at.
- Promotional strategy: What advertising and other promotions are required.
- Distribution channels: To ensure the product reaches all potential customers efficiently.
- Customer support: Support required prior to and after sales.
- Selling: The physical direct selling to the immediate customers.

Identifying the market

To survive an organization must first clearly identify its market:

WHO buys the product?
WHY do they buy a particular product?
WHERE do they do their buying?
WHEN do they buy?
HOW do they buy?

Market classification

Consumer goods:

- Convenience good: Items bought every day, which the customers do not expect to shop around for, e.g. milk and bread should be available in any supermarket.
- Shopping good: Items for which a customer will look at what is available in several retail outlets, e.g. clothing and shoes.
- Speciality good: Items for which the customer expects speciality advice and for which they are prepared to travel to purchase, e.g. ski and professional level sports equipment.

Industrial goods:

- Raw materials: Basic raw materials which will be processed within the organization, e.g. steel bar and plate.
- Capital equipment: Items which are purchased for long-term use within the organization, e.g. machine tools, vehicles and computers.
- Components: Items which go into other products with minimum work being done on them, e.g. nuts and bolts and electronic components such as resistors. These could be sub-assemblies such as radiators for cars.
- Supplies: Items which are used by the organization, but are not incorporated into their products, e.g. electricity, paper, overalls.

These markets differ in how they operate and the trade cycles they experience. In the industrial goods markets, suppliers are completely dependent on other organizations' success in buying and selling their products.

Determinants of demand

Socio-economic factors:

- Population: Groupings by age, number in households, etc. and trends therein.
- Population location, e.g. are they scattered or grouped in one location.
- Net disposable income after tax and basic living expenses.
- Socio-economic groups, e.g. high earners, unemployed, etc.

Psychological factors:

- Motivation: Why people buy particular items.
- Perception: What people see in product benefits.
- Beliefs and attitudes: What people think about certain factors, e.g. organic production. These do not necessarily reflect reality.

- Personality, e.g. attitude to credit, new technology, etc.
- Reference groups: How certain groups are influenced by what a few icons do, or wear.

Market segments can be forming and reforming continually.

Supply and demand

It is a common economic statement that price is a function of supply and demand and it does to a greater or lesser extent.

- When prices are high, producers are encouraged to start producing as opportunities for profits appear.
- When prices are low, many producers find they cannot make a profit and withdraw from producing.

The converse also applies – when there is a shortage, prices rise until the extra profit to be gained entices others into supplying. This balances out the demand and prices then fall again.

There are other factors which can affect supply and demand apart from price.

Factors changing supply relate to their impact on organizations' decisions relating to costs and profits:

- Weather, natural disasters and civil disruptions: This could be small disruptions due to snow and localized strikes. It can also be due to factors such as flooding and earthquakes.
- Technology: Here we are not only looking at what is available, but the take-up of the technology by both suppliers and customers.
- Taxation and subsidies: Here we mean the impact of these on the decisions of organizations.
- Changes in costs of factors of production: Governmental decisions can impact here, but there are also general movements in the marketplace.

Factors changing demand for particular products (include items such as the net disposable income of individual consumers after paying taxes and covering their basic needs such as housing):

- Weather: Not only are there annual cycles, but many organizations find sales differ considerably depending on the weather. The food industry is particularly affected and this has led to major supermarket groups closing following the long range weather forecasts.
- Prices of other products: This can cause changes in buying patterns – an example would be a steep rise in petrol prices. This would reduce money available to purchase other items such as meals out – it may also cause a switch to public transport (or so the government hopes).
- Fashion: As well as the normal fashion cycles there are other fad products which suddenly arise, and almost as quickly fade away. An example is merchandise attached to movies.

- Credit availability: Loosening or tightening credit terms can impact on decisions on major purchases such as houses and cars.
- Advertising and promotion: Although they say that at least half of the money spent in advertising is wasted, it is recognized that a successful campaign can have significant effect on sales. An example is the Gold Blend series of adverts which is reputed to have increased the sales of this product by 30%.
- Population changes: Although these tend to be slow moving, nevertheless they can impact on profitability of certain niches. An example is the longer life attainment of the population leading to an increase in grandparents spending on children's items.
- Confidence about the future: Perhaps a key decision as this will determine if a person saves against the future, or spends as they earn – and perhaps beyond that by using credit.

Some products show different elasticity in their price:demand ratios.

Elasticity of the price:demand ratio

Some products are highly sensitive to a change in price, others less so. This sensitivity we term elasticity of demand.

- Elastic: When a 1% change in price results in a greater than 1% change in demand. Examples are usually shopping or luxury goods such as cars, etc.
- Inelastic: When a 1% change in price results in a less than 1% change in demand. Examples are usually basic items such as electricity, food.

Note that often these descriptions relate to a relative price and with changes in income, customers may switch between certain products. A good example of customers switching products can be found by looking at the cars bought at different stages in a person's life as their income changes.

Product life cycle

Every product has a selling life cycle as shown in Figure 6.1.1. This is very similar to the human life cycle:

- Conception: The basic idea for a product is conceived. This may be after careful consideration of many factors, or can come as a flash of inspiration. Many ideas come to nothing at this stage.
- Pregnancy: After conception, the product idea must be developed within the organization to become fully fledged through a process of design, research and development. Most product ideas die at this stage.
- Birth: The launch of a new product. Many products fail to have a successful launch.

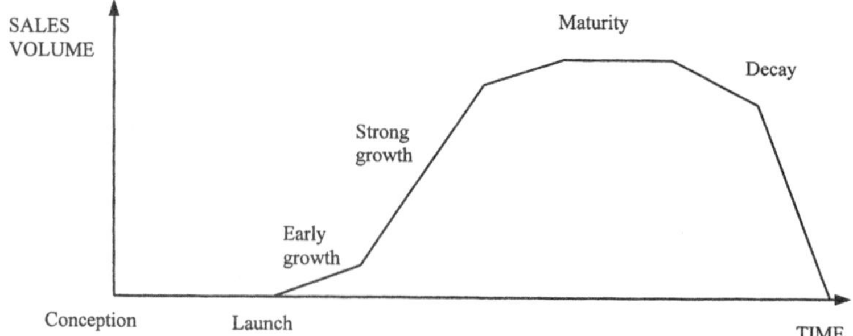

Figure 6.1.1 *The product life cycle*

- Infancy: Growth of sales. During this period there will be a lot of changes and uncertainties.
- Maturity: The product sales have reached a plateau. This may go on for years, e.g. Mars bars – or may be for a short period only, e.g. computers and fashion clothing.
- Old age and decline: Sales start to slow down and decline.
- Death: The product stops selling. Continued support for the product in the form of spare parts may still have to continue. It is not unusual to have a demand for spare parts for over ten years after a white goods item, e.g. a washing machine, has stopped being made.

The main problem with the product life cycle, is that it is difficult to predict what is going to happen in the future. Most product ideas die during the design phase, many fail to break out of the early growth period, whilst others appear to be in a state of maturity for decades.

It may be possible to extend the product life by redesigning it periodically. This can give rise to a surge and extension in the product life cycle as shown in Figure 6.1.2. The car market is a prime example of this with the updating of a named model range after two to three years. It may be fifteen years before bringing out a completely new name model.

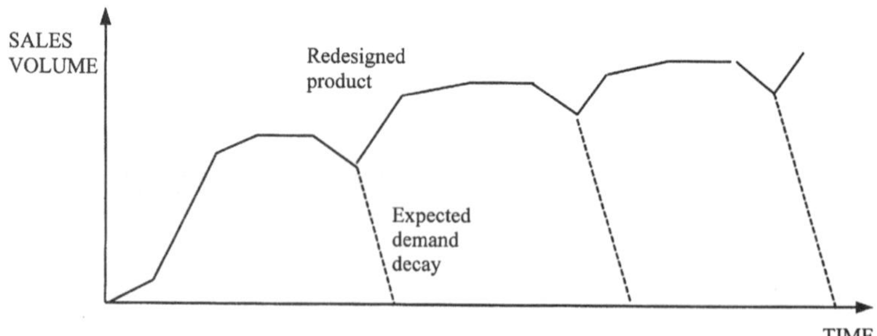

Figure 6.1.2 *Product life cycle extended due to redesign of product*

It is important that an organization maintains a steady supply of new products at various stages of their life cycles to maintain an even cash flow into the organization as in Figure 6.1.3. Many companies rely heavily on sales from products which have been introduced or redesigned within the last five years.

Many products are experiencing shortening life cycles. At the same time, products are becoming more complex and sophisticated. The

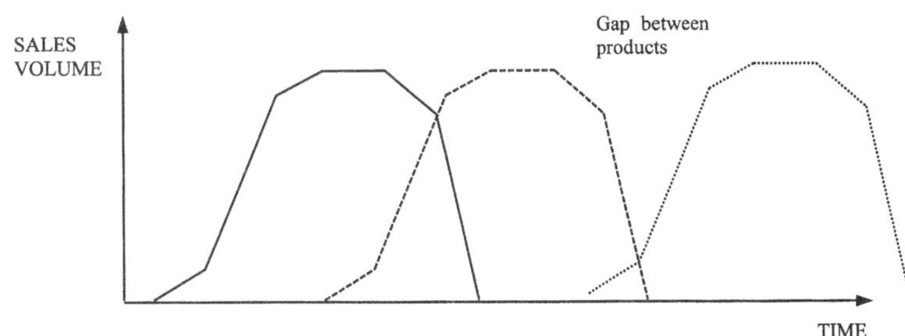

Figure 6.1.3 *Product life cycles for overlapping product launches*

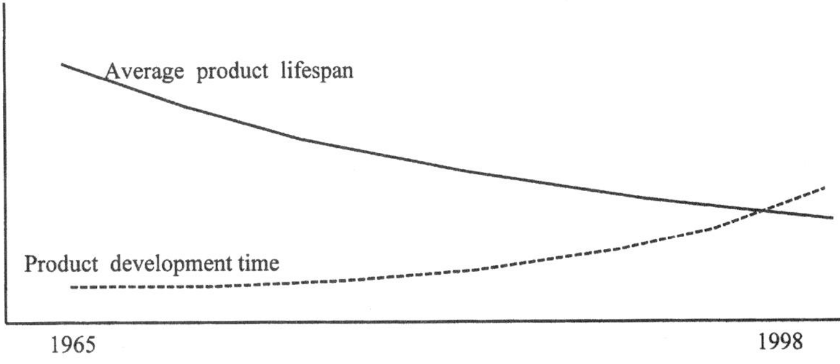

Figure 6.1.4 *Comparison between product life spans and product development time due to complexity in products*

complexity tends to increase development time. Figure 6.1.4 shows this conflicting requirement.

It was once possible to change products into a different part of the market as their sales declined in one region. With worldwide competition this is no longer possible. It is therefore important that several things happen:

- Design cycles are managed to produce new designs in a short time period.
- Launch dates are firmly met.
- Opportunity is taken to maximize sales when at maturity.
- The manufacturing facilities are flexible in product and quantity.

The historic sequence of serial design could no longer be applied because the increased complexity led to jumps in communication channels. The method of organizing the design process had to change from a series set of stages into one where all aspects were being consider concurrently (see Figure 6.1.5).

It is estimated that a late product launch has more affect on profitability than cost overruns of the design stage, or even in extra costs in producing the product itself. An estimation of these effects is shown in Figure 6.1.6.

The reason for the lateness of a product having such an effect on profits is that late products capture a smaller share of the available market. If an organization comes late into a market it finds the competition have established their products and they will therefore gain fewer customers. The product sales will therefore have a smaller enclosed area.

The types of new products are:

- Replacement: For existing own, or competitors', product.
- Extension: To the existing range of products.

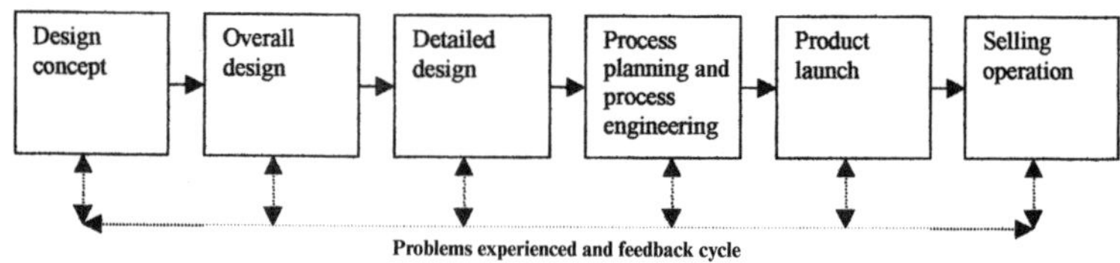

(a) Design process with serial flow

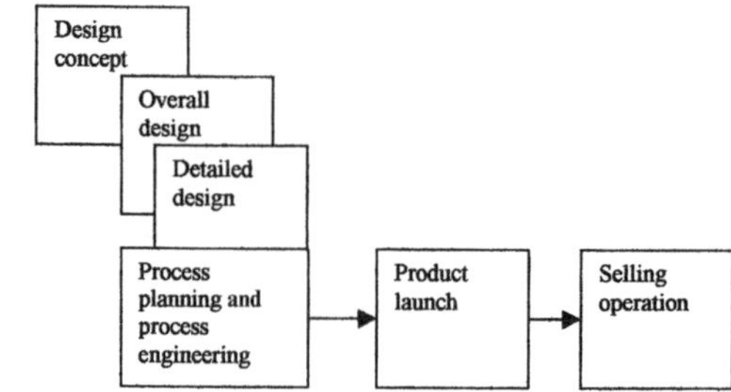

(b) Concurrent activities during the design process

Figure 6.1.5 *The design process – serial and concurrent involvement of different sections*

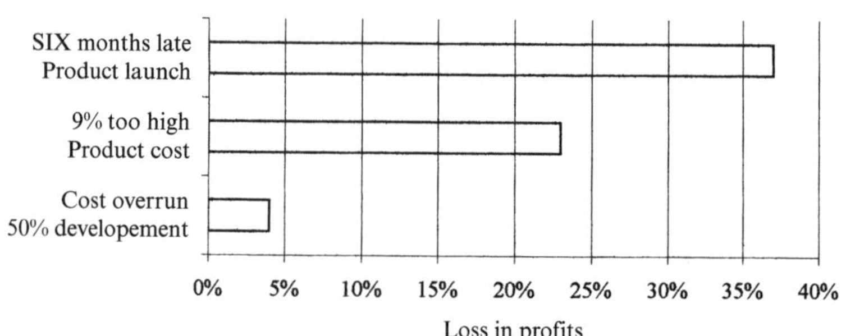

Figure 6.1.6 *Effect on profits of different failures in product development*

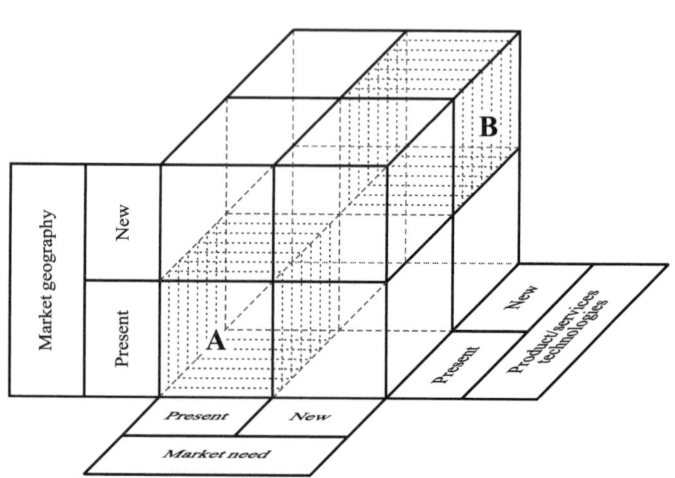

Figure 6.1.7 *Dimensions of the geographic growth vector. (From* New Corporate Strategy, *Igor Ansoff (1988) (John Wiley and Sons))*

- New: Products outside the present range in the present market, or for sale in a new market.

Ansoff's matrix, as in Figure 6.1.7, shows the different type of combinations of products and markets.

In order of increasing risk, i.e. departing from what you know and are good at, the strategies are:

	Current products	*New products*
Current market	Market penetration	Product development
New market	Market development	Diversification

- Market penetration: This means to sell more to existing potential buyers. This can only be gained from competition. An example is Ford attempting to increase sales of its product range of cars.
- Product development: Launching new products to sell within the same market. An example could be Ford introducing the Ka car. The problem here is that the new product may appeal to some existing customers and they buy the new product rather than an existing one. It is hoped that the gain of new customers outweighs the customers who simply switch to a different product from the same supplier.
- Market development: This means to find a new market altogether. Perhaps an example would be Ford's decision to start selling their cars in Russia after the collapse of the USSR. The problem here is that you will be competing with organizations who will have considerable experience in the market and you have to learn about the new market's particular characteristics. You will also have to set up a selling, and perhaps production, facility and cope with new languages, laws and customs.
- Diversification: There are different types of diversification:
  - Forward towards the customer. An example would be if Ford decided to start up a chain of wholly owned car dealers. The problem then is that the organization takes on additional risks and costs which have to be matched to increased profits.
  - Backwards along the supply chain. An example would be Ford deciding to go into the manufacture of windscreen glass. The problem here is that the internal supply has to be limited to selling only to its parent as competitors will be wary of dealing with them.
  - Complete diversification. An example would be Ford deciding it was going to manufacture and sell cakes and similar confectionery.

The organization should build on its strengths, bearing in mind the customer's needs. Most departures from the present products and customers involve the organization having to take on new tasks, buy new equipment and learn new skills. This takes time and money – even if bought in.

Risks with new products

Although new products are important to an organization's future, not every product launch will lead to success. As an organization has invested considerable money, effort and emotion in bringing a new product to the market, it is worth listing some of the areas where launches fail:

Sales may not be achieved due to:

- Competitors' actions – an earlier launch or a better product.
- Lack of market response – not what the public wants.
- Unattractive pricing.

Design shortcomings:

- Lack of technology experience.
- Unreliable configurations.
- Patent infringement.
- Design not manufacturing friendly.

Manufacturing shortcomings:

- Unreliable components purchased.
- Problems with new technology.
- Processing quality problems.
- Initial teething problems leading to late launch, or insufficient output to meet demand.

Servicing shortcomings:

- Insufficient service points.
- Personnel untrained.
- Insufficient spare parts.

Because of the importance of a new product, it is important to get the package right as early as possible in the design cycle as it becomes very expensive to modify the design at a later date as shown in Figure 6.1.8.

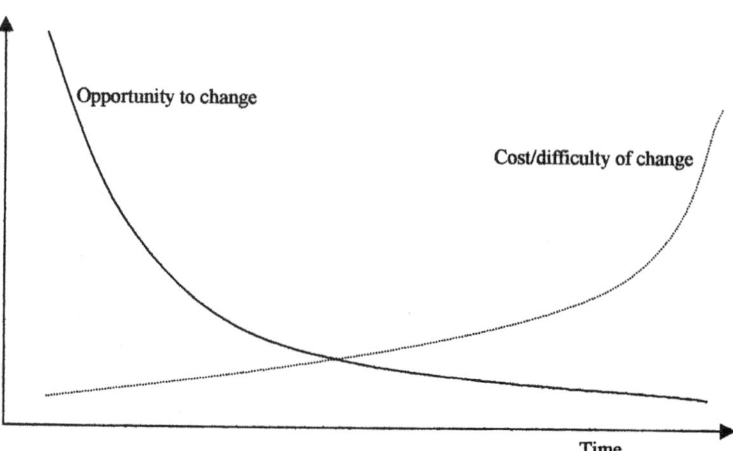

Figure 6.1.8 *Problems in changing project details as time progresses*

Distribution channels

In every market there are channels for products to move through between raw materials and the end user. These channels are important to minimize their effect on the organization, and maximize the organization's use of them. Figure 6.1.9 demonstrates the channels available for electrical components used in telephones.

The distribution channels perform certain functions:

- Transfer of title of ownership.
- Physical movement of the goods.
- Storage of goods.
- Financing of goods at the various points.
- Communication of market trends.

Each organization can decide the degree of control it has over points within the distribution chain:

- Direct ownership: Has the advantage of having full control over the total marketing effort. The disadvantage is that all risks and costs have to be borne by the organization – and all customer contact has to be made.
- Independent intermediates: Has the disadvantage of not having control over the selling effort. It does minimize the cost, effort and risks involved – especially in reaching a large customer base.

Figure 6.1.9 is limited to dealing within a geographic area. In fact the electronic industry is very much a worldwide market in supply at all stages in the distribution chain. This adds a distance, language and cultural aspect into the decision of direct ownership or dealing through intermediates.

In addition there is considerable pressure to reduce the goods stocked in the supply chain and this leads to considerable pressure on production facilities to be highly flexible in being able to produce a range of different products in a very short lead time. Many supermarkets have highly sophisticated computer systems to collect information at the sales point and use this to control their inventory and ordering system.

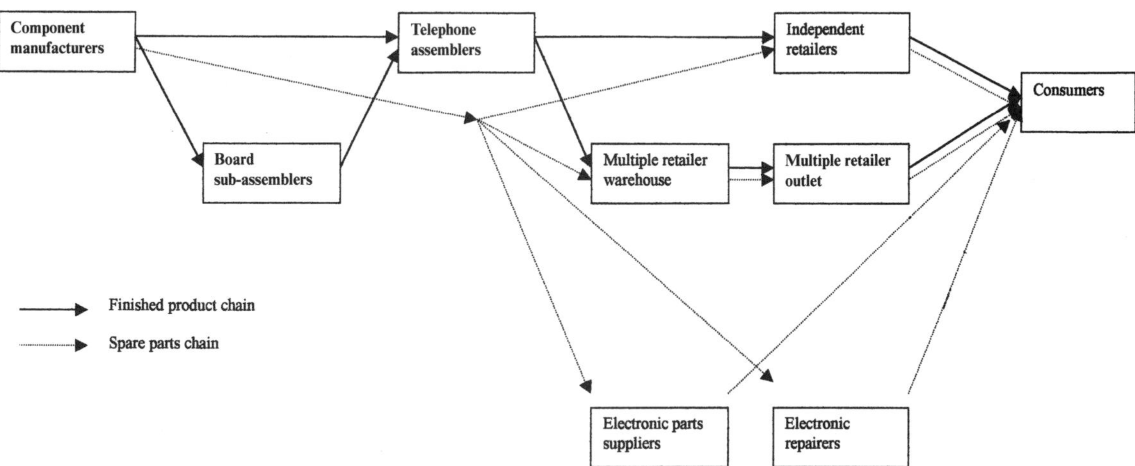

Figure 6.1.9 *Distribution chain for components used in telephones*

Promoting and selling

The different market segments require differing techniques of communication and personal selling to get a customer to purchase a particular product. The stages that a customer goes through and the associated promotional method are:

- Awareness: The customer becomes aware of the products on sale.
- Comprehension: The customer begins to differentiate between the various competing products.
- Conviction: The customer begins to move towards purchasing a particular product.
- Buying: The customer completes the purchase.

It is important that customers complete the route for your product – especially if they already are a customer and have already gone through the process. It is estimated that it costs ten times as much to gain a new customer as to retain an existing one.

Industrial and consumer goods do have differing proportions. Personal selling tends to be more important in industrial selling between organizations.

Typical communication media are (some can be more directed towards particular potential customers):

- Trade and other directories
- Personal contact
- Mail shots
- Direct mail
- Door to door
- Exhibitions
- Telephone
- Sponsorship

In addition there is the mass media, which is more useful for a broadcast approach.

- Television
- Radio
- Newspapers
- Consumer magazines
- Trade & business journals

Each of these has advantages in reaching particular customers, but remember they are individuals with home lives as well as their organizational roles. Reaching them in an out of the organization circumstance may make a greater impact than it would in the organizational setting.

We are not covering the art of advertising in this text, but it is worthwhile examining the above range of media and communication methods to see what they carry in the way of advertisements.

Selling skills

As it is only when a sale is made that the company will obtain an income, we shall therefore briefly look at this process in an organization-to-organization context rather than retailing to broad numbers of customers.

This is particularly important because a small number of successes here can sell a large number of products. The stages to go through are:

Preparation. It is important to ensure that you are prepared for the meeting. This means you have to carefully carry out the following phases:

Research:
- Is this a new prospect or an existing (past) customer?
- Who is the person to be seen – name, title and function?
- How was the call arranged – i.e. cold from your organization, or by invite from the potential customer?
- The company and the industry.
- All the purchase influencers in the organization.
- The competitions' offerings.
- Possible immediate customer needs and future plans.
- Sales potential, now and in the future.
- Buying record.

Preparing for the meeting:
- Dress to match customer expectations – not too formal, or informal. One style does not cover all.
- Have all resources available that you may need – paperwork, price list, technical brochures, etc.
- Ensure up-to-date product knowledge.
- Sales aids which may be needed.
- After-sales support offering that may encourage buying.
- Your business card.

Opening the call. This is important as first impressions do count in any interview situation. A bad start is difficult to recover from:

Importance of the opening:
- Remember from the customer point of view that they are important, therefore consider their needs.
- Orient the discussion to the customer.
- Make a proper greeting, e.g. an interest creating comment; your reason for calling; an opening question.

Agreeing the customer's needs:
Questioning techniques:
- Background questions, not directly related, to establish connection.
- Problem question, e.g. what problems are you experiencing with present supply?
- By implication, e.g. do you feel that you are achieving the best price?
- Establish their needs.

Statement techniques:
- Rephrasing of questions that have not been answered.
- Agreeing needs once established.
- Identifying a need.
- Present dissatisfactions in the supply side.
- Create dissatisfaction with competitors' offerings.
- Factors outside customer's control.
- Acceptable solutions.

Establishing priorities. Normally customer has many, often conflicting priorities. Try to establish what his priorities and working range are on factors such as the following. This may give indications where a trade-off is possible, e.g. higher price for a guaranteed delivery schedule:

- Price.
- Leadtime.
- Delivery reliability.
- Quality.

Presenting your case:
Making ideas understandable:
- Clear and unambiguous statements and offers.
- Visual aids to highlight important aspects of your case.
- Proper structure and sequence in the presentation.
- Jargon free (or use customer's).

Making ideas attractive:
- Stress benefits to your customer, not just the product features.
- Select benefits appropriate to each person's role in making the decision on who to buy from:
  - User
  - Designer
  - Buyer
  - Decider
  - Accountant.

Making ideas convincing by:
- Simple statements.
- Comparisons of different options and competitors' product offering.
- Demonstrating proof of any claims made.

Ensure you involve the buyer by obtaining and giving feedback.

Handling objections:
Techniques for handling objections:
- Anticipate possible objections and prepare the case against them beforehand.
- Listen to exactly what is being said and how it is said.
- Pause and think before replying.
- Acknowledge remarks made in your reply.
- Establish reason/need behind an objection.
- Provide rational answers to match need.

Different types of objections can be raised such as:
- Price – establish value definition and boundary conditions.
- Fear of unacceptable disadvantage.
- Habit, i.e. present supplier.
- Wrong information.
- Conflict with existing systems.
- Previous complaints against suppliers in general, or more seriously your organization.
- Competitors' claims.
- Other interested parties.

Closing is the most important part of any sales meeting:
Obtain customer commitment to at least one of the following:

- A specific order, even if only for a trial.
- Further meeting to hold open possibility of supplying.
- Demonstration of product, either on their site, or yours.
- A formal proposal/quotation to be sent.

It is important to stop selling once your customer decides to buy. Further effort will not only be superfluous, but may appear to be overselling to the customer. Recognizing buying signals such as:

- Direct request to order.
- A request for alternatives.
- The best solution proposed.
- Assumptions.
- Concessions.

Reassure the customer at the end of the meeting by:

- Thanking him for time and attention
- Confirm that their need has been met
- Leave once they appear happy.

Problems 6.1.1

(1) What factors will affect the demand for prints of paintings by a particular artist?
(2) Can you name five products which normally have a selling life cycle of less than one year?
(3) You make hand pottery items, what are the risks attached to deciding to start mass production of similar items?
(4) If you had a hobby of making model ships, identify the possible distribution chain you would enter if you decided to make a business out of it.
(5) If you developed a new process for attaching electronic components to a board, who would you attempt to see in a company making mobile phones.

6.2 Forecasting the future

The ability to foretell what the future holds may be impossible but in business it must be attempted to form the basis for rational decisions which need to be made now. Because of the lead times in design and manufacture you cannot wait until something actually happens before taking action – this will often be too late. This section looks at the reason for forecasting and the techniques available.

Need for a forecast

Forecasting is the starting point for:

- Capacity requirements: Determining what factories you will operate and the mix of labour, machinery and equipment inside them.

- Production planning: Deciding on how much of each product you will make and the dates to produce it.
- Raw material: Deciding on what to purchase – the quantity and timing of this.
- Personnel: Deciding how many people are needed, and their skills and training requirements.
- Budgeting: Determining the cash budget requirements and whether sufficient profit will be made.
- Technology: Determining the changes in the process and the products which will be affected by changes in technology.
- Marketing campaigns: The best time to launch a new product so that it naturally replaces an existing one.

If you make a pessimistic forecast, i.e. you set out to make less than you can eventually sell, then you will have stockouts and irritate your customers – perhaps even to the extent that they will go to your competitor. You may have to react by the expensive use of overtime or sub-contracting to meet customer demand.

If you are overoptimistic, it can also be expensive. You can have high inventories and underused equipment. With high inventories you tie up money and carry the risk of obsolescence, or delaying new designs to use up these inventories.

The main problem with a forecast is that it will never be exactly accurate. There are too many factors which are outside the control of an organization, or even a government. What industry requires is a fairly accurate forecast – which means a probable range rather than a single figure with an indication of the risk attached to it.

Time horizons in forecasting

Each type of industry has different time horizons due to their nature. The chemical and other basic industries, because of their high capital requirement and slowly changing products, tend to think in the relatively long term. Industries such as newspapers tend to think in relatively short terms.

Short Range. One week to three months:

- Purchasing material: Placing, or confirming final details of components and quantities.
- Job scheduling: Determining the final production schedule of what will be made when.
- Working pattern: Short-term alterations to working hours such as overtime working.
- Job assignments: Allocating actual named people to particular tasks.
- Cash flow: Ensuring money is available as required.
- Customer servicing: Arranging the schedule of service personnel's visits.

Medium range. Three months up to one year:

- Sales planning: Detail of visits to be made, and targets to be achieved.

- Production plan: Semi-finished plan of what products will be produced.
- Procurement: Many items have a delivery lead time that requires orders to be placed several weeks before desired delivery.
- Shift pattern: Finalize working pattern of individuals and equipment.
- Personnel level: Finalizing recruitment and redundancy plans.
- Training: Finalizing training schedules of courses and match of personnel.
- Product launch: Finalizing details of product launch.
- Advertising: Finalizing advertising schedule.

Long range. One to five years:

- Strategic plan: To cover whole business.
- Capital investment: Allocating capital spending in conjunction with strategic plan.
- Capacity plan: Make any forward arrangements to open up or close plant including forward arrangement for personnel and equipment placement.
- New products: Determine the product development which will be concentrated on.
- R&D expenditure: Determine the lines of enquiry to be followed to develop new products and processes.
- Facility location: Decide on the opening up or closing down of retail outlets, warehouses and offices.
- New markets: Determine what new product or geographic areas the organization wishes to investigate.

The longer the time horizon involved the greater the risk of a forecast being significantly wrong. When factors such as competition and technology enter the picture this increases the risk involved.

Therefore when considering the medium- and long-term picture, an organization has to take into account what governments may do (in, for example, inflation or unemployment policies); demographic changes in population concentration and age dispersion; general education levels; special grants; etc.

Basis of forecasts

Forecasts tend to be made up from a variety of sources both internal to an organization and external. Some of this will be hard data such as previous sales, other soft data such as an opinion.

Management goals

The aspiration of the top management is often the initial starting point as this decides the length of time being considered in paying back investment of time or money.

Sales force

The sales people should be in touch with the customers and see directly the impact of the competition and contribute substantially to any forecast.

Where the sales cover a large area, or several countries each sales point should initially make forecasts of how the market is growing/shrinking, the changes expected in market share, the changes in technical specifications desired, competitors' offerings and tactics, etc.

Area forecasts can be consolidated into a regional one, then an overall forecast. It is usual on an international scale to find distinct differences between customer demand on products in different locations.

At the same time the organization may be a newcomer into some market areas and much of this information has to be especially gathered. Where they are established, the present sales force should be a good source.

Leading indicators

There are often indicators available locally or nationally which can give some guidance to the overall direction of the economy within a country:

- Changes in inventories levels. These can indicate when the market is slowing down (increase in inventories) or quickening up (reduction in inventories).
- New manufacturing orders: An increase or reduction in these can be especially important to organizations supplying capital equipment such as machines.
- Changes in employment levels: An early indicator here can be changes in the number of jobs advertised. This is of special interest to organizations supplying luxury or shopping goods as it can indicate a possible general trend towards particular products.
- New plant and equipment orders: These indicate how confident industries are in their future trading prospects.
- New buildings started: Of interest to organizations in the white goods or furnishing markets.
- Demand for money: The availability of credit is useful in shopping and luxury goods markets, e.g. cars.
- New government contracts: Those organizations supplying capital goods and in civil engineering are interested in these.
- Elections or government changes: Each party has particular areas they prefer to target, or to diminish, spending on. There may be a rise in spending when an election is about to take place – and perhaps a cut-back afterwards.

Therefore it is possible to come to a reasonable forecast without analysing past sales figures, but often the past does give an indication about the future, especially in seasonal changes in demand.

Forecasting techniques

There are many mathematical forecasting techniques in use within industry. The main problem is that no technique will produce a guaranteed accurate figure and all must be treated with an attitude of partial disbelief. There are too many factors outside the direct control of an organization that can affect both the total market movement and the organization's share of that market. Therefore any technique used must be modified by a good knowledge of the market itself.

Time series patterns

In addition to market fluctuations, there are normally five components of demand (see Figure 6.2.1). There can be several of these in action at the same time.

- Random: Unfortunately in any set of data there are normal variations. Some of these have undefined causes and these are impossible to predict. Even those with identifiable causes are difficult to predict as the causes themselves require predicting as well as their effect.
- Trend: A straight or curved line showing an increase or decrease over time. Normally only reliable for a limited period and may vary.
- Curved, e.g. the product life cycle.
- Short-term cycles: Often seasonal but can be over a day, e.g. traffic density, a week or a year. Can be associated with a single annual event, e.g. Christmas.
- Long-term cycles: Over the longer term there appears to be natural business cycles. The exact prediction of these is extremely difficult.

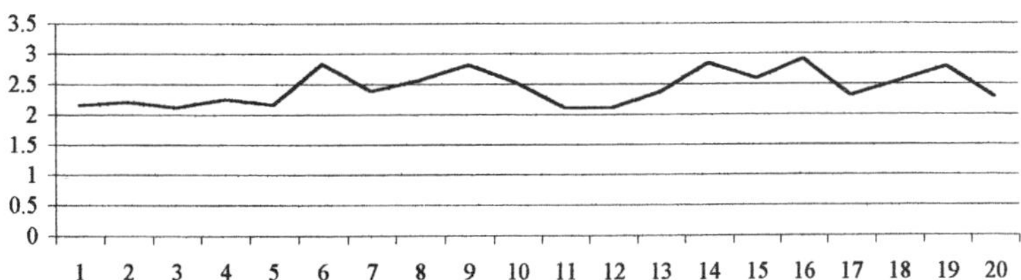

(a) Random variation

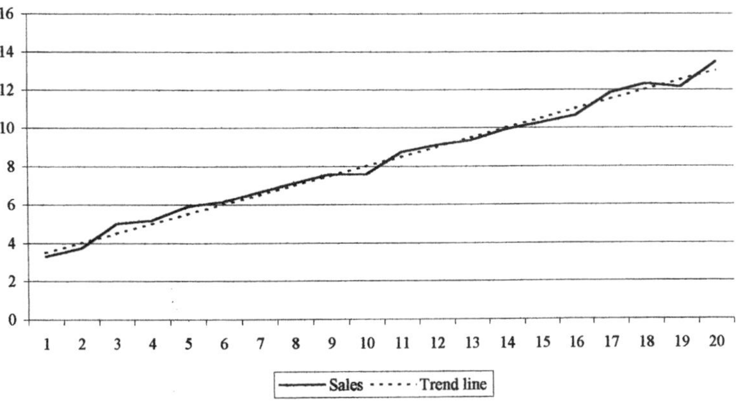

(b) Random variation with a linear trend

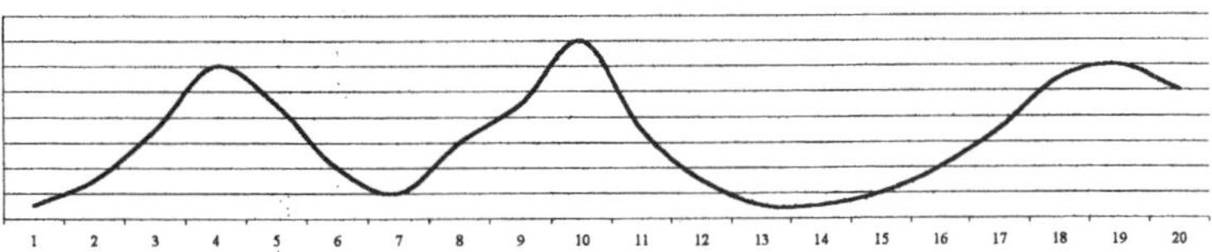

(c) Trade cycles

Figure 6.2.1 *Some sales patterns*

It is these components that make accurate forecasting unlikely – therefore a range of figures should always be given instead of a single number. All quantitative methods rely on a historic database – but remember, history does not always repeat itself!

Quantitative methods

In order to test the quantitative methods, we try them out on past data, i.e. as if they were being used on that data. They are run as if the old data was new, i.e. the future values cannot be seen.

Figure 6.2.2 shows a graph of the historic demand pattern. On first looking at the graph, there appears to be an upwards trend but there is also considerable variation from period to period. The variation may be random, or due to definable causes such as a new product slowing gaining acceptance, seasonal factors, supply problems or competitors' actions.

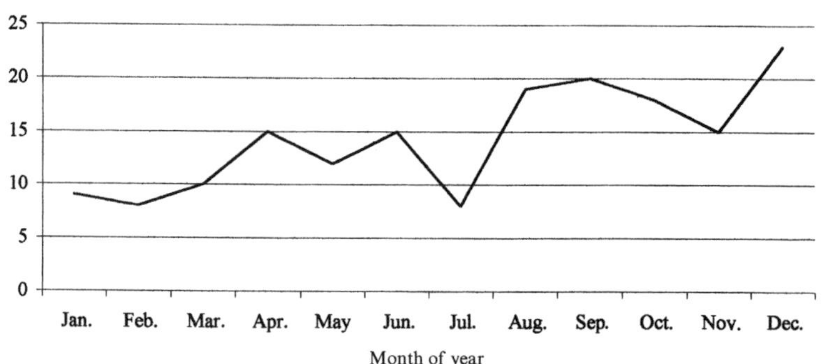

Figure 6.2.2 *Graph of demand over year*

Where there are definable causes then they must be allowed for. It may be, for example, that period 7 was an annual holiday shutdown that resulted in less product being available.

In this particular example, we will assume there are only random variations and make no other assumptions regarding any of the other possible causes, except to say there were no supply problems.

There are a number of error analysis techniques available which we can use to judge the effectiveness of a forecasting technique. Only the BIAS and MAD (mean average deviation) figures are calculated and shown here.

- The BIAS gives an indication of whether the forecast is on average too high (positive) or too low (negative). It can be used to adjust the forecasting process.
- The MAD value gives an indication of the overall accuracy of the forecasts – it is the average error. It can be used to give a probable range to a forecast.

To calculate the BIAS, take the actual sales as the base. Find the error between the forecast and the actual sale, i.e. include the plus and minus sign. Now add up these values and find the average. If the answer is positive the forecast is, on average, too high – if the answer is negative, the forecast is too low.

To calculate the MAD, repeat the process as for the BIAS – but this time take the absolute value of the error, i.e. ignore the sign. The average figure signifies the expected error in a forecast – it can be used to give a range from a central prediction. Note it is an average and therefore 50% of errors can be larger than this. To ensure that all errors are within this may give an extremely large range.

An organization has to find a method of controlling large sales variations, or a method of coping with them – e.g. carrying inventories.

Simple forecasting method

The first method, one can say, is an educated guess.

It consists of examining the immediately past data to guestimate (a hybrid word made up of joining guessing and estimating) the future demand. A graph of historic data, such as that shown in Figure 6.2.2, can give a good guide to what will happen in the future – but only if the future is similar to the past. Any known factors that may cause changes should be taken into account.

Table 6.2.1 shows a series of simple forecasts made and the corresponding actual sales made. Figure 6.2.3 shows the graph of these forecasts against actual sales.

Table 6.2.1 Actual sales against simple forecasts

Period	*Sales*	*Simple forecast*	*Error*	*Absolute error*
1	9			
2	8	9.00	1.00	1.00
3	10	9.00	–1.00	1.00
4	15	9.00	–6.00	6.00
5	12	10.00	–2.00	2.00
6	15	11.00	–4.00	4.00
7	8	14.00	6.00	6.00
8	19	12.00	–7.00	7.00
9	20	15.00	–5.00	5.00
10	18	19.00	1.00	1.00
11	15	19.00	4.00	4.00
12	23	17.00	–6.00	6.00
			BIAS	MAD
13		20.00	–1.73	3.91

As can be seen in the graph in Figure 6.2.3, this method does tend to err some, especially in reaction to sudden changes in demand, but can give a reasonable forecast. It produces BIAS of –1.73 and a MAD of 3.91 respectively that are reasonable in comparison to some of the other forecasting methods, as we shall see later.

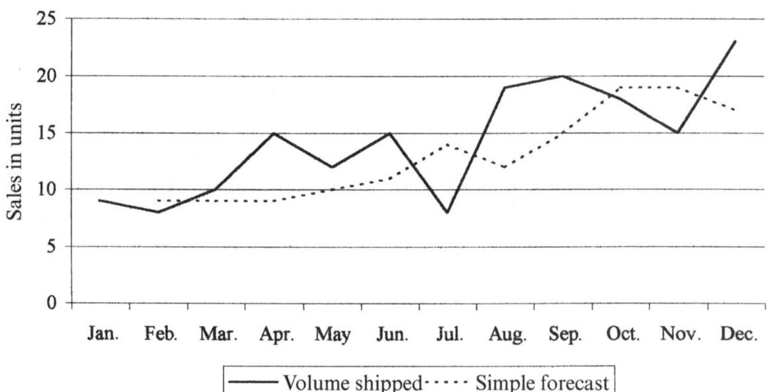

Figure 6.2.3 *Comparison of simple forecasts and demand*

Linear trend

It is often worthwhile to develop the linear regression of the historic demand, to give a guide to a trend. This can then be projected forward to give the forecast demand. The basic regression formula is the formula for a straight line:

$$F_x = a + b \times X$$

where F_x = the trend forecast of demand Y for period number X
a = the estimated inception on the y-axis
b = the estimated slope of the demand line
X = the period number

The equations used to determine a and b are:

$$b = (\Sigma(x \times y) - (n \times \bar{x} \times \bar{y}))/(\Sigma x^2 - (n \times \bar{x}^2)$$

$$a = \bar{y} - b \times \bar{x}$$

where $\Sigma(x \times y)$ = sum over all periods of X times Y for each period
Σx^2 = sum of the X value squared for each period
$\bar{x}$ = the average of the X values
$\bar{y}$ = the average of the Y values
n = number of periods of data used

In this particular case, looking at Table 6.2.2 where we have calculated out the values we have:
We have:

$\Sigma(x \times y) = 1274$
$\Sigma x^2 = 650$
$\bar{x} = 6.5$
$\bar{y} = 14.333$
$n = 12$

$$\begin{aligned} \therefore \quad b &= (\Sigma(X \times Y) - (n \times \bar{x} \times \bar{y}))/(\Sigma X^2 - (n \times \bar{x}^2) \\ &= (1274 - (12 \times 6.5 \times 14.333))/(650 - (12 \times 6.5^2)) \\ &= (1274 - 1118)/(650 - 507) \\ &= 156/143 \\ &= 1.0909 \end{aligned}$$

Table 6.2.2 Calculations and forecast for linear trend projection

Period x	*Sales* y	x^2	$x \times y$	*Linear forecast*	*Error*	*Absolute error*
1	9	1	9	8.33	–0.67	0.67
2	8	4	16	9.42	1.42	1.42
3	10	9	30	10.52	0.52	0.52
4	15	16	60	11.61	–3.39	3.39
5	12	25	60	12.70	0.70	0.70
6	15	36	90	13.79	–1.21	1.21
7	8	49	56	14.88	6.88	6.88
8	19	64	152	15.97	–3.03	3.03
9	20	81	180	17.06	–2.94	2.94
10	18	100	180	18.15	0.15	0.15
11	15	121	165	19.24	4.24	4.24
12	23	144	276	20.33	–2.67	2.67
Σ 78	172	650	1274		–	
Av: 6.5	14.33				0.00	2.32
Period 13				21.42	BIAS	MAD

and $a = \bar{y} - b \times \bar{x}$

$$= 14.333 - (1.0909 \times 6.5)$$
$$= 14.333 - 7.0909$$
$$= 7.2424$$

$\therefore$ $F_x = a + (b \times X)$

$$= 7.2424 + (1.0909 \times X)$$

If we substitute for any period number for X in this equation, we find the resultant forecast. For example, if we want the forecast for period 6, the answer is:

$$\text{Forecast} = 7.2424 + (1.0909 \times 6)$$
$$= 13.79$$

Table 6.2.2 shows in a column the forecasts made using this projection. The table also shows the errors from the actual demand against each forecast. On the error analysis, the BIAS shows up as zero as the trend is calculated from this actual data. The MAD analysis shows that the average individual error is 2.32 against the average volume shipped of 14.333 equal to +/–16%, which appears quite large, but is not unusual where significant random variations in demand are present.

The graph in Figure 6.2.4 shows these forecasts against the actual. It includes a range based on the MAD projected from the single figure forecast to give an average error range. As can be seen, some of the actual sales values fall outside this projected error range. This is because the MAD represents an average error hence some must be larger than it – to include for all possible errors would mean the application of a much wider range.

There are more complex mathematical methods such as multi-regression analysis which can be used, but these are not covered in this textbook.

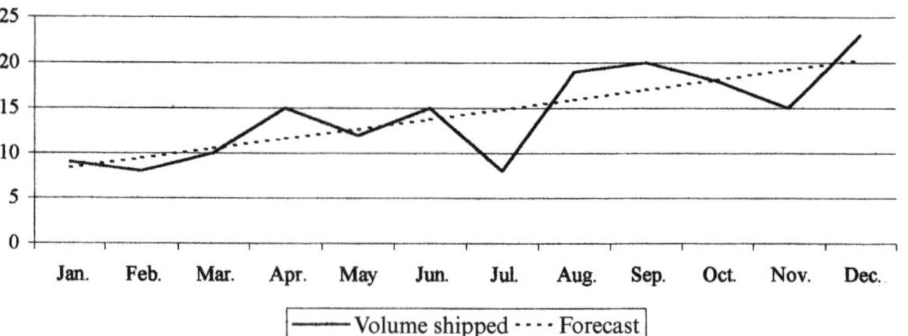

Figure 6.2.4 *Comparison between linear regression forecast and demand*

Moving average

One could use a simple average of all the past data to estimate the future demand. If the market changes over time, as all do, this method could be highly inaccurate.

If instead of taking all past data, the average of only the immediate past was used to project the future then it should be more accurate.

We calculate this 'moving' average by:

$$F_{t+1} = (D_t + D_{t-1} + D_{t-2} \ldots D_{t-(n-1)})/n$$

where t = Period number for current period
D = Demand in period
n = Number of periods included in moving average calculation

Table 6.2.3 shows the values calculated when producing a three-period moving average. For example, the forecast for period 5 is reached by:

$$\begin{aligned} F_5 &= (D_4 + D_3 + D_2)/3 \\ &= (15 + 10 + 8)/3 \\ &= 33/3 \\ &= 11.00 \end{aligned}$$

Table 6.2.3 A three-period moving average forecast compared to actual sales

Period	*Sales*	*Moving average forecast*	*Error*	*Absolute error*
1	9			
2	8			
3	10			
4	15	9.00	−6.00	6.00
5	12	11.00	−1.00	1.00
6	15	12.33	−2.67	2.67
7	8	14.00	6.00	6.00
8	19	11.67	−7.33	7.33
9	20	14.00	−6.00	6.00
10	18	15.67	−2.33	2.33
11	15	19.00	4.00	4.00
12	23	17.67	−5.33	5.33
		Average	−2.30	4.52
13		18.67	BIAS	MAD

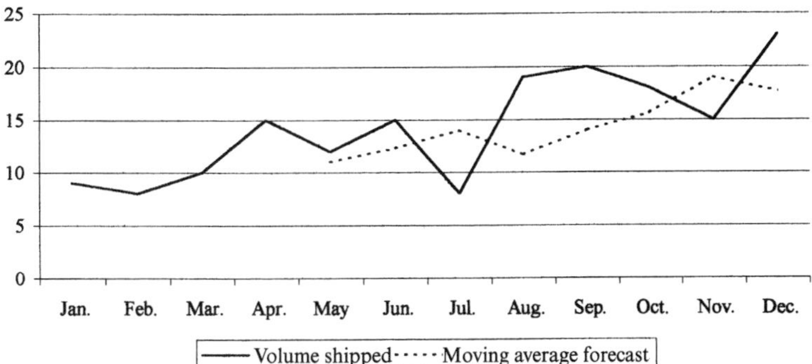

Figure 6.2.5 *Comparison between three-period moving average and demand*

As can be seen in Table 6.2.3, using this method, the BIAS = –1.81 and the MAD = 5.96. Figure 6.2.5 shows a graph comparing this forecast against the actual sales. This shows that, on average, the forecast is low (by 12%) with a large average error of +/–40%.

The use of past data averaging will have a BIAS where there is a trend (positive or negative). Using weighted moving averages can reduce the BIAS because later values have a higher relative value.

Exponential smoothing

We can produce a forecast based on past data giving preference towards either the latest data, or even that from earlier periods by using weighed average methods. The problem remains that in any moving average method any data before the period used in the calculation is ignored.

Another technique called exponential smoothing considers all past data, and modifies its forecast based on how close the forecast was to the latest actual figures. We can decide what is more important – total history or recent history by adjusting the value of a smoothing constant, α, in the formula:

$$F_{t+1} = (\alpha \times D_t) + ((1 - \alpha) \times F_t)$$

The higher values of α give a forecast that reacts much in line with the latest demand. Lower values of α give more preference to the previous forecasts, i.e. a longer historical period. The result of this is that the higher values of α tend to be more effective in following trends but it does not produce a smooth forecast. For a smooth forecast you require a low value of α.

Forecasts for period 5 in Table 6.2.4, for the differing values of α used, are:

$$\text{For } \alpha = 0.1, \quad F_5 = (0.1 \times 19) + ((1 - 0.9) \times 11.44) = 12.20$$
$$\text{For } \alpha = 0.5, \quad F_5 = (0.5 \times 19) + ((1 - 0.5) \times 10.00) = 14.50$$
$$\text{For } \alpha = 0.9, \quad F_5 = (0.9 \times 19) + ((1 - 0.1) \times 9.84) = 18.08$$

If we examine the graph in Figure 6.2.6, we can see:

- α = 0.1 gives a very smooth forecast, but it is not following the trend of the demand well, i.e. it is lagging the trend.

Table 6.2.4 Values of exponential forecasts using differing values of α

Period	*Sale*	α = 0.1			α = 0.5			α = 0.9		
		F'cast	*Error*	*Absolute error*	*F'cast*	*Error*	*Absolute error*	*F'cast*	*Error*	*Absolute error*
1	9									
2	8	9.00			9.00			9.00		
3	10	8.90	−1.10	1.10	8.50	−1.50	1.50	8.10	−1.90	1.90
4	15	9.01	−5.99	5.99	9.25	−5.75	5.75	9.81	−5.19	5.19
5	12	9.61	−2.39	2.39	12.13	0.13	0.13	14.48	2.48	2.48
6	15	9.85	−5.15	5.15	12.06	−2.94	2.94	12.25	−2.75	2.75
7	8	10.36	2.36	2.36	13.53	5.53	5.53	14.72	6.72	6.72
8	19	10.13	−8.87	8.87	10.77	−8.23	8.23	8.67	−10.33	10.33
9	20	11.01	−8.99	8.99	14.88	−5.12	5.12	17.97	−2.03	2.03
10	18	11.91	−6.09	6.09	17.44	−0.56	0.56	19.80	1.80	1.80
11	15	12.52	−2.48	2.48	17.72	2.72	2.72	18.18	3.18	3.18
12	23	12.77	−10.23	10.23	16.36	−6.64	6.64	15.32	−7.68	7.68
	Aver.		−4.89	5.37		−2.24	3.91		−1.57	4.41
13		13.79	BIAS	MAD	19.68	BIAS	MAD	22.23	BIAS	MAD

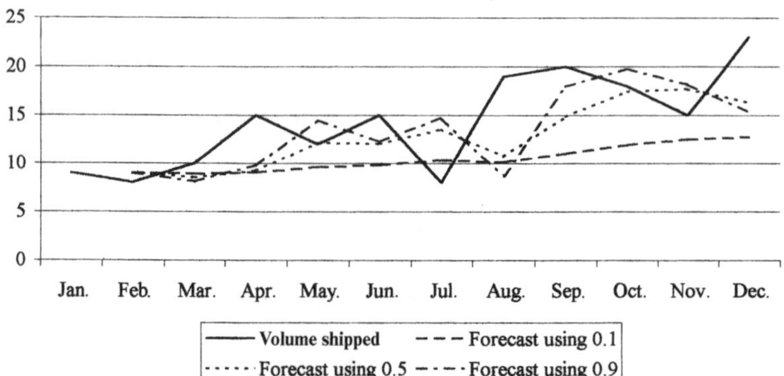

Figure 6.2.6 *Comparison using various values of α against the demand*

- α = 0.9, there is a good correlation to the trend but the forecast is almost the same as the previous period's actual demand and hence is fluctuating almost as much as the actual demand, but one period behind.
- α = 0.5 demonstrates in between results from the low and high values of α.

A low value of α is useful to smooth out fluctuations, but should only be used where there is little trend, as it quickly lags behind the trend.

Note from Table 6.2.4 that α = 0.1 gives a BIAS of −4.89 and a MAD of 5.37 against a BIAS of −1.57 and MAD of 4.41 for α = 0.9. The lower α value gives a smoothing out of the high fluctuation of the actual demand but produces the higher BIAS and MAD because the forecast is lagging behind the actual trend. The best overall value for MAD comes when α = 0.5.

Comparison of forecasts for period 13

Table 6.2.5 brings together the end result of all the methods, including a forecast for period 13 and the BIAS and MAD values. Note that

Table 6.2.5 Comparison between different forecasting methods.

Method	*BIAS*	*MAD*	*Forecast for period 13*
Simple	−1.73	3.91	20.00
Three period moving average	−2.30	4.52	18.67
Linear trend	0.00	2.32	21.42
Smoothing constant, α = 0.1	−4.89	5.37	13.79
Smoothing constant, α = 0.1	−2.24	3.91	19.68
Smoothing constant, α = 0.1	−1.57	4.41	22.23

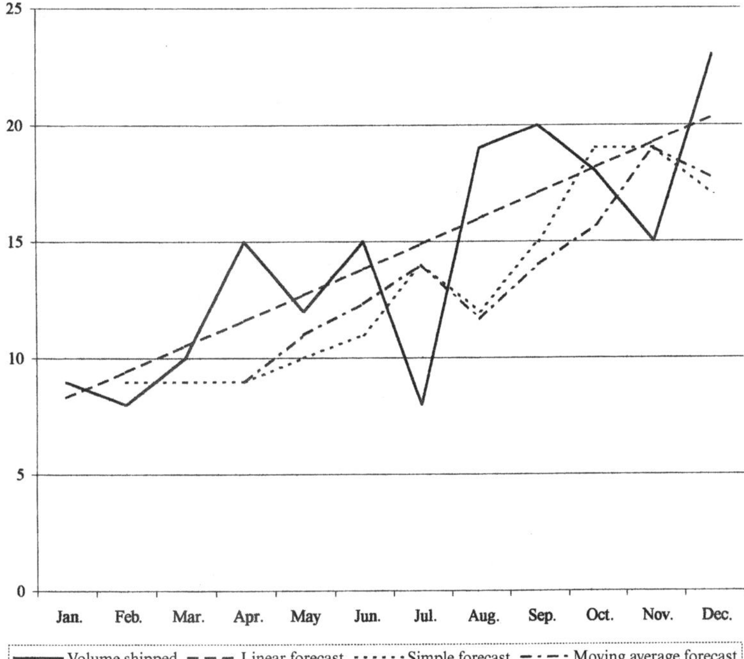

Figure 6.2.7 *Graph showing all technique forecasts*

although each technique is a valid one, they produce answers from a low of 13.79 to a high of 22.23. They are graphed together in Figure 6.2.7.

Which one gives the correct forecast?

The answer is that they are all mathematically correct but probably all inaccurate! For example, the figures for demand may represent a normal year's figures and the actual for period 13 may be similar to period 1.

Only time and trial application of each can tell how useful any of the techniques would be.

Determining cyclic components and outside influences

As referred to in page 273, there are often cycles within any demand pattern. These can be over:

- The day, e.g. traffic volume.
- The week, e.g. food sales pattern.

Table 6.2.6 Daily sales for a food item over four weeks

Week	*Mon.*	*Tues.*	*Wed.*	*Thurs.*	*Fri.*	*Sat.*	*Sun.*	*Total*
1	112	140	105	210	530	180	160	1437
2	98	150	120	250	510	250	110	1488
3	55	195	115	240	490	210	145	1450
4	110	135	125	230	475	195	150	1420
Total	375	620	465	930	2005	835	565	5795
% age	6%	11%	8%	16%	35%	14%	10%	

- The month, e.g. spending on luxury items reduces as the month advances.
- Year, e.g. sales of electric fans peak in the hot summer.
- Particular events, e.g. Christmas, Guy Fawkes (fireworks), new car registration letters.

Where they have existed for a number of cycles, they need to be identified so that production schedules and inventory can be planned to service them. Table 6.2.6 shows the daily sales of a particular food item in a supermarket.

By adding up the sales on each of the days and calculating the percentage they are of the total sales for the four weeks, we can arrive at an expected daily factor that we can apply to any projected week's sales. These factors are shown graphically in Figure 6.2.8.

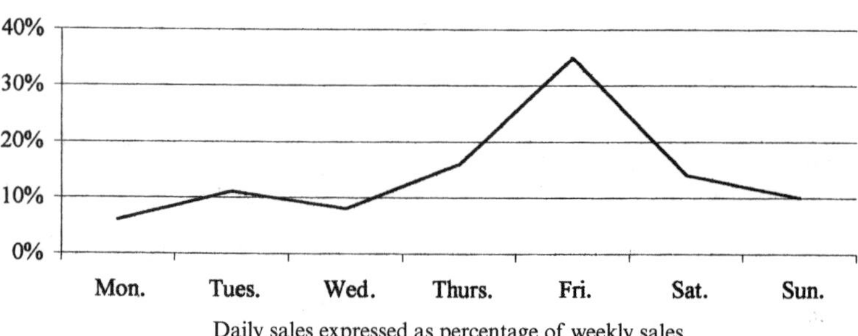

Figure 6.2.8 *Daily sales factors*

In addition to a daily pattern, we could find other patterns and correlations for this item:

- One Monday sales were much less than the other three Mondays – this was put down to a Bank holiday – and this reduction could be built into any future forecast.
- There will probably be an annual pattern for the item as people's eating preferences change over the year.
- There could be a correlation to an outside influence, e.g. ambient temperature, if a popular local football team is playing at home or away, perhaps even if that team wins or loses.
- There could also be a correlation to special offers for this product, or perhaps a negative reaction to a special offer on a similar product.

These all can be built into past data to provide a guide towards the future, but no matter how accurately the recording – all the influences probably have not been captured. The future will always be different, we just hope not too much is unpredictable.

Foretelling the far future

If it is difficult to forecast the immediate future, then how do we go about forecasting even further ahead? Some organizations, such as utilities, have to make decisions now, as the lead time of developing new plant is years.

This does not involve using crystal balls but assembling a group of acknowledged experts and recording their views on what changes are liable to take place, especially in the technology aspects. This can be by focus groups or using the Delphi Technique where each expert records separately their predictions and these are then discussed and put in order of probability.

Other ways are to carry out:

- A full PEST analysis (see page 1) to extend present trends in the political, economic, sociocultural and technologies aspects into the future.
- A five-force analysis (see page 304) to determine foreseeable, or possible, changes in the competitive forces operating.

These long-term forecasts often act as the input to scenario analysis (see page 306).

Problems 6.2.1

(1) What happens if you use a value of $\alpha = 0$ in an exponential smoothing forecast?
(2) What are the problems associated with a highly seasonal product such as fireworks in the UK?
(3) If in the example you have been following in this section the actual sales in period 13 were 10% higher than month 1, what would you use as forecasts for periods 14 to 24?
(4) How would you build the launch of a product revision into a forecast?
(5) If you are supplying the retail market with a range of similar products, what effect do you think a sales promotion on one item has on the remaining items in the range?

6.3 Marketing research

For an organization to make a decision relating to its market, it requires as much information as possible about what people want and how much they are prepared to pay for it. The aim of marketing research is to provide that information with as much accuracy as possible.

There are four main areas that marketing research is involved with:

- Product research: Concerned with the design and range of existing and possible products from the organization and its competitors.

- Customer research: Concerned with the buying pattern of potential customers and trends in their spending and changing criteria for selecting products. Can be consumer or industrial.
- Sales research: Investigations both into internal unit and competing units in sales outlets and sales patterns.
- Promotion research: Investigations into the effect of advertising and other promotional efforts.

Some of that information will be gained from just operating in a market from personnel such as salesmen. However, this is from a narrow base of experience and can be highly personalized and biased. It cannot possibly cover the whole potential range of customers, unless that is very small.

The research programme

There are many ways of gaining useful information from the marketplace. Some of these are:

- Experiments
- Questionnaires
- Telephone surveys
- Mail surveys
- Personal interviewing
- Direct observation
- Group discussions
- Research of governmental and other data sources.

Before deciding on which method to use, marketing research should go through a design sequence to ensure that the total research gives a reasonably valid answer. The phases are similar to that seen in Chapter 5 on project management.

Stage 1 – research brief

Initial discussions about the situation to be investigated need to be held to discuss the extent and limits of the project, especially the population involved, and the degree of accuracy required.

For example, initially it may be suggested to research into the reasons for selecting an intended higher education subject area amongst 16–18-year-olds. But that raises a number of questions, amongst which could be:

- Why do we want to know the answer?
- Is it nationwide (UK), limited to only certain geographical regions or even should it cover the EC?
- Should it cover all people in this age group, even though many of them do not have qualifications to enter higher education?
- Should it be limited to those qualified and intending to enter higher education?
- How broad should the subject categories be?
- How many would be representative of this population?
- How will the survey be conducted?

Once there appears to be agreement about the subject matter, we move onto:

Phase 2 – research proposal

The proposal will contain:

- Clear statements about what is being studied.
- Limits of investigation.
- Methods to be employed.
- Expected accuracy.
- Expected bias and method of coping with it.
- Cost and timing.
- Experience of research team.

Once this is accepted, we can move onto:

Phase 3 – data collection

The main part of the research. Careful records need to be kept of how data was actually collected, the responses received and any significant variations to plan.

Phase 4 – data processing

This is the editing, coding and tabulation of the data collected to determine required response rates and formulation of the results.

Phase 5 – presentation of report

The end of this research project may give management enough information on which to base its decisions, or it may still leave room for intuition, or indicate a further study is required.

Target of the research

Whatever research method is used there is a problem associated with it. How do we identify the group(s) which the survey has to cover.

Typical segmentation models used are:

UK National Readership Survey (NRS)

A	Upper middle class	3% of population
B	Middle class	10%
C1	Lower middle class	24%
C2	Skilled working class	30%
D	Working class	25%
E	Other	8%

ACORN Life Style Segmentation

Household composition: Single; couple; family; shared.
Age structure: Child; young adult; mature; established; retired.
Possessions: Car; telephone; etc.
Pastime activities: Sports; music; reading; etc.

Industry. In some ways easier than consumer as there tends to be a concentration of buyers, unless the target is the small businesses. First place to start is to quantify the actual group that we are interested in, e.g.:

- Electronic
- Computing
- Automobile
- Ship-building
- Construction
- Forestry

Secondary data sources. These are non-obtrusive, i.e. do not require soliciting responses and are termed desk research.

- Internal data, e.g. sales records, complaints records.
- Government publications: Good guides here are *Guide to Official Statistics* (HMSO) and *Regional Statistics* (HMSO). Copies can normally be found within public libraries – the local librarians will tell you what detailed reports they hold in stock, or will arrange inter-library loans for many of them. UK government statistics include social trends, regional trends and family spending.
- Trade press: Again consult your local library or one of the particular industry bodies.
- Business Links and local chambers of commerce.
- Commercial data: there are many subscription services such as Mintel Market Intelligence Reports and Financial Times Business Information Services.
- On-line services: Teletext (Ceefax and Oracle), Viewdata and many of the internet providers. Many of these are subscriber services.
- Libraries: In addition to official and commercial information sources, they are often a contact point for local interest groups and clubs.

Using these sources, a picture can be built up of potential customers, or users, who it may be of interest to carry out a survey on. These sources are initially identified at the research proposal stage, but a more detailed picture is required before we finalize the design of the data gathering phase.

In the industrial market, it is important that we follow the distribution chain to identify all possible uses and potential buyers. Following such a chain will produce a list of contacts and their positions within all the organizations.

Primary sources

- Internal: People directly involved in sales, design and procurement can be a good source of what is happening in the market as they meet and talk to customers.
- External: These are the people involved in the purchasing decision:
  - Designers and buyers within an organization.
  - Retailers and wholesalers.
  - Consumers – buyers and users.

Sampling plan

Once we have the target group, we now have to decide how to collect the data and how much we need. The main constraint on any research is cost. The aim is to collect as much meaningful data as possible within that constraint.

Ideally we would like to solicit information from everyone who may be involved. The cost of doing so would be prohibitive, except in very narrow markets. This forces us into sampling rather than full coverage.

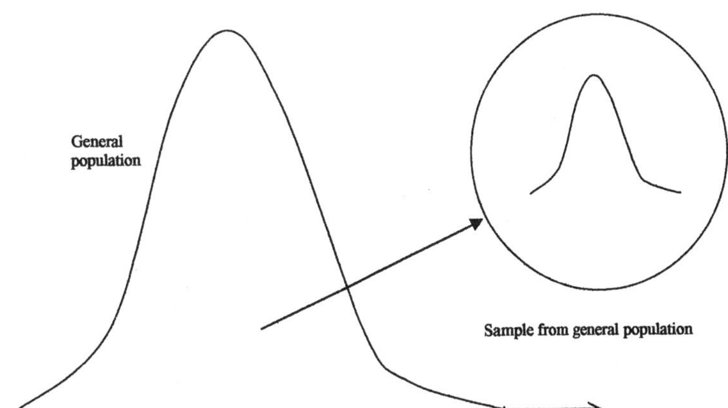

Figure 6.3.1 *Sample taken from the general population*

The basics of sampling are that if a representative group (see Figure 6.3.1) is taken and examined then any statistic of the sample group will be reflected in the total population. The mathematics of sampling is not covered here – just a few words about sampling itself:

- Who is to be surveyed: A clear picture is required of all the different groups that require to be surveyed.
- Sample size: Basically, the larger the better to improve the probability that their response will echo the population we are interested in. Unfortunately accuracy does not follow a straight linear correlation, but depends on the square of the number involved, i.e. to double accuracy, we require four times as much coverage. Note that the larger the survey, the more it will cost to carry out and analyse.
- Sampling procedure – how to choose the respondent: Here one must make sure that a truly representative cross-section of potential customers and decision makers is reachable and is actually covered during the survey. Otherwise we could have a misleading bias in the answers.
- Contact method: Linked to the necessity to reach the respondents. Methods include:
  - Telephone: Does access most of the population, but not all, e.g. ex-directory at one end and poorer households at the other. Times of availability of different consumers can be problematic, i.e. many family members will be out during normal working hours.
  - Postal: Will reach a good spread, but heavily reliant on recipients responding. In many surveys a 10% response is considered good. However, this self-selecting group may not be representative of the whole population.

  - Interviewing: Expensive method, but does allow more information to be collected. Has problems in reaching balanced audience and in the questioner influencing the answers by body language.

In addition to surveys, we may wish a more detailed response, although the actual sample used will be smaller to prevent excessive cost. This is achieved by:

- Observational research: This is non-obtrusive and involves recording reactions such as stopping to look at displays and resultant action, e.g. enter shop – make a purchase. Can be expended into analysing facial responses.
- Focus groups: Here one selects a small group of intended target and conducts detailed enquiry, e.g. taste panels are used by most food groups as a test/intermittent activity during product development.
- Experimental research:
  - Can be similar to observational research, but using a selected group.
  - Test marketing is another form.
- Test marketing: Here the product, or service is offered in a limited geographic area to test response to the product, or advertising campaign. This gives an opportunity to alter the offering before a nationwide launch.

Each method has problems ensuring responses, and a post analysis is required to ensure spread of respondents matches the initial target.

The questioning technique

This is the key skill area in any marketing research and therefore potentially the most error prone. Care has to be taken that questions are not leading, especially in a sequence which may guide participants towards particular answers.

We have to decide what information we want and the best way to structure interviews to get it:

- Facts and knowledge, i.e. about particular products or services.
- Opinions: Attitudes towards products and services – may also require depth of feeling.
- Motives: What makes people do things.
- Past behaviour, e.g. buying patterns, brand loyalty, etc.
- Future behaviour, e.g. intentions to purchase.

To collect the information, there is a variety of ways we can frame questions:

- Closed-end: Very quick to collect and analyse, but forces a direct choice, which may not reflect reality.
  - Two choice: These are typically of the yes/no type. May be agree/disagree. Need to collect number of non-responders, i.e. 'don't know'.
  - Multi-answer choice: Here a choice is given of several probable answers as in Figure 6.3.2. Can be single or multiple choice. Need to leave room for 'others' and don't knows.

Tick which of the following newspapers
you read regularly

The Times ☐
The Telegraph ☐
The Guardian ☐
The Mail ☐
The Express ☐
The Mirror ☐
The Sun ☐
Other (please name) ____________

Figure 6.3.2 *Multi-choice answer*

How much do you like baked beans:

Not at all |——|——|——|——|——| Very much

Figure 6.3.3 *Scalar question*

Why did/are you attending this course?

____

____

____

____

____

____

____

Figure 6.3.4 *Open-ended question*

- Scales: Here responder has to quantify answer as in Figure 6.3.3:
  - ♦ Bipolar, e.g. indicate between two opposites which is nearer.
  - ♦ Importance straight judgement required.
- Open-end: Here the actual words used by responder to a question or stimulus are recorded. Ways of framing question/answer can be:
  - Unstructured answer: Normal invite is by use of the word 'why?' as in Figure 6.3.4.
  - Word association: Immediate answer to a word – normally a trade name or similar.
  - Sentence, story and picture completion can be used to give stimulus to open up inner feelings.

Note open-ended type questions can be severely influenced by the questioner leading. They also need much more careful analysis.

Several problems in collecting data are:

- Refusal to respond.
- Bias in question, e.g. 'You don't think . . . do you?'
- Questions too long, e.g. asked to select several from a list of twenty items.
- Wording of question may be unclear, or can be interpreted in different ways.
- Response is based on what responder thinks is the 'correct' one.
- Responders may, on reflection, not carry out stated intentions.
- Researcher misinterprets answer.

Using the results

The answers from a market research project can be used to:

- Determine the potential market for a new product at different selling prices.
- Form a competitive picture of a product against its main competitors.
- Identify and rank customer needs/wants for a new design specification, i.e. first stage of QFD (quality function deployment).
- Crystallize an advertising campaign.

There have been notable occasions where a market research has apparently been correctly carried out and acted on – with disastrous consequences. The most recent famous failure was that of the 'new coke' recipe by the Coca-Cola Company.

Indications were that its launch would be successful which persuaded the company to make a complete changeover from its previous recipe. Instead it cost Coca-Cola substantial market share which was only addressed by bringing back the traditional recipe.

Product image

By making a profile of the customers' responses against selected attributes or images associated with a product, we can see a product profile as in Figure 6.3.5. This enables a company to see the overall image of its product and gives areas for possible improvement or to stress in advertising as we can build on existing images.

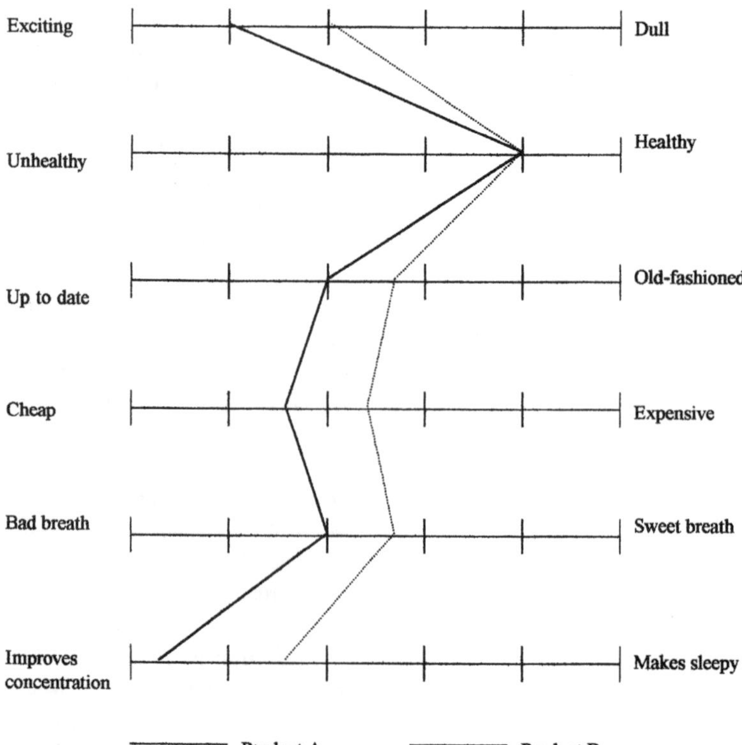

Figure 6.3.5 *Product profiles following market research exercise*

Product attribute correlation

These matrices can be used to form a series of cross-correlation between attributes of a product, and its competition, against stated customer desires. For a sports car, the competing desires could be:

- Speed:comfort
- Price:finish

It is not unusual to find that different market segments will not only put different desired values to these attributes, but they will also score models differently. Being able to differentiate what the different segments value and how they score can guide targeted advertising campaigns in what is portrayed and where different messages should be published.

In the next section (6.4), we shall examine an overall company profile where comparison on a range of supporting attributes is made. Marketing research is the basis for this.

The customer matrix

It is possible to amalgamate all the attributes when measured into a particular segments customer's perceived use value and then compare this with the customers' perception of the product's price to determine an overall competitiveness.

- Identify all the attributes that the customers value.
- Assess the importance of each attribute to the customers – i.e. score them to give a value weighting.
- Rate each competitor and own products by customer.
- Amalgamate the attribute scores into a perceived use value for that product's attribute, i.e. multiply the attribute rating for a product by that attribute's customer value weighting.
- Average out the different attribute values into an overall customer perceived value (PV) for each product.
- Determine the customer perceived price (PP). This may not reflect actual price differences as the customer may have only an initial price in mind but an amalgamation of life costs.
- Graph each product onto the matrix as in Figure 6.3.6.
- Analyse the matrix as Table 6.3.1.

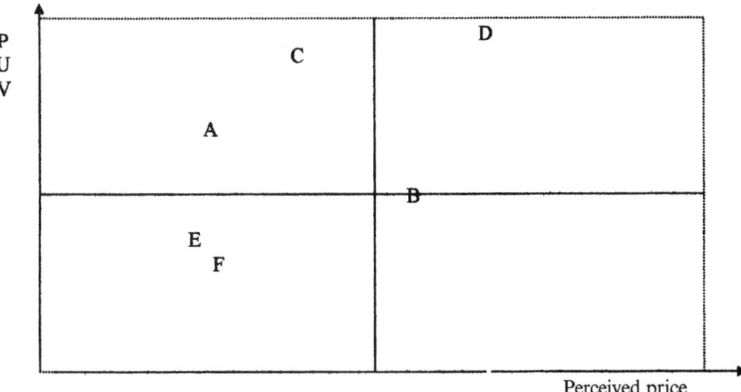

Figure 6.3.6 *Product placement on the price:value matrix*

Table 6.3.1 Matrix of sectors within perceived price and perceived value

		Perceived value	
		Low	*High*
Perceived price	Low	In line with market	Highly competitive
	High	Uncompetitive	In line with market

We should find a reasonable correlation between the perceived value and the perceived price across the range of products.

Where a product appears to be highly competitive these perceptions must be maintained. Conversely if a product is seen not to be competitive, then action for improvement must be made, or perhaps a withdrawal from the market.

If the matrix indicates that the product is in line with the market then by concentrating on improving the attributes that the customer values, we should be able to drive up the customers' perceived value. This will increase the sales, or allow us to set higher prices without affecting sales.

It is more difficult to alter the customers' perception of the price, except by changing it. The basic question is how much of a change in price is required to change an impression of being highly priced. Will that change be judged as being achieved by cheapening the offering? Perhaps a price revision needs to await a new opportunity arising from a new product.

Problems 6.3.1

(1) How many people do you know? Out of these, how many own a Ford car less than three years old? From your answers, estimate how many Ford cars a year will be sold in the UK in the next calendar year? How accurate do you think this answer is?
(2) Using the NRS and ACORN models, describe yourself and determine the best method to gain a good response rate from your group.
(3) If you wished to open a shop specializing in fishing tackle in your home town, what basic information do you think you will need to form an estimation of potential customers? Where will you find it?
(4) Can you think of several questions where your answer may be swayed by a desire to be seen as being politically correct?
(5) Using the matrix in Table 6.3.1, categorize a number of restaurants.

6.4 Market strategy

For an organization to be competitive it must have products that the market wants, at the right price and in the right place backed up by effective promotion.

It must satisfy all its customer's needs to win orders by critically examining its competitive strength. It must then determine which market it can best serve, if need be by forming partnerships with others. These are examined in this section.

The marketing mix

In marketing terms, there are said to be four 'P's which lead to a customer purchasing a particular product:

Product Place Price Promotion

As these are what makes someone actually buy the product, it is worthwhile examining them in detail – especially in relation to what your competitors are offering.

The products

As anyone who has watched advertisements should realize, a product is more than just a physical mixture of features. Each product has many connotations and images attached to it, which are worthwhile examining separately.

Benefits

Designers of products, especially in engineering, tend to concentrate on the features of a product as that is the physical matter that they are creating. When you buy a product you are actually buying benefits rather than features. Benefits are everything that a customer gets from owning the product, over and above the physical product itself. These are the needs that a product should be designed to meet. They are what our salesmen should be stressing to potential customers.

Intangible benefits. These are the deeper underlying reasons why people buy particular types of products. There can be more than one intangible associated with a product. Examples are:

- Perfume and cosmetics – to achieve increased appeal to others.
- Sports car – to achieve increased appeal plus excitement.
- Washing powder – to show that you are a proper parent.
- Fashion clothing – a sense of belonging and being up to date.

Tangible benefits. These are less deep than intangible, but could be said to be the more apparent need that the product is meeting. Again there can be several tangible, and intangible, needs that a product could satisfy:

- Cars – transport, speed, safety.
- Vacuum cleaner – cleanliness.
- CD player – sound (music).
- Mobile phone – communications.
- Soft drink – satisfy thirst.
- Table lamp – giving light.

Feature attributes

These relate to the physical product itself and can be easily perceived by the customer, sometimes even before purchasing. They are normally easy to compare to the features of our competitors' products externally. They are therefore important in being pleasing and what the customer expects in a product.

- Size: Could relate to any measurement such as overall length or internal width of a car, or the screen size of a television.
- Shape: Customers associate a certain shape to a product – that is why it can be difficult to sell a technically superior product when it comes in a shape that the customer does not associate with the product.
- Quantity: Could relate to the content of a package, or the seating capacity of a car.
- Service: Could relate to the time to be served in a restaurant, the number of trains each hour between two towns.
- Taste: Could be related to sweetness or bitterness of a soft drink.
- Odour: Could relate to the smell of a perfume on someone's skin – or even how long that smell lasts.
- Colour: Could relate to the range of colours that a car comes in.
- Power: Could relate to a car engine or the wattage of a CD sound system.
- Efficiency: Could relate to the kilometre per mile performance of a car or the amount of water used by a washing machine.

Signal attributes

These are more aesthetic and aimed to appeal to the senses, although they are also closely linked to the intangible benefits above. Sometimes they are used in the selling environment, rather than in the product itself.

- Odour: Smell of baking in a supermarket.
- Material: Not only of the product itself, but often of the packaging of items such as jewellery and cosmetics.
- Style and design are very important here.
- Sound: Background music in a clothing retailer.

Product range

As not all customers want exactly the same product, we have to produce differentiated products so that we can attract a variety of customers. This differentiation causes problems within manufacturing – so our designs have to be careful to use as many common parts as possible, whilst being able to show real differences. There are two measures here:

- Range: The number of different product lines available. Examples include the different named models from a car manufacturer or the different toothpastes produced by the same company.
- Depth: Depth within each product line. This could be the various models (can be over twenty) from a car manufacturer under the same name. Could also refer to the different size of packaging containing the same product as in soft drinks or breakfast cereals.

Packaging

Packaging has the prime function of ensuring that the product is protected in transportation and storage so that it reaches the customer in good condition. However, it can often contribute to any of the above attributes:

- Intangible by display of a brand, or containing images.
- Tangible by ease of use.
- Features such as reuse, or easy to recycle.
- Signal by design and colouring.
- Product depth by single use packs, combination packs.

Place (distribution)

This relates to ensuring that the product is in a place where potential customers can see and access it, so that they will make the purchase. This must be done efficiently and effectively.

- Distribution cover – how many potential customers are reached and the channel used.
- Physical distribution:
  - Transport used, e.g. road, rail, postal service, specialist carriers. Speed of delivery may be important as in the newspaper industry.
  - Service – covers matters such as number of deliveries per day/week and reaction time to a variation in order pattern.
  - Inventory – how much and where is it stored.
- Nature of outlets – the product should match expectations of the customer in where to find it.
- Direct sales via websites and newspapers/magazines.

Promotion

This is how the organization communicates with its customers along the distribution channel. There can be a variety of reasons for communicating, e.g. special offers or a new advertising campaign can be directed at retailers. There are a number of avenues used:

- Advertising: The common recognized means of drawing people's attention to a product offering (see page 266 for range of aims and means).
  - Focus: Can be directed to particular channel members or individual groups of end customers.
  - General: Direct to all members of channel or general public.
  - Publicity: Although more a general broadcast, this can be tying a specific product to an event, such as Formula One racing, or getting sport celebrities to wear, or use, particular products.
- Sales promotion: This relates more to direct inducement:
  - Special offers to retailers, or customers, e.g. two-for-one.
  - Tokens collected to 'purchase' items.
- Sales: This relates to the direct selling:
  - Sales system. How the actual transaction takes place.
  - Personal selling. How many staff deal with customers. In supermarkets no direct selling takes place.

- Public relations: More general publicity around an organization to improve particular aspects of its image, e.g. by support of a special charity event or a local organization.
- Packaging: Although primary function is protection and appeal, it also offers an opportunity to send messages to customers.

Price

Although price can be very important, we have to remember that it determines the profit after deducting all the costs. Less than cost price can be used as a marketing tool, but only where long-term profit results, e.g. to penetrate a market, or as a loss leader to entice customers to buy other profitable products. It can also be used to target particular customers who may be useful either through the bulk purchases they bring or through future sales. We have to be careful about our pricing structure in relation to our other customers and our competitors.

Allowances and discounts. These are used to pass on some of the savings that result through large commitments. Usually varied in relation to quantity or order value, but can also be used to entice early payment of invoices, e.g. 2% reduction if paid within seven days.

Pricing gap. This is between all the costs and the price. Can be used to set allowances and discounts (see Chapter 5 for margins).

Customer valuation of benefits. This is where real profits can be made. If there are criteria such as quality or on-time delivery that the customer values highly, they may be willing to accept a higher price.

Service extensions

In addition to the four Ps, there are other criteria that customers use in deciding where to trade, and who to trade with. These could be said to be another three Ps.

People

There is a variety of your people that your customers come into contact with when trading with you. It is important that staff are trained as being representatives of your organization and show the expected attitude towards your customers. This includes their appearance and behaviour.

Customer contacts of special importance are:

- Telephone operators: Almost certainly the first contact point. How they greet your customer and handle their calls make a lasting impression.
- Delivery personnel: These people represent your organization at the customer's premises. Their manner and attitude will be noted.

Order process

The most important transactions with the customer are the placing of the order and the associated payment. The procedures should be friendly, efficient and error resistant, even when chasing late payments.

Automation may appear to save money, but can be a turn-off for your customer. They like to deal with friendly and efficient people, not machines. Information technology can assist your people to be fast and efficient – when the systems are correctly designed. Credit control procedures are especially important here.

It is possible to increase customer involvement as in self-service in supermarkets and on the Internet, but remember to have staff available as a back-up to answer customer queries and problems.

Physical surroundings

The customer expects certain environments when they are doing business. You therefore have to examine all areas that the customer may visit and ensure that they meet the customer's expectations. Key areas are the approach to your plant/offices (i.e. look at them as if you were approaching them for the first time) and the impression of the reception area and meeting rooms. Do not forget the factory and despatch areas – your customer may visit these.

Market qualifying and order-winning criteria

For every product, the customers not only examine the product offering, but have a further set of criteria in their minds when examining which of a number of competing products to purchase. Terry Hill of the London Business School laid down the following in his book *Manufacturing Strategy*:

- Market qualifying criteria are attributes that the customer expects from all competing suppliers. Some are more critical and are termed order-losing sensitive qualifiers.
- Order-losing criteria are those considered to be critical to the customer and a failure to demonstrate them will exclude the supplier from consideration.
- Order-winning criteria are those attributes that enable a company's offerings to gain an advantage over their competitors in the eyes of the customers. These eventually win the actual order.

Over time, order winners will tend to change to market-qualifying or order-losing criteria and new order winners take their place. We must therefore continually identify what these criteria are and ensure that our internal processes can deliver them to the customer and ensure they are satisfied.

Criteria can be either related directly to the product itself, or they can be more service oriented. A company, by focusing on those criteria in which it does well, will often gain a competitive advantage. Criteria looked for include:

- Speed of design process: Being able to quickly give a customer an idea of how you can meet his stated need. Increasingly important in conjunction with customer's redesign process. Component and sub-assembly suppliers are increasingly working hand in hand with their customers on new designs. Being able to communicate electronically can be key here.

- Speed of quoting: As many customers are looking for quick deliveries of the product itself, the speed of replying to their requests for price and delivery can be crucial in gaining the order.
- Quality: This covers several aspects:
  - Conforming to the customer specification for use.
  - Number of defects in a batch. Especially important as more customers are moving towards having no incoming inspection and putting incoming components straight onto the manufacturing line. This can apply to information as well as tangible products.
  - Reliability, i.e. the maintenance of good functional operation over time. Important to all customers, but more so to some, e.g. aeronautical and defence. May include aspects of ease of maintenance.
- Delivery on time: Being able to give a reliable time frame for a delivery is becoming more important as customers move towards minimal inventory of raw materials and are geared up to JIT (just-in-time) operations. Being able to meet an advertised product launch date is especially important in maintaining an organization's customer image.
- Quick delivery: In many situations being able to give the customer immediately what he wants when he asks for it is critical. In this case, we either need extremely efficient manufacturing or need to carry a stock of items that may be wanted. The latter involves storage facilities, opens up the risk of obsolescence and ties up our money. Many items, however, are expected to be available ex-stock, especially in convenience goods. However, we do need to be able to give all customers their delivery within the time they need it. The P:D ratio can be critical in winning orders (see Figure 6.4.1).

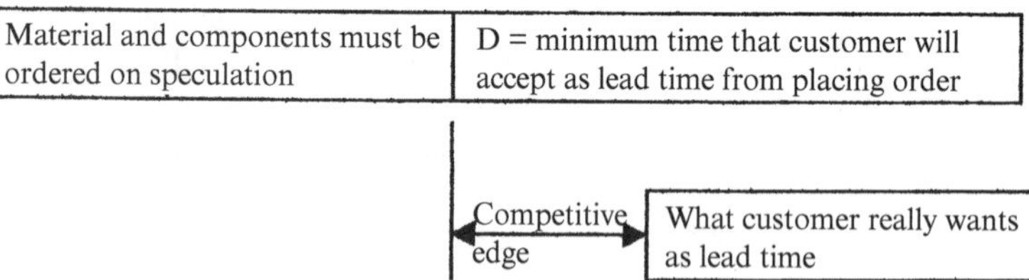

Figure 6.4.1 *P:D ratio showing where a competitive edge can be gained*

- Flexibility in volume: Being able to react in tune with the quantity needed by the customer in volatile markets can be attractive.
- Flexibility in design: Being able to quickly customize standard designs to meet the customer's particular needs. Modular design is advantageous here.
- Technical information: As many customers will not carry the technical expertise which should be within your organization, you may need to be able to quickly and accurately answer technical queries. This can be pre-sales or after-sales.
- After-sales service: Often the product's relationship does not finish with the sale – there can be a large degree of follow-up service

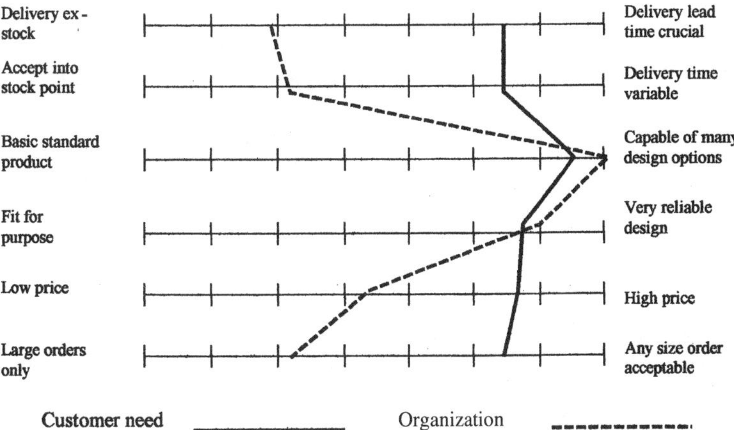

Figure 6.4.2 *Profile comparison between customer need and organization's performance*

required, e.g. service engineers, spare parts, technical information about operations or maintenance.
- Ethical behaviour: As worldwide trading expands, concerns such as child labour, a fair price for indigenous materials and other human rights concerns have to be addressed to ensure that different customer concerns are being meet.
- Environment: As this is another area of concern, processes have to be environmentally neutral by producing a minimum of waste and pollutants. Recyclability and reuse may be important.

Figure 6.4.2 demonstrates a profile comparison between what the market requires and an organization. Where the two profile lines harmonize the organization is matching the market's needs.

Where there is a large difference, this is an indication of either of two things:

- We are not being sufficiently efficient in matching what the market criteria are. We therefore need to improve this weakness in our internal processes and facilities, or find a different market to supply to.
- We are overefficient in meeting the market criteria. We can either:
  - Reduce our effort and expense in meeting this criteria without a loss to the customer, or
  - We can make use of this excess by finding new customers, or attempting to increase this aspect in our customers' perceptions.

Case study

Client's requirement

The basic brief was to increase productivity at a minimal capital cost. Time available was an eight-week BESO assignment.

PROFILING

A competitive profile derived for the organization is reproduced in Figure 6.4.3. Time was too short to interview many customers and the market requirements were based mainly on discussions with

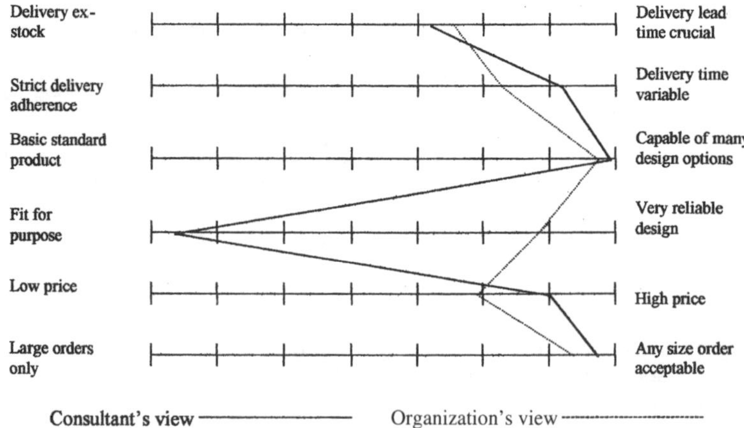

Figure 6.4.3 *Company profile analysis showing difference between consultant's view and organization's view*

the organization's marketing personnel. The company's achieved performance was arrived at by observation and discussion.

This clearly showed that the organization's competitive advantage lay in its design skills. However, it also showed several weaknesses, namely:

Lead time, due date adherence, price, quality.

Initially the identification of quality as one of the company's major weaknesses was not believed as they had a reputation of producing high-grade products for their customers.

Effect of quality problems

The starting point in persuading personnel that internal quality problems were the most important and were behind the other symptoms started with demonstrating the consequence of quality faults as per Figure 6.4.4. The important items that were identified were the following internal failure costs.

Figure 6.4.4 *Relationship of poor quality and reduction in profit*

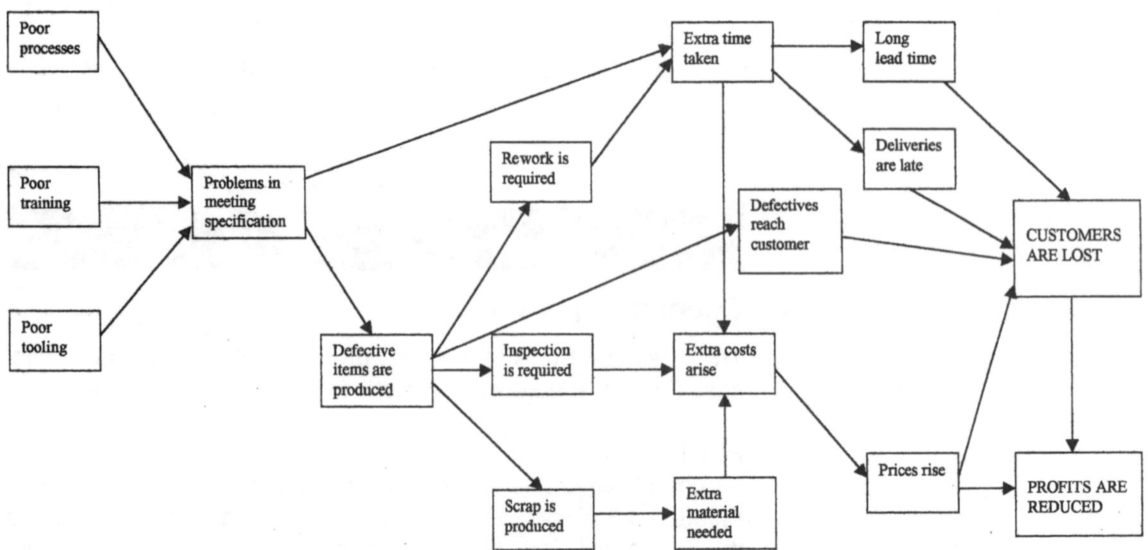

- Scrapped products – extra costs and delays in producing replacements.
- Reworked products – extra costs and delays.
- Inspection – highly skilled labour used in non-productive work.

The personnel within the organization, by not identifying root causes, were too often demonstrating their skill in rectifying defects rather than in preventing them reoccurring. By eliminating much of the quality problems they could produce items quicker and with less costs.

The competitive profile

You need to closely assess your own position within the marketplace, not only against your customers' criteria, but also in how your competitors are matching these.

A polar chart (see Figure 6.4.5) or a profile comparison can illustrate the difference between you and your competitors in meeting the customers' criteria. This demonstrates their and your relative strengths and weaknesses under a variety of attributes.

Weaknesses have to be corrected, or protected. Strengths have to be built on – especially if they are superior to the competition. In Figure 6.4.5, we have a weakness in new equipment, but a strength in cash reserves. The strong cash reserve could be used to purchase up-to-date equipment that may convert the present weakness into a strength. It could be used to support price reductions, mount a large advertising campaign or even to buy over a competitor.

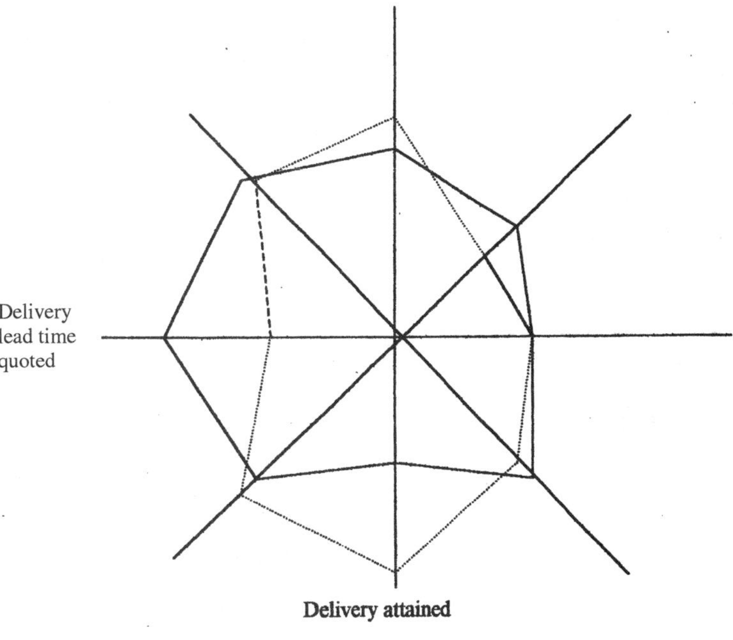

Figure 6.4.5 *Polar chart comparison of organization and major competitor. The profiles are similar except that the organization is performing better in quoted delivery but worse in reliable delivery*

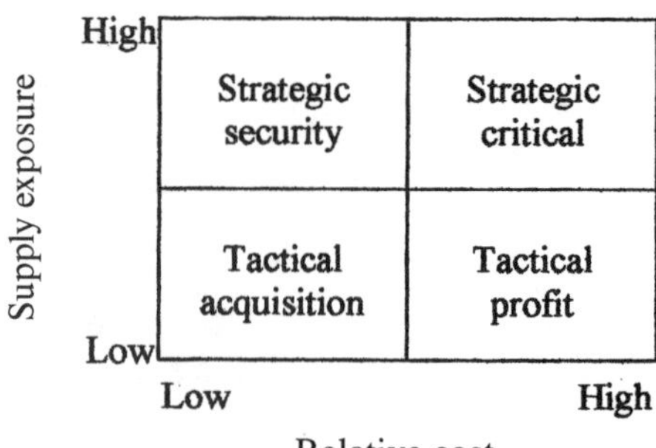

Figure 6.4.6 *Supply positioning*

We also need to examine our entire supply chain for existing, or potential, risks and problems. We are especially concerned with our reliance on either particular suppliers or customers.

Suppliers

Here we are looking at our vulnerability to problems in supply from happenings such as strikes, takeovers, fires, etc. Question how dependent on each supplier you are, and how important you are to them in turn. Indicators are:

- What proportion of the requirement for any item does any individual company, or small group, supply?
- What proportion of their output do you take?
- What other sources of supply are available?

Where we are heavily dependent on one, or a few suppliers we must take strategic measures to ensure maintenance of supply. This may involve seeking other possible suppliers.

The problem here is that relationships with suppliers can be crucial if they have expertise, or are willing to be flexible in their dealings with you. Spreading purchasing can damage long-term relationships that have been built up.

Looking at the supply position matrix (Figure 6.4.6) shows the four extremes:

- Exposure low – low cost item: Buy tactically – play off suppliers against each other to give us special terms on delivery, etc.
- Exposure low – high cost item: Buy when prices are low, or demand is low.
- Exposure high – low cost item: Arrange a guarantee of supply, perhaps using a few key suppliers.
- Exposure high – high cost items: Serious if vulnerable – secure long-term supply agreement.

Customers

Here we are looking at our strength in relation to our customers. What happens if a major customer stops buying from us? In other words, how important we are to each customer and how dependent they are on us. Questions to be asked are:

- What proportion of your output goes to your main customer, or even the top three customers?
- What proportion of your customers' needs is met by you?
- What alternative sources of supply are available?

When we are heavily dependent on a few customers we are vulnerable not only to the possible loss of the trade, but to pressure being put on us by the customer, e.g. to reduce prices, accept late payments, reduce lead time, etc.

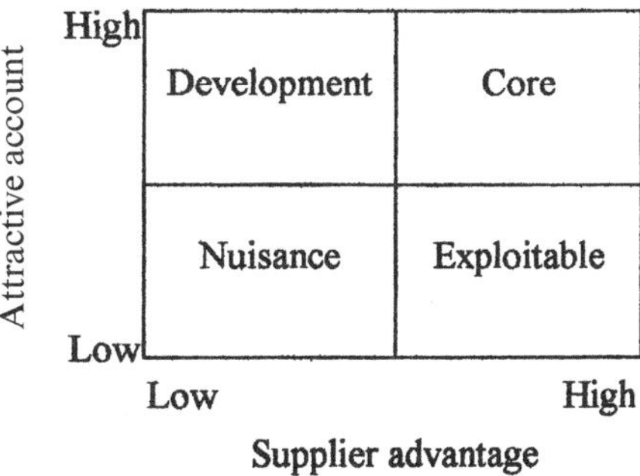

Figure 6.4.7 *Supplier preference*

Again we have a matrix to guide strategy – the supplier preference (Figure 6.4.7) where we examine our competitive advantage against the attractiveness of the customer.

- Low profit – no competitive advantage: This is a nuisance, and we should examine it to determine marginal profits involved in dropping it.
- Low profit – high competitive advantage: This should be exploited to maximize revenue, bearing long-term relationship in mind.
- High profit – no competitive advantage: We need to develop long-term supply arrangements, being careful what we are willing to give in return.
- High profit – high competitive advantage: This is our core business and we need to ensure that we do not lose it.

Threats of new entrants to market

Even when we a do have a strong position in a particular market, we are not necessarily secure there in the future. If it is a profitable area, this will attract new producers to move into the industry. We need to ensure that this is difficult by considering:

- Economy of scale: There are breakeven points for manufacturing different quantities. We have to ensure that we maintain these for our own facilities which will make it difficult for new entrants to get enough orders to cover their costs. We also have to guard against unnecessary centralization as that adds complexity and makes it easy to get out of touch with customers.
- Capital requirements: Often costs are contained by using expensive processes. We have to ensure that we maximize this through investment and keeping capacity filled.

- Customer loyalty: The cheapest customer to get to buy from us is one with whom we have already built up a relationship. We should ensure we keep our customer happy.
- Learning curves: As an organization builds up experience in design and manufacturing, it discovers ways to improve output at reduced cost. We have to ensure we maintain our own learning and improvement and keep ahead.
- Sophistication of technology: We must ensure that we keep up to date in both product and process technology.
- Distribution Channels: The complete channels give access to a range of customers. We have to maintain good relationships all along the supply chain – both into us and from us to the end user.

Where entry barriers are weak, we need to strengthen them by building cost and relationship advantages.

Threats of substitute products

All products have a natural life. The ideal situation is that the end of that life is within your hands and you can replace products readily. To maintain that position we need to be continually researching the market to determine what substitute products already exist, and researching to see how easily they could be developed.

These researches should also feed back into your own development programme for new products. Better that you bring them to market, than your competitors.

Existing competition

We have to continually be aware of our competitors and what they are doing. We also need to be aware of factors that may lead to intense competition developing:

- Numerous, or equal sized companies: Difficult to maintain a dominant position.
- Slow market growth: As capability improves, we normally need to expand sales to maintain our return on investment. Slow market growth means that all the suppliers intensify their activity to gain whatever growth there is.
- Lack of product differential: Where all suppliers offer the same basic product, then other non-feature aspects are used to try to gain customers. These include advertising campaigns, price cuts, special offers, etc.
- Capacity utilization: Where spare capacity exists either in the industry as a whole, or in a particular competitor, then special efforts will be made to fill that capacity. Often the only way to do this is to capture customers from others.

Looking at the influence of each of these four factors together leads to what is termed a five-force analysis – the fifth factor being the resultant rivalry.

Market strategy and planning

It is important that an organization carefully plans out its future strategy in product development.

The Ansoff matrix (see page 263) explored potential areas of growth. The General Electric matrix (Figure 6.4.8) attempts to denote an organization's market competitiveness. The circle size indicates the size of the market and the segment the market share. The length and direction of the arrow indicate direction and speed of the expected change.

Competitive position

GE MATRIX		**Strong**	**Medium**	**Weak**
Market attraction	**High**			
	Medium			
	Low			

Figure 6.4.8 *GE matrix showing relative market size and direction of market trend*

Competitiveness position is determined by a mixture of market share, strength of product line, relative cost/price structure, distribution facilities, brand profile, technical skills, production facilities, raw material sources, strength of sales staff, etc.

Market size, market share, profit margins, capital investment needed, competition, technical synergy, customers, risk, growth, etc. determine the market attractiveness.

From an analysis of where we are and where we can/wish to go, we can draft detailed marketing plans:

- Determine relevant markets and their segmentation.
- Identify characteristics of the potential segments.
- Evaluate segment attractiveness and the organization's competitive position serving each segment.
- Select target market segments.
- Set objectives and strategies for each selected segment.
- Develop full marketing mix for each selected segment.
- Implement marketing plan with feedback and control.
- Decide degree of diversification.

An organization's ability to turn these plans into reality is determined by their capability. It is important therefore to be able to clearly establish what the organization has in resources, i.e. its staff and their skills; techniques; processes, etc. They can then ensure that the market plan utilizes their strengths, or that action is taken to rectify any weaknesses. A full SWOT (Strengths, Weaknesses, Opportunities, Threats) analysis needs to be carried out looking both internally and externally.

Co-operation with other organizations

At sometime, most organizations have to co-operate with other organizations. The idea behind co-operation is to make use of the separate skills, experience and contacts in a way that each of the separate organizations are unable to do by themselves. The co-operation may be due to:

External forces:
- Globalization: In order to gain access to markets it is necessary to have a very wide presence. This can be expensive and difficult.
- Share technology: Because of the cost of researching and developing new products and processes and then attended risk, many organizations share the cost and benefits.
- Economy of scale and scope.
- Trading in other sectors can counterbalance turbulence in the organization's normal sectors.

Internal forces:
- Competence weakness in skills, technology.
- Capital involved may be difficult for one organization to fund.
- Risk limitation or sharing.
- Speed of development – joint forces increasing effort.
- Defence against predators.

Type of co-operation:
- Alliance: A temporary informal joining together, normally to tackle a particular problem.
- Consortium: Coming together to make up a wider skill base – such as bidding for the building and equipping of a power station.
- Joint venture: Forming a separate legal entity to exploit a particular market or product development. Often the only legal way to work in some countries is a joint venture with a local company or individual.
- Merger: A full joining into one organization.
- Acquisition: Buying over another organization – may be hostile (unsupported by the board of the bought company) or friendly (recommended by the board).

In all cases there will probably be problems arising through several different cultures and management styles competing in the new organization or relationship. It is noticeable that over 50% of mergers end up with a smaller benefit than envisaged.

Making contingency plans

The difficulty with the marketplace is that it is always changing and a company must be ready to react to these changes. One way is to follow up a forecast of the possible future changes in the market or the environment by devising plans to cope with these possible changes.

The way that this is achieved is by scenario analysis. This is based on the organization's competitive analysis and then applying the principal impact factors from three scenarios, derive the likeliest possible severe impact on the organization. By looking at this impact and its probability the need to develop a contingency plan to deal with the impacts will be prioritized.

As an example take a university offering engineering degree courses. The scenarios are:

(1) No change from present trends.
(2) The government increases funding to home students.
(3) The government reduces funding.

The principal areas of concern are:

(A) The level of costs.
(B) Nature of competition.
(C) The strength of demand.
(D) Difficulty in attracting staff.

Table 6.4.1 Scenarios

Impact factors	*Scenario*		
	1	*2*	*3*
(A) Costs	Slow rise	Fast rise	Pressure to fall
(B) Competition	Steady	Slow rise	Fast rise
(C) Demand	Slow rise	Fast rise	Fast fall
(D) Staff attracted	Slow fall	Fast rise	Fast fall

The evaluated events are then placed on a scenario matrix as in Figure 6.4.9. Using the letter and number, e.g. A.1, is the cost factor change under scenario 1. The axes are 'probability of occurrence' and 'strategic importance'. Quadrant one, i.e. top right hand, is where the likely strategically important factors will congregate. In this case, strategic contingency plans must be made for events A.2, C.2 and D.1.

Further outline plans may be considered for those just outside this quadrant, i.e. events within the dotted lines, e.g. A.1, B.2, C.1 and A.3.

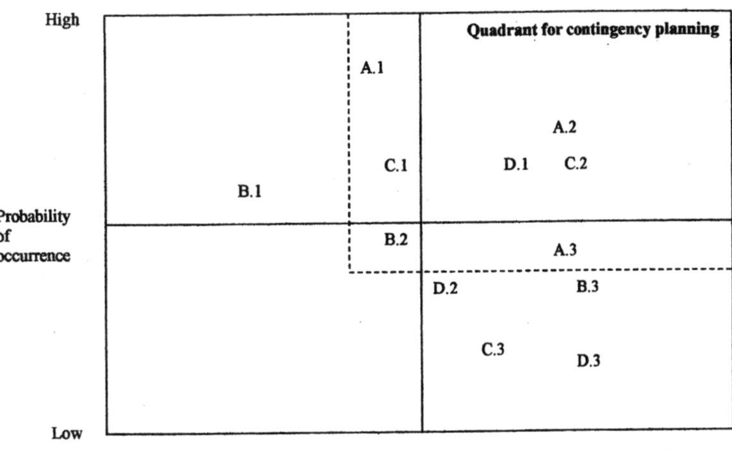

Figure 6.4.9 *Scenario matrix*

Case study

Situation

A small company supplied specialist jams to a few major retailing groups. Six months previously a new competitor appeared who was making inroads into the company's market. This extract is from a consultant's initial report.

1. Introduction

The following are the steps to be taken to analysing the company situation and develop proposals for action in the short and long term.

2. Measurement of key result

The following standard financial measures should already be available, or could be calculated quickly. They need to show at two points – prior to entry of new competitor and changes since.

- Profit and loss.
- Sales in units and money.
- Production costs and gross margin.
- Administrative expenses and net margin.
- Cash flow.
- Inventory levels.
- Dispersion of profit – tax, dividend and retained.
- Borrowing and resultant leverage.
- Return on capital employed and net assets.

These will give an indication of the seriousness of the present situation.

In addition the following need to be evaluated for the same time points:

- Customer base – number of, sales per unit, location, etc.
- Production, capability and capacity utilization.
- Quality indicators such as rejects and returns.
- New products introduced and in development cycle.

The final items are to establish the market in my mind – especially the strength, or weakness of the company and its competitors that will guide any recommended action.

- History of the company.
- Major changes in key personnel in last two years.
- Age and condition of equipment.
- Product life cycle and seasonal effects.
- Supplier base and production cycle time.
- Recent advertising and other promotional campaigns.
- Competition analysis – especially the new entry.

From all the above, the trends involved can be established to show time available for action and the direction the company may move in both the short term and longer term to (re)establish good results.

3. Possible improvements

Once the initial evaluation has been carried out, there may be several ways in which the company should proceed. Which particular one, or combination of several, will be determined by the analysis. The following are not listed in any preference order.

3.1 Reduce outgoings

Although this is a common approach, it easily results in future problems if not approached correctly. However, there should be no problems in ensuring that the company maximizes its present use of resources:

Supplier base. Ensure that they are well served by suppliers – high quality; delivery as and when required, value for money, quick reaction, etc. Should be minimum number if possible, consistent with security of supply.

Production base. Ensure processes operating to maximum efficiency through well-maintained machinery, well-trained and motivated operators, good planning, low reject rates and minimum waste, including effluent.

Design. Ensure design cycle matches customer needs with minimum failures. Ensure process of changeover from development to production is well organized beforehand. Severe cutbacks in this area will damage future viability.

Buildings. Ensure that premises match requirements in space and minimal energy losses.

Overheads. Ensure these contribute towards the survival and effectiveness of the organization.

Systems. Ensure these are geared towards customer requirements in most efficient manner, e.g. quick and accurate order taking. IT must be fully supportive.

Management. Do they have skills and experience required?

A main problem arises if analysis shows that there are too many staff to support envisaged operations and a reduction is indication. If handled poorly this can result in low staff morale and send a message of weakness to our competitors and suppliers.

Targets and controls should be set up in each area.

3.2 Increase sales income

There are several ways of increasing sales – dependent on the present customer base, and loyalty, and the strengths of the existing, or potential, competitors.

ANSOFF'S MATRIX	Current products	New products
Current Markets	Market penetration	Product development
New Markets	Market development	Diversification

3.2.1 Market penetration: This is basically attempting to increase the company's share of the existing market. There are several ways in which this can be done:

(i) Advertising and other promotions: This can be successful, but does not give a guaranteed result. It can be expensive and can be reacted to quickly by competitors. If competition has greater access to reserves or other funding the advertising battles can be exhaustive.

(ii) Meeting customers' order-winning criteria: In the medium term this will pay off. Main problem is establishing these criteria and ensuring they are met or exceeded. Criteria include – low delivery time, reliable delivery, high specification, low price, high quality, quick reaction to change in customers' demand be it volume, or new products.

(iii) Increased selling effort – reaching customer and persuading them to place orders: Can initially be expensive and have to be careful not to make sales pitch too hard.

3.2.2 Market development: This is attempting to find new markets for the company's products. These could be:

(i) Greater cover to reach all potential customers within the present sales area (if any missed out).

(ii) Going beyond present geographic area into other areas of same country, or into export market (if not already there) – or intensifying effort if already doing so. UK food goods still have a good name in foreign parts, especially within the commonwealth – and full advantage of this has not been made. As distance widens, however, so does the complexity of control – decisions such as opening own sales offices or dealing through agents need to be made.

(iii) Finding new customers different from present ones. For example, it may be supplying restaurants, or high-class department stores or sales at venues such as airport duty free shops.

3.2.3 Product development: A way to find different products that can be sold to the existing customer to meet their other needs. Again there are several ways to do this:

(i) Customizing existing products to meet particular customer wants.

(ii) Developing, or buying in, complementary products that can use present contact points.

(iii) Widening product mix – not just in variety of jams but also in variety of packaging, e.g. present type packages.

3.2.4 Diversification: A way of finding different products to sell to an entirely different customer base – in effect a combination of 3.2.2 and 3.2.3.

The competitiveness position is determined by a mixture of market share, strength of product line, relative cost/price structure, distribution facilities, brand profile, technical skills, production facilities, raw material sources, strength of sales staff, etc.

From an analysis of where they are and where they can/wish to go, we can draft detailed marketing plans:

- Determine relevant markets and segmentation.
- Identify characteristics of potential segments.
- Evaluate segment attractiveness and organization's competitive position.
- Select target market segments.
- Set objectives and strategies for each selected segment.
- Develop full marketing mix for each selected segment.
- Implement marketing plan with feedback and control.
- Decide degree of diversification.

These will prove necessary to the long-term survival of the company but need to be developed now. They all require a reservoir of resources, experience and skills that may not be available within the company.

4. Capability and resources
The company's ability to turn plans into reality is determined by staff's capability. We must establish what the organization has in resources – people and skills, techniques, processes, etc. They can then ensure that the plan utilizes their strengths, or that action is taken to rectify any weaknesses.

Weakness can be met by retraining, importing new skills or co-operating with other organizations to gain from that organization's strengths.

The type of co-operation can be alliances, consortium, or joint ventures. Each of these has advantages but possible problems in lessening control.

Problems 6.4.1

(1) What are the product features and attributes associated with a wrist-watch?
(2) Why do order-winning-criteria tend to change to market-qualifying criteria over time?
(3) If you were in the home delivery service, what factors would you consider worthy of considering in a scenario analysis?
(4) If you were a university, what factors should you consider in a competitiveness analysis?
(5) What partners could a university consider in the UK and further afield and what would be the best relationships to be formed?

7 Information technology and electronic commerce

Summary

Information technology is increasingly important in everything we do. This chapter should indicate to you what skills you will need to have, and what you will use them for, in order to use IT effectively.

It begins with a survey of the changes that IT is producing at the beginning of the twenty-first century, and the social and legal results of these changes. There follow sections on the use of the most common operating system functions followed by the major applications used for text, number and general data processing, and applications more specifically used by engineers. The chapter concludes with a section on the technology and use of the Internet and the growth of e-commerce.

Objectives

By the end of this chapter, the reader should be able to:

- recognize and respond to the effects of IT in changing the organization;
- use a computer, particularly a PC, efficiently;
- use common 'office suite' applications effectively;
- get the most from engineering applications;
- exploit the power of the Internet, particularly e-mail and the Web;
- understand the potential for the use of e-commerce.

7.1 IT and change in organizations

IT (which is sometimes called ICT, information and communications technology in the academic world) has produced and is producing vast changes in the way companies and organizations function. This section describes the main IT influences and their impact.

- Large payments can now be made instantly and securely.
- The time taken to process a document – respond to an order or pay a bill – has decreased from days to hours or less.
- Next day delivery is expected. Without it the order is often lost.
- Work analysis processes are removing slack time from work schedules (far more effectively than time-and-motion studies ever did).
- Sophisticated data analysis systems allow much more efficient targeting of direct mail (already the most cost-effective form of advertising).
- Internet processes threaten to wipe out many middlemen, wholesalers, retailers and brokers by allowing direct sales of products and services.
- The legal framework in which a company operates changes frequently as the law tries to keep up with new issues created by the technology.

To master the IT field requires a raft of skills to be acquired or at least to be understood well enough to direct others (and the old advice not to direct others to do something you cannot do yourself still holds in the e-commerce age). But do not get bogged down with details. Whichever software application you are using, the next version is likely to function slightly differently; with an uncluttered head you will be able to find out how and convert.

The effect of the speed of change

People have difficulty coping with exponential change. Explosions cause damage! In the two decades between 1980 and 2000 computer technology, particularly PC technology, has changed dramatically. The speed of processor chips, typical memory sizes (both working RAM and backing disk storage) and communications speeds have increased 1000 times or more. This exponential growth of the power of the technology (see Figure 7.1.1), with the number of transistors per chip continuing to obey Graham Moore's law of 1985 and double in less than two years, makes prophecy about the capabilities of IT a very inexact science.

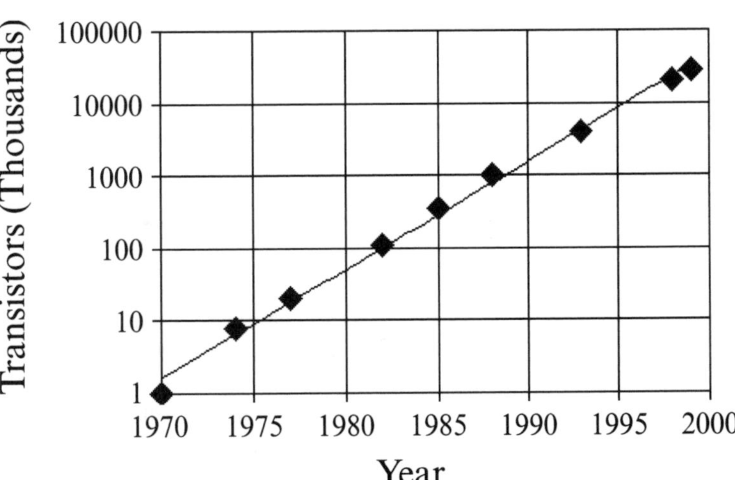

Figure 7.1.1 *Graph of Moore's law describing chip complexity (data from Intel website)*

Whole industries need to rethink their structure. When data could be communicated at 300 bit/s, the digital data on a CD would have taken nearly a year to send from Hollywood to London. Prevention of copyright was a matter of keeping tight control of originals, and of seizing poor copies of copies 'in the interests of preventing the customer suffering from an inferior product'. With cable modem internet connections at 1 Mbit/s, a single bootleg copy of a music album, or video of a new film can be broadcast round the world in about an hour, seriously disrupting the neatly phased marketing strategies and income generation of the entertainment industry.

Activity 7.1.1

Imagine you have a piece of paper that you fold in half, stacking one sheet on the second. Imagine that you repeat this twenty times. Without calculating, estimate how thick the piece of paper would be? Did you imagine a sheet of paper about 0.2 mm square and 100 m high? The brain has difficulty in dealing with exponential processes of this sort, seeming to imagine twenty folds and not 1 million, and getting the height totally wrong.

Working smarter not harder?

The implication of much IT advertising is that if you buy more computing power people will be able to compete more effectively whilst working no harder. In practice the effort saved is illusory, and the expectation is that this time is available for equally productive work. Some concern has already been raised about this. In Japan the phenomenon, called 'karoshi', has been known for some years, and in 2000 the first Japanese family was compensated for their son dying through overwork (*Financial Times*, London, 24/06/00). The problem is not restricted to Japan.

The solution is not easy to find since karoshi affects all walks of life. With IT systems in place, work time may be measured far more exactly than could ever previously be done, and everyone is encouraged to compete with the best, until they 'burn out'. Continuous change devalues experience and the opportunities for the older slower brain to rest awhile and let the young and thrusting forge ahead until they meet an obstacle that only the experienced have met are diminishing.

Data mining

Large organizations have decades of archive data, once simply kept for the historical record, for 'just in case'. With hindsight it is possible to see uses for this data that were not perceived at the time, or simply not possible at the time due to lack of number-crunching power. Data mining reuses all this old data to answer useful questions.

One very serious question was the discovery, on old tapes of meteorological records of the Antarctic bases, of baseline data from which the rate of development of the hole in the ozone layer could be established and its link to the growth in CFCs established. One interesting feature of this was the difficulty the scientists had in reading the tapes. These had been archived properly, in secure, dry conditions, but the tape readers had become obsolete and only a frantic search revealed the existence of a working model.

Engineering museums might be necessary as reservoirs of technological artefacts in the future.

Direct sales

Commercial organizations also have large repositories of information, and use it to target their customers with direct mail, from which they claim to get a good response despite the amount of such paper that is thrown away daily. (To put this in perspective, 3% is regarded as an excellent reply rate for direct mail.)

Big supermarket chains remain quite open about the value of their store cards as marketing tools. They could go much further and link the store card data with their sales data, building up a detailed profile that would permit personalized letters reminding about particular products that, for example, had not been bought for at least two weeks.

Consumer groups have declared this intrusive and some supermarkets have publicly denied considering doing it, or that their computer systems have such capacity at present. The Data Protection Act will also limit the ability to use such personal data.

Activity 7.1.2

Take some junk mail that you have received. Try to work out how they found your name and address. Did they use the electoral roll, or have you responded to some previous advertisement? Try adding a (false?) middle initial when completing unimportant forms, and then notice when it appears unexpectedly.

The legal framework

The law affects computing in several ways. It protects the buyer from shoddy goods; it protects the producer from being undercut by a copy from whose sales the development costs do not have to be reclaimed. It protects everyone from the misuse of personal data. It discourages those who would maliciously damage software systems.

The sale and supply of Goods Act 1994 and the Sale of Goods Act 1979

These laws state that items sold must be fit for their purpose as described or recommended by the seller, provided the customer is using it according to instructions. So it is possible to take that computer program back, even with its broken seal, and demand your money back if you were led to believe that it would do more than it did. It is difficult, however, since imperfect performance seems to be tolerated much more in the IT industry than in other industries, and proving that you did not take an illicit copy is impossible.

Intellectual property: overview

Legal protection of intellectual property, the right to exploit ideas that someone thought of first, has a long history. Charles Dickens campaigned for the first Copyright Act to protect his income as a writer. Patents originated in the 1623 English 'Statute of Monopolies' that gave inventors a monopoly on their ideas if they agreed to publish them. In the UK the Copyright, Patents and Registered Designs Act 1988 and the Trade Marks Act 1994 are most recent. In the US there is a Digital Millennium Copyright Act 1998 among others.

Copyright

Copyright originally applied to written materials only. The definition has been extended to cover recorded music and video, broadcast media and computer software. Essentially it is illegal to copy any of these unless the author's permission is given, but there are many exceptions and the length of time for their protection varies from 15 to 70 years.

Patent

Patents were originally designed to allow inventors to make money from their ideas by preventing illicit copying, with the additional benefit to society of having their details published and thus promoting innovation. They largely apply to innovatory things and processes that are useful in the real world. Some overlap with copyrights may occur in a product design, where a new software idea might be patented, and its documentation and 'look and feel' copyrighted.

Trade mark

Trade marks apply to the names, logos and other devices – such as color, sound and smell – that are used to identify the source of goods or services and distinguish them from their competition.

Application of IP laws

Application of these laws to IT and computing continues to require modifications to original definitions, and has become much more complex with the speed with which ideas can be spread across the world. In software the value of the merchandise is entirely in the thoughts of its creators and in what it can do, and many person-years of work can be copied in a few seconds. For this reason a lot of effort has been expended by FAST, the industry-sponsored Federation against Software Theft (website: http://www.fast.org.uk) to investigate and act against software piracy.

The laws also vary from country to country. US law often predominates because the US has the largest market, and currently dominates the Internet, and so IT users have to comply with US regulations.

Commercial pressure from major multinationals is slowly forcing countries to enact similar laws in order that, eventually, no one will be able to buy a $500 software package legally for $20 as has been possible.

Details of US law can be obtained from the US State Department's website, or one of many legal advice sites, e.g. http://www.nolo.com.

Possible abuses of intellectual property protection

Some trade marks are potentially problematical, e.g. Microsoft™ have trade marked Windows™. Fortunately commonsense prevents this becoming a problem for double-glazing firms.

More seriously, there is an increasing number of patents (applied for and even granted) which if enforceable could have far-reaching effects. In the computer-based learning domain, for example world patent WO9822864 has been granted:

> **COMPUTER EDUCATION AND EXECUTION INTERFACE OPERABLE OVER A NETWORK**
>
> A computerized, multimedia education and execution interface system and method for educating and entertaining a student user according to a teacher user's options. The system incorporates the education techniques of video segments, on-line student and/or teacher activity, written instructions, student's activity monitoring, student rewarding and education and entertainment program content filtering. The environment uses various input devices including a mouse and a keyboard. Compact disk drive with main memory and disk drive store the various programs and education data. An audio interface and video controller and screen with a central processing unit communication over a bus complete the system.

Such a broadly defined patent, if enforceable, would cover a huge slice of computer-based learning and make some corporations extremely wealthy. Its enforcement would enrich many lawyers!

Shareware and freeware

Authors who do not wish to market their software conventionally, either because their product is too similar to another, which is already well marketed, or because their target audience is too small/specialist to make marketing practicable, can use the shareware option. In this they retain copyright, but allow free copying of their software with the proviso that after a trial period a token sum is sent to the author, often in exchange for an up-to-date version or manuals. Though many people do not pay up, a sizeable income can be obtained from marketing a good product in this way.

Authors with altruistic tendencies can distribute their work as freeware, whereby they retain copyright but allow free use of the software. There is a whole sector of the computer industry involved in production of products for LINUX (the freeware version of UNIX originally developed by Linus Torvalds), and sharing ideas about development and bug elimination.

Shareware and freeware authors must be careful not to infringe the patents of others. Unisys, who currently own the patent for the type of data compression used in the popular GIF graphics files, now ask for a fee before allowing authors to use this type of file. The sum required is enough to deter anyone unsure of any profit, and has added impetus to the development of a freeware PNG portable network graphic standard.

It is quite possible to make money from freeware! Red-Hat and Caldera, among others, have built businesses by packaging up LINUX and its tools and applications, freely available from the Internet but distributed in many places, and selling them on as (relatively cheaper than on-line charges) CD-ROMs.

Activity 7.1.3

If you have a PC at home, assess how careful you are to use only software that you are entitled to use.

Data protection

The Data Protection Acts 1984 and 1998 protect people from the electronic equivalent of gossip, malicious and otherwise. Before 1984 it was possible to be told you were a 'bad risk' for a loan, and to be unable to do anything about it. Now those who have personal data on their systems have to reveal its existence by registering with the Data Protection Registrar, only to retain it for the purpose for which it was gathered, and protect it from illicit access. We have rights to find out whether data is held about us, what data is held, and to have it corrected where there are errors. Everyone should be aware of the eight basic principles (from the Data Protection Registrar's website, http://www.dpr.gov.uk), either as a data user or as a data subject:

> Anyone processing personal data must comply with the eight enforceable principles of good practice. They say that data must be:
>
> 1. fairly and lawfully processed;
> 2. processed for limited purposes;
> 3. adequate, relevant and not excessive;
> 4. accurate;
> 5. not kept longer than necessary;
> 6. processed in accordance with the data subject's rights;
> 7. secure;
> 8. not transferred to countries without adequate protection.

Personal data covers both facts and opinions about the individual. It also includes information regarding the intentions of the data controller towards the individual, although in some limited circumstances exemptions will apply. With processing, the definition is far wider than before. For example, it incorporates the concepts of 'obtaining', 'holding' and 'disclosing'.

The Computer Misuse Act 1990

The Computer Misuse Act was designed to discourage tampering with computer systems by hacking, either software or data, or by the writing of malicious programs, e.g. viruses. There are three types of offence:

(1) Unauthorized access to computer material (i.e. hacking and attempted hacking into a system).
(2) Unauthorized access with intent to commit or facilitate commission of further offences (the further offences are unspecified, but the penalty is higher than in 1).
(3) Unauthorized modification of computer material (this is interpreted to include both hacking in and changing something and the writing of viruses).

There have not been many prosecutions under the Act, since evidence is very hard to acquire and action is hard to take. It would be a brave bank that helped prosecution of a hacker for successfully breaching their security, since such revelations would seriously damage its own credibility. And though viruses regularly appear it is difficult to pursue someone living across the other side of the world.

Activity 7.1.4

Use an internet browser to access the Data Protection Registrar's site and find out more about the rights and responsibilities of data users and of data subjects under the Data Protection Act.

Health and safety

Organizations have a major responsibility when it comes to health and safety. Fortunately most computer systems are relatively safe (against electrocution, for example), but it is still possible to break the law with trailing power leads or jury-rigged network cables.

VDU radiation

Concern is still shown by some as to the effect of low-level radiation from computer monitors. No real proof has come to light. Unborn children were thought to be particularly at risk and pregnant women advised to work elsewhere. However, both mother and child are currently thought to be more at risk from restricted movement due to poor sitting arrangements.

Posture and general ambience

Working long hours sitting slumped over a computer terminal with poor lighting is the most likely cause of workplace related illness for users of IT. Regular eye tests are prescribed for employees working at a terminal for several hours per day. There are fairly stiff guidelines for positioning of monitor, keyboard and chair, and for window blinds and types of lighting, to try to eliminate the major effects of eyestrain, back problems and repetitive strain injury (RSI). RSI is a serious problem for those who must input large amounts of text or data. These guidelines should be followed, together with the advice to take regular breaks, if you have to take on such a task, even for a limited period. See the Health and Safety Executive's website at http://www.hse.gov.uk for more information.

Activity 7.1.5

Use the internet URL given to find out the rights and responsibilities under the Data Protection Act and the Health and Safety (Display Screen Equipment) Regulations 1992.

Problems 7.1.1

(1) What does Moore's law say about the growth in power of the microchip? Why does this make the effects of change in IT difficult to predict?
(2) Estimate the number of bytes required to store the text of an encyclopaedia (any large book will do – count the characters per line including spaces then multiply by the number of lines and the number of pages). How many such books would fit on a CD-ROM of 600 MB? How long would it take to transmit such a book across a network at 100 Mbit/s (roughly 10 Mbyte/s)?
(3) You have received a shareware program from a friend. Which laws protecting intellectual property affect your use of it?
(4) What is RSI? How can you protect yourself against its effects?

7.2 Using computer applications efficiently

Applications are the programs for which computers are bought, in order that letters can be written and edited, CAD drawings modified, databases accessed, etc. Every application has its own set of methods and processes, but there are many common ones that apply. You need a portfolio of skills in the use of these applications. You will find it most profitable to notice the features common to all (most?) of the applications (these, like cut and paste, only need learning once). Then get a grasp of the main processes in particular types of application (these, like mail merge in a word processor, differ in detail only with manufacturer and version, and a little investigation of a new package can usually make it work).

Application skills: to be able to:

- produce documents using a word processor or desk-top publisher;
- analyse data and model processes using a spreadsheet;
- organize and extract data efficiently using a database package;
- handle e-mail effectively;
- search the Internet and other databases effectively;
- obtain new skills using a computer-based learning package;
- present information to clients and the outside world.

This chapter is too short to teach you all these skills, and certainly not in detail, especially since the precise detail varies between packages, and also between versions. The aim is to outline what you should try out on the software that is available to you most abundantly whilst studying so that you can be confident when starting work. Above all you need understanding and an overall grasp. The detail of the software changes, from source to source, and from version to version. What you can do with it is much more constant.

Operating systems processes

The operating system is the software that runs the basic processes of the computer. It permits storage of computer data in named *files*, and allows those files to be stored systematically in *folders* (also known as directories) and sub-folders. It further allows such files to be loaded and saved, created and copied, moved and destroyed, printed and executed.

File naming conventions

Filename rules vary. Some early PC operating systems, e.g. MS-DOS, were limited to 8 letters, plus a 3-letter extension (the 8.3 convention, e.g. MYLETTER.DOC), but most allow much longer names, with longer extensions. It is even possible to have a space in a name, e.g. MY LETTER.DOC, though this is best avoided since it minimizes compatibility across platforms. Some computer users join words with an underscore as in My_letter.doc, or use the increasingly popular CapitaliseEachWord convention, producing MyLetter.doc. The extension, e.g. .doc, is often used by the operating system to link the file with the application that produced it, allowing it to be opened more easily.

Some operating systems, e.g. UNIX, are case sensitive, and will treat MyLetter.doc, myletter.doc, and Myletter.doc as three different names. You will have to learn when to be careful.

Graphical user interfaces

Operating systems processes are all defined so that applications programs, the ones you actually buy the computer to use, can carry out these basic processes in a straightforward fashion by using pre-defined routines. To provide some manual control there was generally a command-line interpreter so that such processes as copying a file (i.e. loading it and saving it somewhere else) could be carried out using a typed command. Most modern systems, e.g. Windows^{TM}, have added (or even replaced it with) a graphical user interface or GUI, where such processes can be carried out by moving a mouse to control a pointer on screen, and pushing icon buttons, or even by dragging and dropping the files where required. The UNIX/LINUX family of operating systems has its own X-Windows equivalent.

GUI processes are preferred to command-line processes by most users since they are found to be more intuitive and easier to operate. Occasionally still there are processes that can be carried out better by command line, particularly where a series of processes can be executed in sequence using a batch file.

Operating system skills: to be able to:

- safely start up the computer and shut it down;
- copy files, creating back-ups, or copies usable elsewhere;
- delete files, lest your disks become full of outdated work;
- reorganize files by moving and renaming files and folders/directories;
- write-protect files;
- find files in a typically tree-structured set of folders;
- execute files that are programs;
- install new software and remove old software from the hard disk safely.

Detailed knowledge helps, once basic principles are understood. For example, in most mouse-driven environments, the left mouse button picks objects, operates programs, and uses them to open files.

However, in Windows^{TM} you also find that:

- It is possible to swap left and right buttons, or change the sensitivity of the mouse.
- The right button gives access to properties of the object, allowing it to be made read-only, for example, or to actions related to the item letting it be copied easily.

- <SHIFT> with the right button may allow better control of those actions, e.g. opening a Word^{TM} document with a different package from the desktop.
- The <F1> key gives access to context-sensitive help.

Exploring the use of such options and processes can greatly increase your productivity (and hence salary).

Help!

Modern software generally comes with an on-line help facility that can be actioned simply with a keystroke or two. Help files are a preferred means of documenting features of the program, especially where it would take up an inordinate amount of paper.

Learning to find out what help is available, as documentation or as examples, is a skill that will enable software to be learned more quickly. Many computer users read the manual as a last resort. Use the help facility more frequently.

If you prefer books there are a plethora of new ones produced every time a popular package is launched or updated. You may find that even *The Complete Moron's Guide to X* has useful information for beginners. If you have to buy an expensive book, check out the 'sale' shelves for guides to a previous version; the basic software may not have changed much.

Networks

Most computer installations are networked, even if only by modem to the Internet, in order that outside resources, e.g. websites or software archives, can be accessed. Their use requires you to be able to connect to and disconnect from the resources you require, to understand and operate the password and other security requirements of the system, and to have some grasp of the complex data structures that a large network houses.

Network skills: to be able to:

- log on to the network;
- change the password where possibly compromised;
- understand how to get to and use your home workspace;
- understand how to get to and use any shared workspace;
- give others limited access to your files and directories;
- find and execute appropriate system and application software;
- log off the network.

Auxiliary hardware: peripherals, connections and storage

A peripheral is an extra piece of equipment attached to the computer and controlled by it, such as a printer or scanner. Network connections and disk storage systems used to be peripherals, but are now generally housed in the main processor box and considered as part of the main computer system (except for handheld computers, and this may yet change).

You need to know how to service these items, preparing them for use by replacing disks, paper, ink or leads, and cleaning them when required. When working with a complex system it is useful to know how to troubleshoot (and to know when to call in the experts!).

The mouse is a simple example, particularly when heavily used in a greasy or dusty environment, when the pointer action it produces becomes erratic. Cleaning the accumulated grime from its rollers by opening it up and swabbing them with a small amount of methylated spirit is not difficult. Nor is replacing it with another mouse costing a few pounds; the latter may be more cost-effective.

Hardware skills: to be able to:

- insert and extract removable data storage of whatever kind your system has, e.g. floppy disk, Zip drive, DVD;
- prepare floppy disks (and other magnetic media) for data storage by formatting;
- write-protect magnetic disks and tapes, and keep them away from damage, both mechanical (mainly dust and finger grease) and magnetic (usually electromagnetic fields such as those controlling a VDU);
- look after CD-ROM and DVD media, distinguishing the various types;
- replace paper (and disposable parts such as ink heads) in printers;
- check network, modem and sound cables (where appropriate) when they could be causing a malfunction.

When things go wrong

Fault-finding

Each part of the 'Standard' PC, the processor, the circuitry, the operating system, each peripheral, and each application program, has probably been worked on by thousands of people. It is a miracle that it works at all, and hardly surprising when things go wrong. When it does you have two choices – to send for the repairers or to fix it yourself. Either way you will save a lot of time by being able to describe what is wrong clearly, using correct terminology. The computer, just like, say, the motor car, gives off audible and visual signals as it works, and changes to these (particularly smells!) can be significant.

Fault finding skills: to be able to:

- name the major parts of your system;
- find a list of the manufacturer and model number of relevant parts;
- describe what happens to the computer as it 'boots up';
- find a record of occasionally used passwords in a safe place;
- describe what happens when applications are executed;
- recognize any changes in the usual sequence of operations, especially when these occur soon after changes to the system;
- use correctly the reset button and important combinations of keys like CTRL-ALT-DEL on a Windows™ system;
- react appropriately if the PC starts up in a default 'safe' mode.

Disaster recovery

Even though these may amount to several thousand pounds the most valuable part of many computer systems is neither the hardware nor the software but the data that it contains.

Consider the simple data input costs and how they mount up. A good typist might manage 50 words per minute, for 8 hours per day, for 250 days in a year. At an average of 6 letters per word this is 600 000 characters/bytes or 0.6 Mbytes. Now estimate salary and on-costs of £20 000 per annum (this is £10 per hour). Each £1 of salary buys 30 bytes of data.

So a 3.5-in floppy full of text data could hold about £40 000 of data, and a 600 Mbyte CD-ROM of documentation could cost £20 million just to type in! (For real data this would be an exaggeration, since so much is in graphical form, and a scanned photo or other graphic could easily take up 0.6 Mbytes, and take less than 5 minutes to produce. At £10 per hour this might cost less than £1 to produce, plus copyright costs of course.)

Of course the value of data stored, and of its organization into searchable and usable form, may be many times the cost of its production. When this is taken into account, the need to keep the data secure is clear. At least one large (computer!) company, and doubtless many small ones, has been brought to bankruptcy by the loss of its data in a fire.

Fire is reckoned to be the major cause of serious loss of data. Fireproof safes are available, designed to keep magnetic media protected for 30 minutes, giving time for data to be rescued.

Secure operation skills:

- retrieve originals of software in case of disaster;
- find relevant back-ups of data;
- restore software and data back-ups where practical.

Activity 7.1.6

Take out the mouse ball (of your *home* PC) and clean the rollers. Gently vacuum your keyboard to remove dust. Clean your VDU with a damp cloth.

Find the originals of all your software and store them carefully in a safe place. Make a back-up of all the important data files you have created, or at least, calculate how much storage space you would need to back them up and make a decision that you cannot afford it!

Problems 7.2.1

(1) What problems were there for users of the old 8.3 filename convention? Why does the use of spaces in long filenames sometimes produce problems? What alternatives are there to make long filenames easier to understand?

(2) List some of the advantages and disadvantages of using a networked computer. When is it essential to change your network password? What potential problems are there in keeping the same password for a long time?

(3) List some of the problems that dust can cause in a computer system. What practical steps can be taken to reduce the potential damage to the computer?

(4) What is involved in a disaster recovery plan? List the software and the data that you would need to install if your computer had to be replaced after a fire.

7.3 Using the major office applications

There is a wide range of applications available for computer users, ranging from the general word processor to specific industry packages, e.g. lighting design.

Initially, each application, being written by a different team with their own perspectives, had its own set of keystrokes and mouse actions to drive it, often linked to the computer operating system on which it was first produced. With the increasing predominance of a small number of operating systems, most notably the Windows™ GUI, a standard set of processes has been enforced, both across different vendors of the same type of package, and across different types of package. The preponderance of Microsoft Office in the PC world is due to its having a headstart in this, since they also owned the operating system.

General application skills

There are some processes, often related to the GUI, that are common to most applications. Practise them in each! This will enable you to learn a new package more quickly, and, generally, to integrate your usage of the packages. For example, in Windows™ text can be moved inside a document by:

- highlighting it (pointing to the start with the mouse, holding the left button down and wiping over the chosen area);
- cutting it (holding ⟨CTRL⟩ key down whilst tapping the 'X' key;
- moving the mouse to the required destination; and
- pasting (holding down ⟨CTRL⟩ and tapping the 'V' key).

(Note: there are alternative ways of carrying out each operation in Windows™, and similar techniques in every other operating system.)

Once you have learned to copy and paste inside Word™, say, you can copy inside WordPerfect™, or from Word™ to WordPerfect™, or from Notepad to Aldus PageMaker, or from Paradox to Lotus 123, etc. The result is sometimes different from what you might expect, but there are generally alternative options if you do not agree with what the programmer thought was the 'obvious' outcome.

Graphical user interface skills: to be able to:

- resize, minimize, maximize and close a window;
- move using Cut and Paste;
- copy using Copy and Paste;
- move or copy using Drag and Drop with ⟨CRTL⟩ and ⟨SHIFT⟩ to force Copy and move;
- highlight using ⟨SHIFT⟩ and cursor keys, home and end;
- swap between Insert and Overtype mode using the ⟨Insert⟩ key;
- use Save As to make a second copy of work done for back-up;
- use keyboard shortcuts rather than mouse processes when these are more efficient.

Document production technology

In the beginning was the text processor, devised by programmers to enable them to code their programs, and as a spin-off to write and (above all) to edit text documents. Then someone invented word wrapping and the word processor was born. Some time later the desk-top publisher appeared, as a means of taking the relatively crude text from a WP and merging it with pictures, to produce, for example, a newsletter or brochure, and making a lot of skilled typesetters redundant at the same time.

Currently standard word processors can do anything a desk-top publisher package could five years ago, and have largely usurped their old market in small business advertising. Desk-top publisher packages have become specialist publishing software handling colour separation (for printing with yellow, blue, magenta and black ink), image editing, chapters, indexes and any other publishing features that would slow down a standard word processor if it were included. The text of this book was produced using word processors. The vast majority of the layout was decided by the publisher, who was supplied the content as plain text.

Word processing

However, until you begin to publish commercially, your documents will normally be produced using a word processor. Since you may be using it (or its derivative – the e-mail editor) rather a lot, it is worth getting to know its main features, because you will want to use bold type, different fonts, and many of the other features available. If you do not, your documents will be outshone by more eye-catching competition. You will certainly have to comply with (company) document standards (heading position, fonts, etc.). You will also want to ensure that you do not lose work and that you can keep track of changes.

Elementary WP skills: data input and editing

There are some skills you probably already have or can pick up easily once you can activate your chosen word processor. These allow you to produce a basic document or message, and are similar to those required for e-mail messaging.

Basic WP skills: to be able to:

- begin a new file with a pre-defined template, and to use the template efficiently;
- save and then reopen the file in order to continue editing;
- type as well as possible. There are many typing tutors available to enable you to practise. Speech recognition systems that do not require any typing skills are yet to be developed;
- copy, move and delete blocks of text;
- embolden, underline and italicize text;
- activate the spell-checking system;
- to print the document.

Intermediate WP skills: formatting your document

There are many examples of inferior products winning the battle for market share by better marketing. Some would argue that the PC is a case in point. Similarly ideas which are more smartly expressed often win out too; as a sixties guru, McLuhan, said 'It is the medium that matters, not the message.' An attractively laid out document is more likely to be read. Only the really dedicated, such as those marking exam scripts, feel the need to read scrappy documents.

What makes a document 'attractive'

Over the years some basic principles have emerged. There is no reason why you should not break these rules; indeed slavish adherence to any system leaves the way open for the iconoclast to attract attention by being perverse. But at least be aware of why they are there, so that you break them for a purpose.

- Keep the number of fonts (the word most programs use for typefaces) on a page to a minimum to avoid the 'busy' or 'muddled' look. The same typeface may be of different sizes, underlined, italicized, etc. Serif fonts (with tails, like Times Roman) allow for easy reading and are useful for body text. Sans-serif fonts (without tails, like Arial or Helvetica) are believed to slow the reader. These are best for headings (unless you want your reader to read more slowly of course).
- Typescript is more difficult to read than text printed with proportional fonts. Reserve fixed-width fonts like Courier only for data that has been structured by spaces, e.g. directory listings. Computer programs are conventionally laid out in fixed-width fonts.
- Use the Styles feature of your WP to control the appearance of your text. Then if you wish you can modify the appearance of each paragraph or of each type of heading in one hit. When you have hit upon a suitable set of styles, create your own template so you can use it in further documents.
- Use plenty of white space. It may be environmentally more friendly to pack the words on the page, but it can overface the reader. Break your paragraphs up with subheadings summarizing their contents to improve readability. Headings should be separated from their paragraphs by half a line, and blocked paragraphs (i.e. with no indent on the first line – normal in reports) separated by a full line. Headings of most formal documents need numbering in some systematic fashion.
- Size matters! 10 point text (a point is 1/72 of an inch – an old printing measure) is as small as you should use in multi-page documents for readability. 12 pt is easier for most people to read.
- Colour is excellent for emphasis, decoration and graphics, provided you can print it. But colour photocopying is not cheap, and if your document is copied in black and white your careful emphasis will be replaced by greying out!
- Columns may be useful in newsletters. Newspaper columns are designed so the reader can read a line with minimal eye movement and hence can scan quickly down the page.
- For structured data either:
  - use the tab key and left, right, centre and decimal tabs set in the ruler to line up the text at absolute positions, or
  - use the very flexible table feature of your word processor (turn off the borders if necessary). In tables, individual cells, columns and rows can be formatted separately to give the desired effect.

Intermediate WP skills: to be able to:

- change font and font size, the paragraph alignment and line spacing, etc.;
- set up styles for headings, paragraphs and other items to simplify formatting;
- number headings and paragraphs;
- use columns to make long narrative sections of text readable;
- format structured data in tables;
- create specific document templates.

More intermediate WP skills: including graphics

Many documents include graphics, either as decoration or to display information in an easily accessible form. It is useful to know how to insert the graphic, perhaps by cutting and pasting it from elsewhere. It may need to be cropped, to remove extraneous detail, and then must be positioned at an appropriate point in the narrative, with text perhaps flowing round it. It will need to be labelled so that it may be referred to.

Activity 7.3.1

Add a graphic to a block of text. Add a caption. Change the wrapping options so that the text flows round the graphic. Move the graphic right and left and decide its best position. Try the graphic boxed and unboxed.

Advanced WP skills: document automation and management

The beauty of preparing documents on a computer is not speed – most could be produced faster by hand – but in their editing and reuse. A document may have a table of contents and index added automatically. It may then be reformatted with a different number of columns, different size of paper and font, and yet have all the page references changed correctly.

It may be rehashed into a handbook, developed into a presentation or published as a web page. It may even be automatically translated (not perfectly but intelligibly) into a foreign language.

A document may have to be saved in a different format, ready for a different processor, even in text-only format. If available in text-only format some judicious use of Search and Replace facilities may be needed before paragraphs wrap correctly.

Mail merging is another example of automation, linking one document to another. A letter is linked to an external data file, and instead of actual names and addresses, a reference to different parts of the name and address is placed appropriately. Then when the merge process is executed, a series of letters is printed, one to each recipient. A merged document need never actually exist.

This produces an interesting problem for those who keep records of *everything*. How is the creation of such a letter recorded? If the source letter and/or database are subsequently modified for another letter, the only possible record is a merged file, which is subsequently archived.

Modifications to files at least may be recorded. Where a source file is edited, perhaps by more than one person, a modern word processor may be set up to highlight the additions and deletions in different colours, at least until the latest version is agreed and the highlights then removed. The problem of keeping records of files in various stages of development has to be solved by archiving them to some permanent storage periodically. This may be done as an addition to the back-up process, as described later.

Advanced WP skills: to be able to:

- import a text file and reformat with unnecessary paragraph marker's removed and headings correct;
- create a table of contents using suitably defined heading styles;
- create an index from labelled words in the text;
- set up a mail merge letter and data file and execute the merge;
- use suitable file developmental stage labelling as a document evolves.

Activity 7.3.2

Set up a personal contacts database (see 'Databases', page 337) and then create a mail merge letter, link to the database and merge to a file (you do not have to print!).

Financial and numerical technologies

Some say that it was the spreadsheet, first VisiCalc on the Apple II and then Lotus 1–2–3 on the first IBM personal computers, that really sold the PC to business. Prior to that keyboards were nasty things that typists used, and managers didn't want to handle. Suddenly it was possible to set up whole business scenarios like 'What would happen to our current situation if we reduced our spending on computers?' in a couple of hours, rather than go begging the data processing department for the use of one of their programmers. (Doubtless the possibility of doing it without anyone else having to know also appealed.)

Spreadsheets are the all-purpose tool for setting up numerical models to answer such questions (and many other non-financial queries as well). There is also a range of specialist software available, particularly for accounts. Each of these has their place in the modern business.

Spreadsheet introduction

A spreadsheet is a rectangular array of boxes or cells into which numbers and letters can be placed. It is initially two dimensional, rather like the layout of a book-keeping page. These may be linked to make a workbook.

The columns are labelled with letters, from A to Z, AA to AZ, BA to BZ as far as IV. Rows are labelled with a number, from 1 to 800 or more. Each cell has its own name or cell reference, made from the column letter or letters, and the row number, for example, A1, B23, BZ235.

From one set of data it is possible to produce a second that depends on it by known factors. Hence it is possible to use this year's figures, with an estimate of possible inflation factors, to predict likely future performance.

As well as giving the possibility of setting up arrays of figures and providing rapid totalling of the rows and columns, spreadsheets have facilities for simple production of graphics from the data.

Column → *Row No.*	*A*	*B*	*C*	*E*	*F*	*G*
1						
2		Year 2000	Jan	Feb	Mar	
3		Gas	37.32	29.24	19.23	
4		Water	56.23	29.88	32.73	
5		Electric	17.23	18.34	22.42	
6		Rent	75.00	75.00	75.00	
Using the SUM() function, automatic totalling can be added. In the row below, C7 contains =SUM(C3:C6), and F6 contains =SUM(C7:F3)						
1						
2		Year 2000	Jan	Feb	Mar	Total
3		Gas	37.32	29.24	19.23	85.79
4		Water	56.23	29.88	32.73	118.84
5		Electric	17.23	18.34	22.42	57.99
6		Rent	75.00	75.00	75.00	225.00
7		Total	185.78	152.46	149.38	487.62
Inflation may be allowed for in the coming year by making a copy of the table in rows 12 onwards, modifying the labels, and adding a cell with the relevant inflation value.						
Replacing the 37.32 in C14 by =C3*(1 + C12) puts the estimated new value in C14.						
Copying this formula to each data cell (not SUM cells!) gives a block of estimates. These may be replaced by actual values when the time comes.						
12		Inflation Value	10%			
13		Year 2001	Jan	Feb	Mar	Total
14		Gas	41.05	32.16	21.15	94.37
15		Water	61.85	32.87	36.00	130.72
16		Electric	18.95	20.17	24.66	63.79
17		Rent	82.50	82.50	82.50	247.50
18		Total	204.36	167.71	164.32	536.38
19						

Figure 7.3.1 *A simple spreadsheet example*

Cell entries

From Figure 7.3.1 we can see that a cell in a worksheet can store one of the following:

- Text, useful for labelling cells.
- A number or value stored typically to 18 decimal place accuracy.
- A cell reference so that B2 might contain =C2 so when C2 changes so will B2.
- A function, often referring to a cell reference, e.g. SUM(B2:B5), adding up all the values in cells B2, B3, B4, and B5.

Spreadsheet general principles

Distinguish between the data and its appearance. For example, the number 36 526.6 may appear as £36,526.60, the date 01-Jan-2000 (using just the whole number), the time 14:24:00 (using only the fraction 0.6 of a day), as a percentage 3 652 660.00% or even as an improper fraction 36 526 3/5. These are interchangeable by reformatting the underlying number.

Do not leave unnecessary gaps between rows or columns of data. Add any extra rows and columns, to separate data and make the printout clearer, last of all. Extra rows often get in the way of efficient formula copying, among other irritations. Extra columns, added simply to separate data, are unnecessary; widen the columns instead. It is relatively simple to change fonts, add borders, and to shadow areas of the sheet for extra clarity.

Activity 7.3.3

Work through the example given in the Table in Figure 7.3.1. Add a table of inflation values, one for each expenditure item, so they can be changed independently. Add a third year and further predictions.

Spreadsheet graphics

Apart from simple number handling, producing row and column totals with more reliability than most humans, spreadsheets are often used for production of simple graphs, particularly for display purposes. The days of hand-drawn graphs on displays are long gone.

The range of options is enormous, so it is important to have some clear principles in mind. You can break them, just as in word processing, but have a good reason.

(1) Colour aids clarity wonderfully on screen. Ensure that it will print and copy well when that is necessary.
(2) Graphs should be clear and uncluttered. 2D graphs are generally more easily interpreted than 3D graphs, particularly if forward-placed data series obscure others. Resist the temptation to add unnecessary detail such as picture backgrounds unless you are deliberately trying to mislead your audience.
(3) Pie charts are simple statements of different contributions to a whole, usually 100%. Too many slices confuse, so some may have

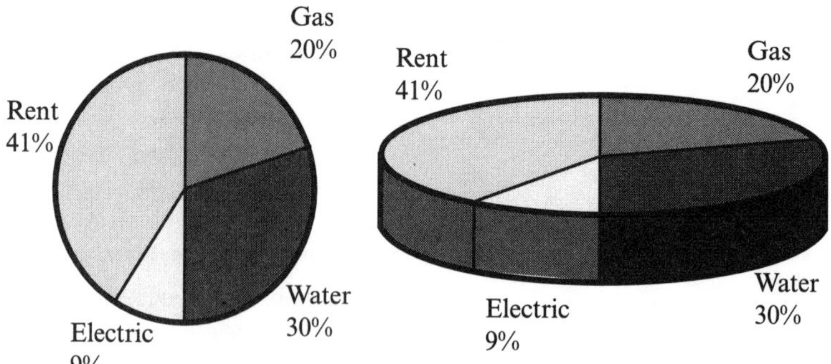

Figure 7.3.2 *3D and 2D pie charts: compare the representation of the same data*

to be aggregated into 'the rest'. 3D pie charts may make it difficult for those without good perspective awareness to compare areas.

(4) Line charts and bar/column charts are used to show change in a variable when a category changes, e.g. salary against employee, pollution level against town. When a line chart is used for X–Y

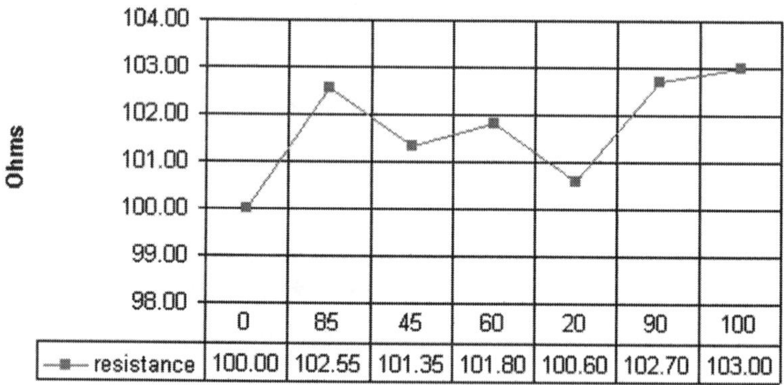

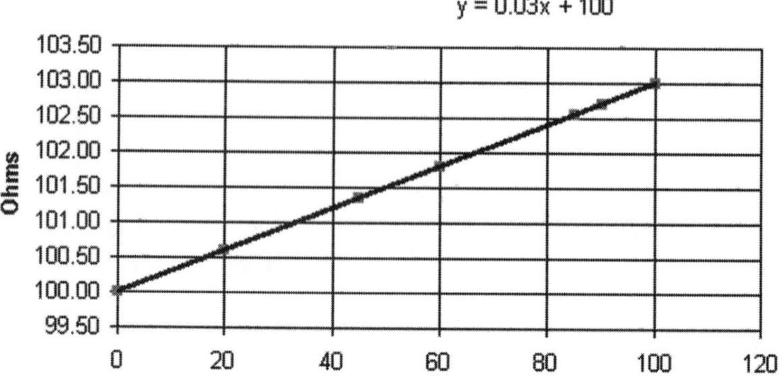

Figure 7.3.3 *Line and XY charts of unsorted data. Notice that the relationship is lost on the line chart, and X values are irregular and not on gridlines*

data the X values, however irregular and in whatever order, are drawn equally spaced; the Y axis does not pass through X = 0.

(5) Bar/column charts tend to be used for items that mount up, especially for discrete data that counts up, e.g. number of days of sunshine, salary, rainfall. Line charts are more useful for data that is continuously variable, like temperature.

(6) Where continuous data is grouped (it may be discrete data measured in tiny steps) and plotted, e.g. number of houses in different price ranges, the gaps in the resultant bar chart may be reduced to zero producing a histogram.

(7) XY chart data may need to be sorted in ascending order of X. Alternatively, if available, plot the points and get the package to plot its best-fit line (and give the equation if Excel 97 or similar). X–Y chart axes have lots of options, including reverse plotting and logarithmic scales. Use them appropriately. Beware of heavy gridlines. At graph-paper densities you will see nothing else.

(8) There are many specialist graph options, such as radar charts. Make sure you know when it is appropriate to use them.

Activity 7.3.4

Use a spreadsheet to create suitable types of graphs needed for your course. Explore the options available for titles, labels and the axis scale. Remove any backgrounds and cut and paste to a word-processed document.

Back calculation: using goal seek to find the answer

One of the most useful features of a spreadsheet is its ability to back calculate. Given that $y = x^2 + 3.5x - 7$ it is not difficult to work out a value for y, given values of x. Much more problematical is finding both x values given a y value. Plotting the graph, we find:

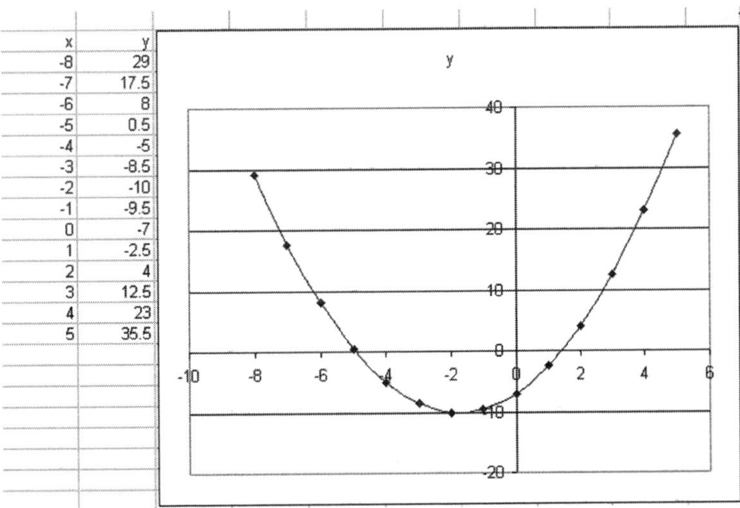

x	y
-8	29
-7	17.5
-6	8
-5	0.5
-4	-5
-3	-8.5
-2	-10
-1	-9.5
0	-7
1	-2.5
2	4
3	12.5
4	23
5	35.5

Figure 7.3.4 *The quadratic equation: values and graph*

To find the solutions of the equation, where $y = 0$ the Tools>Goal Seek function is used on the contents of cells B8 and A8. (Other values could be used but close values are safest – there are two crossing points):

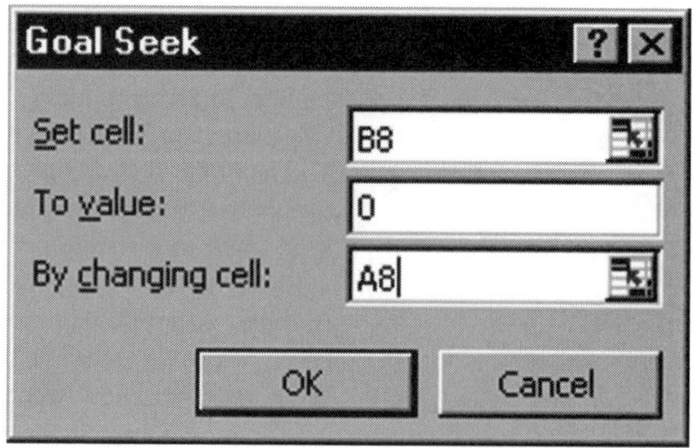

Figure 7.3.5 *Finding the first crossing point*

and the value is given as:

x	y
−4.922 15	9.32E−06

Let us assume that 0.000 009 32 is close enough to zero!

Similarly we can find the other crossing point by goal seeking on A14 and B14, say:

x	y
1.422 038	−0.000 67

This time y is negative and a little bigger in size but probably good enough

An even more powerful tool exists. Remember ploughing through all that calculus, simply to find a minimum value. Tools>Solver will find that too:

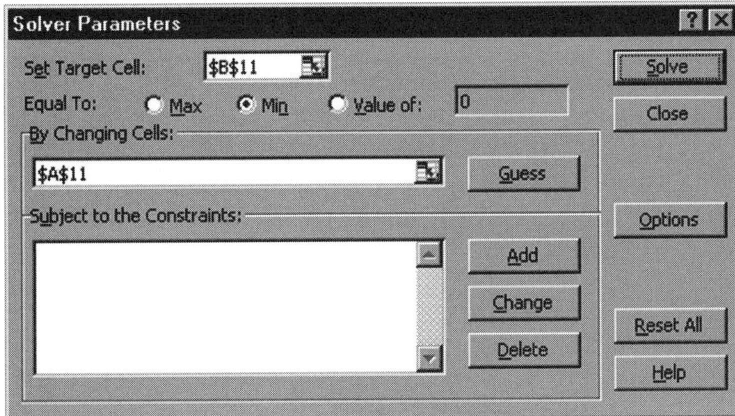

Figure 7.3.6 *Finding the minimum value, with Excel's Solver*

Pressing the Solve button gives the answer that y has a minimum of –100 625 when $x = -1.75$.

x	y
–1.75	–10.0625

Activity 7.3.5

The height of an object projected vertically upwards at 23 m/s under gravity is given by the formula:

$$y = 23t - \tfrac{1}{2}(9.81/t^2)$$

Set up a column of values for t, and a column of values of y, calculated from t. Produce an X–Y graph – it should be a parabola. Now use Solver to find the maximum height reached (y is maximum), and the time it takes to get back down to the ground ($y = 0$).

Databases

Whereas probably more than half the PC screens in use at any one time are showing a document processor, most overall computing time is devoted to database handling. Huge databases of personal and stock-control information are held by the government, utilities, banks, supermarkets, etc. Most engineering and business catalogues come in the form of CD-ROM databases. And then there is the Internet.

An engineer should have some database searching skills for use in, for example, libraries or CD-ROM catalogues. Since most records systems are becoming electronic it is helpful to know some of the basics of setting up a database so that the importance of keeping data in a structured fashion is appreciated.

Database basics

Data is stored as bytes, 8 bits = 1 byte. Modern computers handle data in 1-, 2-, 4- or 8-byte chunks, 8-bit, 16-bit, 32-bit or 64-bit data.

No of bits	*No of bit patterns*	*May represent*
8	$2^8 = 256$	Integers 0 to 255, characters from an enhanced ASCII set, numbers as text, especially when communicating data.
16	$2^{16} = 65\,536$	Integers 0 to 65 535, or –32 768 to +32 767, Unicode characters.
32	$2^{32} = 4\,294\,967\,296$	Integers from 0 to 4000 million or –2000 million to + 2000 million.
64	2^64 = 18 million, million, million	Integers or Floating point numbers between
		10^{10000} and $10^{-10\,000}$, with an accuracy of 1 part in 200 million, million. (Floating point numbers may represent dates and times.)
Note: ASCII, the American Standard Code for Information Interchange, is the default PC mapping of patterns of 7 bits to American English characters. Enhanced ASCII adds a bit and maps a further 128. Unicode can cope with all the alphabetical characters plus Chinese/Japanese/Korean ideograms.		
Pictures, video and sound are constructed from large groups of bytes, many millions in the case of video, called binary large objects, or BLOBS! Database tools to handle these fully, e.g. find a picture within a picture, are still being developed.		

Figure 7.3.7 *Data storage options*

Creating an example database

Databases are built from tables, each of which should describe some individual type of entity, e.g. a person, an invoice, an item of stock, etc. The exercise below takes you through some of the decisions that have to be made in order to create one table in that database.

Suppose a small company wants to set up a stock database. Each item of stock is an example of the stock entity.

The stock entity has attributes that describe it. Notice that it is a bad habit to have spaces in attribute names since, as with filenames, some software treats the space as an end-marker. Notice also the beginnings of an extended database. The supplier is just given a reference to supplier table; this will eliminate address space and, more important, the need to update many records should details like phone numbers change.

Attribute	*Content*	*Length*	*Problem/opportunity*
stock_id	letters and numbers, e.g. 324MPR-9	8 bytes, text	a unique value for each item. It may contain a fixed pattern of 3 numbers, 3 letters, dash, 1 number, in which case an 'input mask' will reduce errors
warehouse_position	e.g. R04C07L02-	9 bytes, text	-row 4, column 7, level 2. This 'coded data' minimizes the space used
name	a name of fixed length	e.g. 20 bytes	if such a length is fixed, short names will waste space. Fixed lengths do make searching quicker
description	very variable, may include data table		need to use a database that can handle variable length fields, such as an MS Access 'memo' field, or restrict the contents
supplier	name and address	8 bytes, text	a full name and address would be repeated for each item supplied. Instead use a supplier_code and separate linked table for a supplier entity
number in stock	integer	4 bytes	not negative. A short integer of 2 bytes might be used but searching may actually be slower
reorder_level	integer (whole number)	4 bytes	not negative
last_date_used	date	4 bytes	could be entered automatically when any change to the database takes place

Figure 7.3.8 *Stock data outline description*

Designing the table

The next stage is to specify the table contents in the package you are using. This is very package specific. Opening an MS Access database and designing a new table produces screens like Figures 7.3.9 and 7.3.10.

In Figure 7.3.9 the field names have been inserted, and the stock_id attribute put in the first field. The key button makes it the primary key

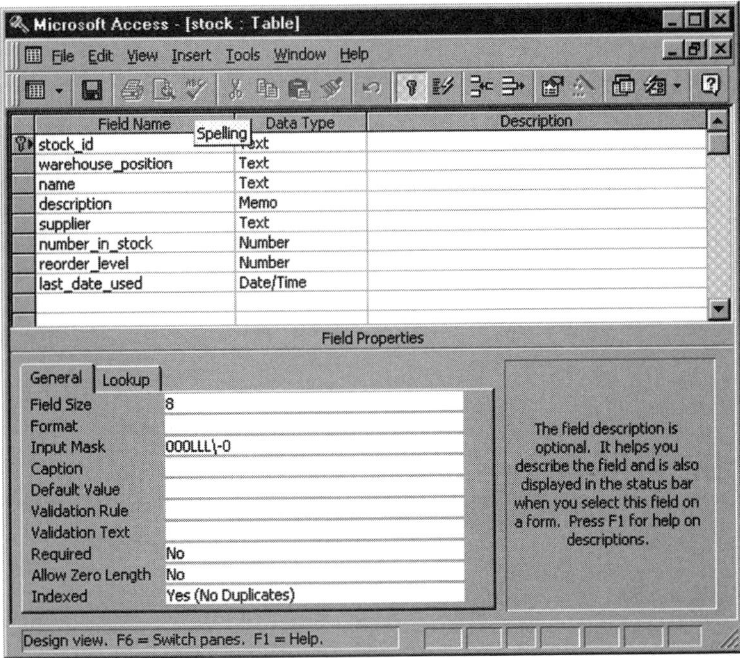

Figure 7.3.9 *MS Access table design, defining the Primary Key*

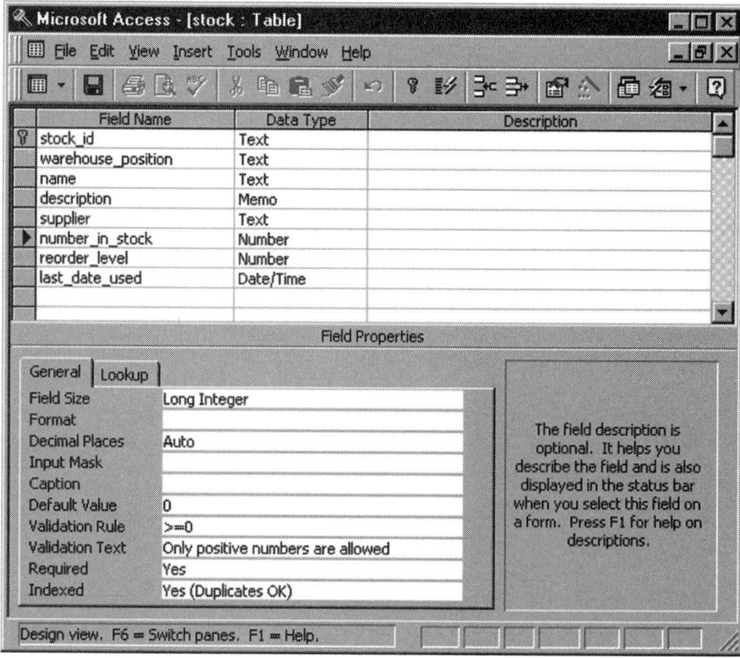

Figure 7.3.10 *MS Access defining a number field with validation rule*

field that defines the order in which the data will be stored. The field size is set to 8, and the mask set to be 000LLL\-0, which means 3 numbers, 3 letters, a dash and 1 number, all compulsory.

In Figure 7.3.10 the number_in_stock field is being defined. The field size is Long Integer (4 bytes for Access) and the validation rule prevents the error of the value entered being negative.

Data input forms

Having defined the fields the data is inserted. The simplest way is to create an input form for the table (Figure 7.3.11).

Most database software has tools to help speed frequent processes. This form was generated automatically from the stock table. It can be edited, and the order of the data input changed.

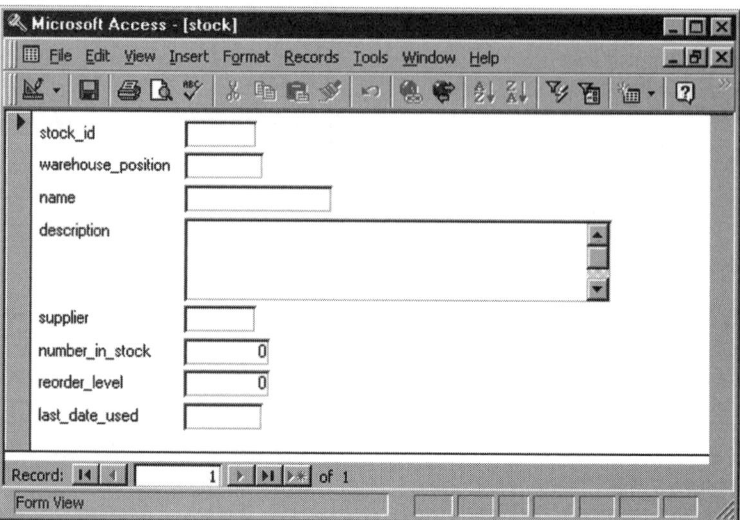

Figure 7.3.11 *MS Access: automatically generated input form, blank*

stock

stock_id 123ABC-9
warehouse_position R04C12L02
name Ring Spanner 10 mm
description 10 mm AF Ring Spanner, Steel
supplier PQ123456
number_in_stock -5
reorder_level 0
last_date_used 12/12/99

Microsoft Access
Only positive numbers are allowed
OK Help

Record: 1 of 1

Figure 7.3.12 *MS Access: data input is checked by validation rule*

The form adheres to rules already defined in the table. In Figure 7.3.12 an attempt was made to input a negative number_in_stock, and a pop-up warning was produced.

Querying a database

The simplest way to query a database is to use QBE, query by example. An MS Access query shown in Figure 7.3.13 will give the data in Figure 7.3.14.

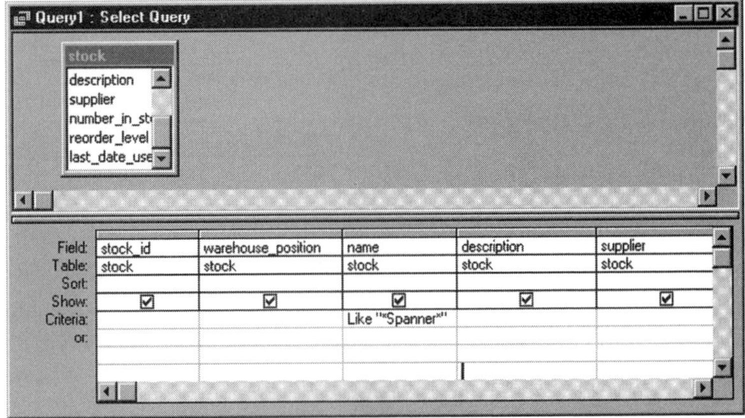

Figure 7.3.13 *MS Access query: find a spanner*

Query1 : Select Query

stock_id	warehouse_p	name	description	supplier
123ABC-9	R04C12L02	Ring Spanner 10 mm	10 mm AF Ring Spanner, Steel	PQ123456
134BGH-9	R02C05L01	C Spanner, 25mm	one inch C Spanner, Steel	AR67897

Record: 3 of 3

Figure 7.3.14 *MS Access query: the spanner is found*

Reports from the data

The object in inputting data is to be able to retrieve it in a more organized form, usually called 'reports'. Access will produce a report from a table, or from a query on the table (see Figure 7.3.15).

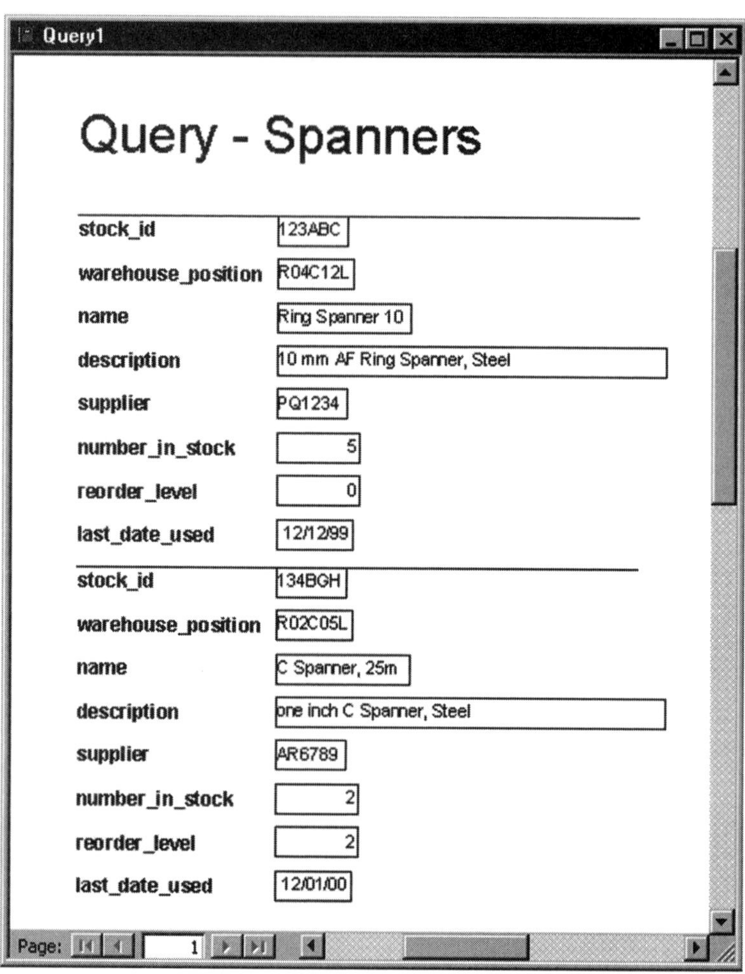

Figure 7.3.15 *MS Access report*

Database conclusions

Databases can become very complex, with hundreds of tables spread across many computers and many countries. A small-scale PC database can, however, handle the stock requirements for a small business, and its use improves skills that may then be used in the extraction of data from larger systems.

Crucial concepts are:

- GIGO: garbage in, garbage out. If inaccurate data is input an excellently presented report is useless.
- Input validation: Trying to prevent inaccurate input by automatic checking of values by rules.
- Refining queries: Trying to be precise about what is wanted (see Finding information on the Internet, page 351).

Activity 7.3.6

Use a database application to create a list of personal contacts, addresses, phone numbers, date-of-birth, etc. Use it in the mail merge activity (see page 331).

Personal information managers (PIMs)

PIMs are specialist databases used to keep track of diaries, contacts, etc. When networked they enable diaries to be shared, times of common meetings to be inserted, and other information to be shared around colleagues in a business. Such information may be kept on portable systems as small as a palm top or Psion and synchronized with a large central database at regular intervals.

Synchronization of data between databases

Where multiple database systems carry the same information there is potential for confusion. In database design a major principle is that any given data item should only be kept in one place. This is part of the process of normalization of a relational database.

However, duplication is inevitable where a company representative carries information away on a portable system, whether it be as contact or sales information or just a personal diary. If as a result of visits the portable data is changed, there will obviously be a difference in the data held. On return to the company and plugging in to the central system, synchronization software must add new information in either direction, and where discrepancies occur (e.g. representative and boss schedule a new appointment at the same date and time), resolve them.

Presentations

The cartoon Dilbert has many jokes about 'Death by Powerpoint'. Despite their overuse, and particularly the overuse of some of the facilities, they are very valid tools for engineers, both in providing rolling displays for exhibitions, and in providing focus for lectures/seminars, etc. Presentations are easy to put together and provided you can guarantee a suitable projector or screen, easy to carry on a floppy disk.

The basis of a presentation is a series of slides that can be sequenced by timing, use of buttons or both. Each slide has:

- a background which needs to be chosen to suit the subject;
- text;
- images which may or may not have their own background;
- sound/video attached;
- printable summaries or slide contents.

The main pitfalls are:

- 'pretty' backgrounds that obscure the text/images. Backgrounds are rarely the point of the presentation;
- excessive amounts of text. These are bullet points. Your talk is much fuller;
- images that are overcomplex. They need to illustrate a single message;
- sound that is intrusive. That wake-up call may be effective first time. After that it becomes annoying;
- video may be better projected separately, rather than in a reduced-size window;
- animation, particularly of text bullet points. Though text can fly in from all directions with a screaming sound does not mean that it should;
- slide printouts. These are great if given out first, since the audience can use them as a basis for notes (assuming that they are legible with a contrast between back- and foreground). If given out as a summary they are probably inadequate, and a two-page summary sheet would be far better.

Activity 7.3.7

Produce a presentation next time you have to share your results orally with the rest of your group. Quite apart from content, get feedback about your use of the features available. Were there any that they found detracted from, rather than enhanced, their understanding?

Problems 7.3.1

(1) What features of a word processor allow you to take plain text and improve it to produce a readable document? List some of the pitfalls of overusing some of these features.

(2) A company can either invest £1000 capital and make objects at £5 each, or invest £5000 and make them at £3 each. By setting up a table of differences between total costs for different numbers of items using the two processes, work out using Goal Seek at what number of items it becomes cheaper to make the greater capital investment.

(3) Estimate the number of bytes required to record the information given when applying for a driving licence. By multiplying by the number of drivers on the country's roads (estimate it if necessary) work out how much storage would be required for the drivers' data file.

(4) List some of the problems likely to be encountered when a sales representative returns to head office and attempts to synchronize the data on his or her laptop with that on the main office machine.

7.4 Industry specific applications

CAD

Every sector of engineering involves the design of new products and has its own computer aided design applications, capable of providing different sorts of drawing models. Some generic packages, e.g. AutoCAD, provide a general drawing environment which may be enhanced with add-ins, for example their own architectural engineering and construction extension that will draw images of 3D walls, insert windows, doors and winding staircases, and give perspective views of what has been produced. Companies like HevaComp have converted their own package so that it works in conjunction with AutoCAD, since they can then set up the required lighting and other services.

There are electronic add-ins to these generic packages, but the pressure to produce a circuit that can be tested electronically (for example, be seen to oscillate or amplify) as well as be converted into a printed circuit board layout, results in a wider range of specific products.

The trends in the CAD world are to:

- draw in 3D if perspective/isometric views are required. This means that there is only one model of the item, rather than plan, front elevation, side elevation and third angle projection to be changed when modifications are made;
- work with libraries of pre-produced, editable and reusable components to save costs. A manufacturer may provide a library of their accessories ready for insertion into a product drawing (thereby locking them in as supplier, of course);
- design not just draw. It is increasingly possible to ask the questions 'Will that type of cooker fit here?', 'How would its temperature change if it was painted black?', and 'Would the circuit performance change if I used a smaller battery here?'
- to be able to interchange data with others electronically, either by standardization on particular packages, or by using some of the drawing conversion processes that are available.

Project management

Good project management is vital if a project is to succeed, and most developments these days are in project form, where a team of people with different backgrounds and skills is put together to satisfy a need. All the project management tools, such as arrow diagrams, PERT diagrams, Gannt charts, etc. are available to display the plan for the project, and to monitor its progress. The computer can track efficiently the usage of each resource (person, tool, and raw material) and measure the effectiveness of the various processes. Resource levelling may be provided in which checks are made (for example, that the plan needs three mechanical engineers at this time and only two are employed; therefore either the job must be rephased or a temporary person appointed). Project management is dealt with more fully in Chapter 4.

Simulation software

As already stated under CAD, computer models can now be used to test circuits before they are built. This process of simulation depends upon accurate knowledge of the behaviour of components and accurate calculations, together with a great deal of testing of computer models against known reality. Once the system has been created and tested adequately, however, it permits expensive systems to be tested to destruction, even nuclear weapons to be designed without explosions. Similarly aircraft pilots can be trained to fly a plane before it is off the ground, and to land many times in hazardous conditions before it is necessary to do so.

Computer-based learning (CBL)

The provision of learning opportunities is expensive. Setting up large study groups in which each student gets the same delivery can minimize its costs. Particularly in specialist subjects this conventionally implies gathering from a large region at a particular time which is not always practicable.

The 'solution' is to put the delivery on a computer – CD-ROM, Internet, etc. – and deliver it to the student in their own time, with some access to tutors but most interaction using computerized systems, e.g. automatic testing and e-mail. There are many people working on providing such CBL, and there is evidence of its working effectively in some areas, particularly where companies want their staff (who will get a pay rise if they succeed) trained in the use of a specific product. But most of the 'Let's teach this by computer' initiatives founder on the lack of motivation of the students. All the enthusiasm the CBL people see and wish to harness in a crowd of disaffected adolescents playing 'Quake' evaporates when they have the option of 'Basic Maths' or 'Solving Kirchhoff's Equations' tuition.

Students of engineering are assumed to be motivated. They also come in relatively small, expensive classes. They may well find their time increasingly being devoted to CBL packages. Their use to replace or supplement lectures is sometimes understandable, particularly where students have difficulties with time and place. Their use to replace hands-on use of equipment is a particular problem if it removes the chance to carry out real operations. An analogy is driving tuition; you may learn the basic ideas practising on a simulator (and it is certainly a safer way of learning to change gear whilst moving), but you are not safe, and would not pass a driving test without significant road experience.

Problems 7.4.1

(1) What are the major differences between drawing something on a CAD system and drawing the same item on a drawing board? (Consider scales, accuracy, speed, editing, flexibility, sharing of information.)

(2) In what circumstances can a simulation be as good or better than the 'real thing'? (Consider detail required, hazard avoidance, levels of supervision required.)

(3) Under what circumstances would you accept computer-based learning in place of tutor-guided learning? Are there circumstances in which you would prefer it, and why?

7.5 Electronic commerce

The power of modern communications, particularly networked computers, is changing the face of commerce. Its effect on engineering is to reduce the time-to-market for products by:

- increasing the speed at which new products ideas spread, thus creating a demand. When a new games console is produced in Japan, publicity about it spreads fast. If sufficient demand is built up by this the distributors may be forced to export more quickly to avoid too many internet purchases distorting their carefully phased marketing plans;
- allowing more rapid dissemination of ideas to a vaster audience. As these ideas spread there is a greater possibility of two being put together to produce an unanticipated development.

The greatest influence on this process is the Internet, the huge sprawling network of connected computers that is the basis of most B2B, business to business, and B2C, business to customer, communications.

The Internet

The Internet began as the US Defense Department's computer network in the early 1970s. At the time there was a small number of isolated mainframe computer networks which could not interconnect because they used incompatible proprietary systems, whose suppliers feared loss of their market if they allowed connection to others. The US Defense Department's Advanced Research Projects Agency (ARPA) set out to create a decentralized network of computers that would still communicate if some of them were hit by nuclear bombs. The ARPA devised a set of rules, called a protocol, designed to be robust. This has developed into the TCP/IP (transmission control protocol/internet protocol) standard that the Internet uses today

In the following decade universities and research institutes began to be linked to this network and it began to develop into the Internet. In 1989 Tim Berners-Lee, working at the CERN Research Institute for Particle Research in Switzerland, proposed a modification of the hypertext systems then being used for document storage as an easier means of access to data. With this hypertext transfer protocol, the World Wide Web was born. As the Internet became easier to use, so more people outside defence/academic research began to use it through an increasing number of ISPs (internet service providers). As more people began to use it, commercial opportunities began to be seen for advertising and sales, and the letter e began to be tagged on to many words, increasingly without the hyphen, and generally lower-case, hence email, ebusiness, esales, even etailing.

TCP/IP and decentralization

The Internet is nowadays a complex web of intelligent communicating devices called routers, with computers attached as sources of messages. It is decentralized. No one controls it or can switch it off. A computer in the UK uses TCP/IP links to communicate with a computer in Hawaii by chopping its message into packets, putting a header on each which gives the (currently) 4-byte address of the Hawaiian computer, and passing it on to a router in the general direction of the Pacific. The packets are passed from router to router until they reach Hawaii, perhaps in short hops, perhaps by satellite link. At the other end the message reaches its destination computer and is put back together, interpreted and a response or acknowledgement sent.

Each packet may take a different route. If a router breaks down, or if a particular channel becomes overloaded, the messages are routed round it (see Figure 7.5.1). Redundancy, and therefore reliability, is built in.

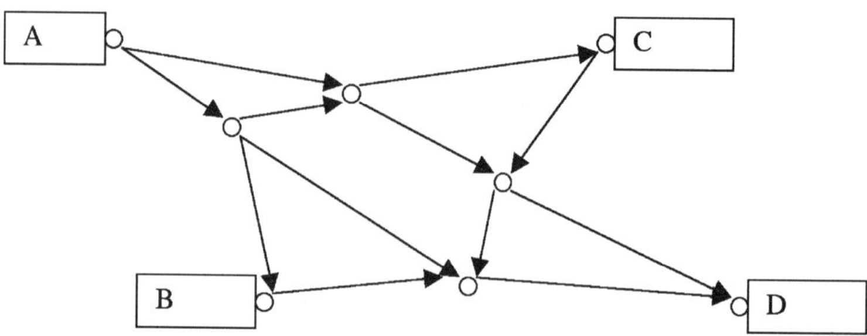

Figure 7.5.1 *Message passing using TCP/IP*

There are several possible routes for sections of messages to pass from computer A to computer D. Two of the routers could break down and communication would still be possible.

Internet terminology

The Internet has spawned a host of acronyms: Internet itself, TCP/IP, ISP, B2B, etc. Any that are relevant you rapidly learn. Most books have a glossary of terms used. Some of the most important ones are:

URL: uniform resource locator. This is a text alias such as www.freeserve.co.uk for what would otherwise be something like: 195.92.249.48. In an address like http://www.bton.ac.uk/main/index.htm the www.bton.ac.uk is the server address; there could be several servers, in which case the address might be http://library.bton.ac.uk/. After the single '/' follows the directory and/or file address, e.g. '/main/index.htm' which the browser will fetch. If no file address is given a default home page is supplied.

hypertext: text with active links, which when picked with a mouse or otherwise activated, takes the user to a different document or part of a document.

http: hypertext transfer protocol. This is the set of rules obeyed by browsers and servers in interpreting the hypertext pages used on the Web. It also makes such activities as downloading of files easier.

server: a computer that provides information in response to requests from another computer.

e-mail address: a text string such as j.k.wilkinson@bton.ac.uk (my e-mail address). The server is the part after the @ symbol; indeed http://www.bton.ac.uk accesses the University of Brighton website. The remainder is unique to me. Anyone else called J. K. Wilkinson would have to be given an alternative address.

com: this is an example of a top-level domain. Others exist like .gov, .edu (and licensing of new ones will happen in 2001). These were originally all US URLs, with other countries adding .uk, .de, .fr, etc., but .com became the symbol of the internet-savvy company, forcing its release to the rest of the world.

Internet access costs

Most people access the Internet using a phone line. This normally costs money, unless their ISP has provided a freephone number. Freephone connections are not free since someone has to pay the phone charge somewhere, the cost is generally recouped on monthly charges or advertising. ISPs who do not charge a monthly access charge reckon to get revenue from advertising.

In principle, dial-up costs can be minimized by accessing at as high a rate as possible, but since on a normal dial-up line the technical limit is 56 000 bits per minute and has been reached by modems available in every new PC, no more improvement can be expected.

Increasing the speed of access

The other advantage of speedy access is that it enables high bandwidth multimedia applications, such as full screen video, to be accessed. 56 kbit/s lines are inadequate for this. The alternatives are:

ISDN: Integrated Services Digital Network, which allows permanent access in both directions at 64 000 bits per second for a significant cost. Extra capacity in units of 64 kbit/s is obtained by using extra parallel phone lines. ISDN is mostly limited to business use because of expense.

ADSL: Asynchronous Digital Subscriber Link, which relies upon a single phone line being able to carry data for short distances (up to 2.5 km) at much greater rates than for longer distances. It assumes that most of the traffic will be to the consumer, and is ideal for services like video where the consumer is less active.

Cable modems: these rely on the much higher bandwidth (carrying capacity) of cable networks than simple phone lines. The capacity does depend on how many users share a cable link to the exchange, however, and will fall as more people use any particular segment of the cable network.

Satellite: satellites have huge capacity, Gbit/s (billions of bits/s) per channel. If used to broadcast data they could provide millions of web pages per second or hundreds of video feeds. This would be adequate for many purposes, provided that the reverse channel via a normal phone line was available to enable the pages broadcast to be selected by the consumer.

Activity 7.5.1

Start to compile a list of useful research websites for your course. Create a web page for the purpose, so all you have to do is call it up, and activate the hyperlink. Link it to your website when you have created one.

Electronic mail

E-mail is the electronic equivalent of 'pigeonholes' into which incoming mail is placed in organizations that do not deliver to desks. It is easier to mass communicate with pigeonholes, since posting memos into adjacent holes is much quicker than giving them to colleagues. However, you cannot guarantee that the mail message is ever received, certainly not that it is read quickly.

E-mail accounts are electronic pigeonholes held on the internet service provider's computer. The account owner has to access their account to read the mail message. If they are on holiday, no one else will see a piece of paper on their desk and act on it instead. And many e-mail users simply read the message and do not acknowledge it, so you cannot be certain that it has been read.

So e-mail is not a panacea for poor communications. However, it is good for rapid communications between people who read it, particularly where the telephone is problematic. Attaching files (reports, pictures, etc.) allows documents to be shared quickly and easily, and edited and returned where necessary. Co-operative working between people separated by continents is quite possible. And distribution of text to hundreds of e-mail accounts is possible where the access software allows groups to be set up.

Some words of warning

Blanketing large numbers of people with unwanted mail is called spamming, and is definitely against the code known as 'nettiquette'. It causes great irritation and the spammer may well be 'flamed' with large numbers of angry e-mails.

E-mail is more or less permanent. Once the Send button is pressed, a copy of the e-mail is stored on the local computer, one with the local ISP, one with the destination ISP, then with the destination. Because all of these are backed up and archived a record of the message is retained indefinitely, and may be read by anyone with the right to investigate. E-mail is not the medium for private discussions about company policy, enquiry about jobs with deadly rivals, or gossip. Do not send an e-mail that you wouldn't sign.

Some companies have the policy of attaching a disclaimer on the end of every e-mail, e.g. from REPP:

****

The information in this internet email is confidential and is intended solely for the addressee(s). Access, copying, dissemination or re-use of information in it by anyone else is unauthorized. Any views or opinions presented are solely those of the author and do not necessarily represent those of Reed Educational & Professional Publishing or any of its affiliates. If you are not the intended recipient please contact Reed Educational & Professional Publishing, Oxford, +44 1865 311366.

****

The legal status of such disclaimers has not yet been tested.

Activity 7.5.2

Familiarize yourself with the e-mail processes of your system. Try setting up groups of users. Set up a system of folders for incoming mail that you wish to keep. Clear out unwanted mail.

The World Wide Web

As explained before, the World Wide Web is a huge network of computers around the world that uses TCP/IP and responds to HTTP requests. This sprawling, rather anarchic network contains a huge amount of information, some for sale, mostly free, with varying usefulness and reliability. If someone wants to set up a website devoted to their own obsessions there is little to stop them, provided that no laws are broken in the country of their ISP. (There is currently some legal argument about the liability of the ISP if the law is broken. Some argue that the ISP cannot be held responsible for pages under the control of individuals; most ISPs play safe and, given a complaint threatening legal action simply disconnect the pages.)

Activity 7.5.3

Familiarize yourself with the use of a web browser. Load pages by picking links and by typing their URL. Save pages and their graphical content on your local or network drive. Use the browser to access local pages.

Finding information on the World Wide Web

With hundreds of millions of pages of information available, how is the useful sorted from the useless? There are various strategies:

- Use a table of useful links set up by your institution, often the library. This is a web page devoted to links. If you find other useful pages, the information is generally gratefully received and posted as well. Or try a professional body (see Figure 7.5.2).
- Know the URL and type it directly. This obviously works. What is not so clear is that you can often guess the URL by topping and tailing the company name with http:///www. and .com (e.g. Microsoft is at http://www.microsoft.com). Some browsers will top and tail automatically. For UK companies try.co.uk instead of .com (e.g. the BBC at http://www.bbc.co.uk). Sometimes you must be a bit devious; there is a phone enquiry company at http://www.192.com!
- Use a search engine. Which one you use depends on what you want to find. Some, like http://www.yahoo.com, have developed a huge directory of categorized services to which pages have been submitted; this makes them best if your enquiry fits their categories.
- Others, like http://www.google.com (see Figure 7.5.4), have programs continually scanning the web and keeping their indexes up to date. There are also programs downloadable from the web, such as

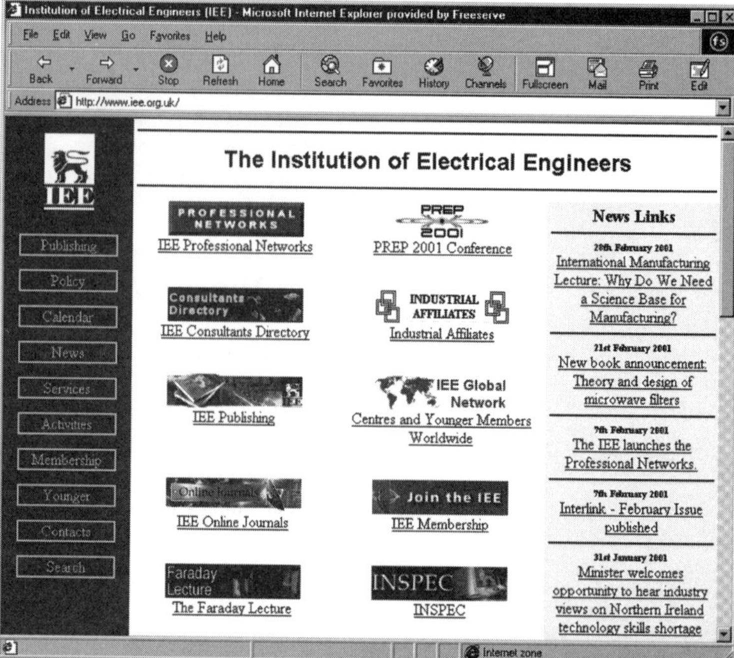

Figure 7.5.2 *A page of links set up by the IEE in the UK*

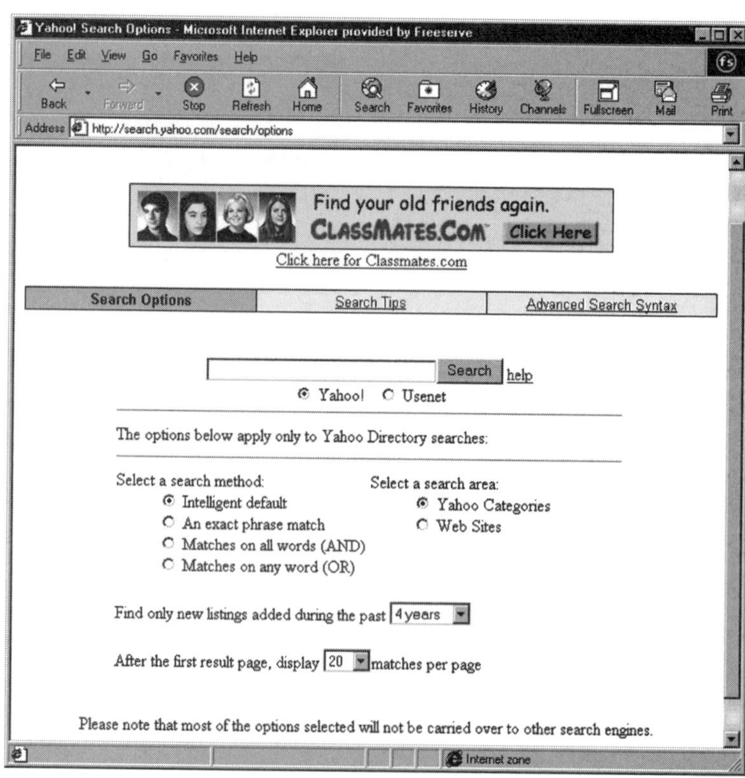

Figure 7.5.3 *Yahoo*

Web Ferret (from http://www.ferretsoft.com) that know about searching many search engines and can send requests to many and at the same time remove duplicates. This increases the chances of finding a particular page.

If you want specific information from a specific country you can try the national equivalent, e.g. http://www.yahoo.co.fr, but be aware that the information is likely to be in French.

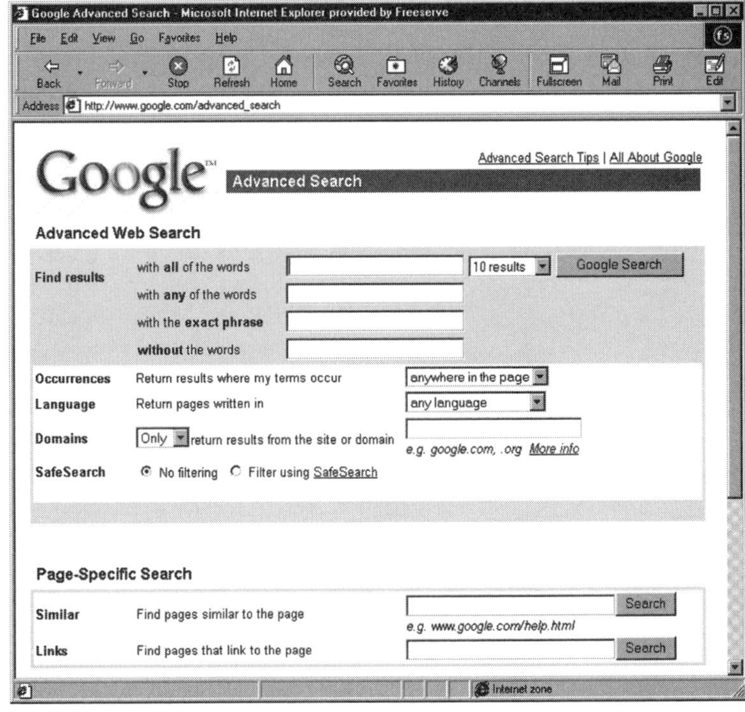

Figure 7.5.4 *Google search engine*

Search Options

Search for [ ] **Quick Reference**

All the words (any order) ▾ | Find !

What catalog?

- ◉ The Web
- ○ Pictures
- ○ Books
- ○ UK & Ireland Sites
- ○ Sounds
- ○ MP3 Files
- ○ Tripod Homepages

What part?

- ◉ Entire document
- ○ Title only
- ○ URLs only

What language?

15 languages to choose from: All ▾

What is relevant to you?

Match all words...	high	Close together...	medium
Frequency of words...	medium	Appear in title...	medium
Near beginning of text...	medium	In exact order...	medium

How to display the results?

Display 10 ▾ Results

Sort by Domains ▾ | Find !

Figure 7.5.5 *Lycos: more complex but focused search options*

Intelligent searching

Internet searching is like searching any other form of large database. Suppose that you are searching Lycos for articles about CAD. With many millions of pages available, a search for computer aided design will give 'hits' on between 10 and 20 pages from millions containing the words computer, aided and/or design. They are often just the last 10 indexed. So you need to be more specific. Try 'computer-aided design' in quotes, to search for the exact phrase.

Intelligent searching requires the use of the Boolean connectives – AND, OR and NOT, with and without brackets. So you might search for ('computer' OR 'electronic') AND 'design' AND NOT 'manufacture' to refine a search. Remember that:

AND means both must items connected must be present
OR means that either one or the other or both must be present
NOT means what follows must not be present

The order of processing matters. Use brackets to change the normal 'order of precedence'. Without them NOT is processed first, then AND, and finally OR. In the example above leaving out the brackets would mean that any article mentioning 'computer' would be returned.

Lycos (see http:/www.lycos.com, and Figure 7.5.5) also allows the operators ADJ, NEAR, FAR and BEFORE, giving hits from 'computer design' and 'design computer', 'computer' and 'design' within the same paragraph, or within 25 words, 'computer' and 'design' more than 25 words apart, and 'computer' before 'design'.

Activity 7.5.4

Try searching for some job information on the Web. Try looking for ('job' OR 'employment') AND 'engineering' AND 'electrical', for example. Try refining your search with such extras as NOT 'America' or NOT 'UK'.

Some Internet problem areas

Journalists love scare stories about how bloody crimes were incited or made easier by internet sites. Information is more readily available from a networked computer than from a public library, and without the danger of observation. Nevertheless the 'Molotov cocktail' petrol bomb and its effects, as so many other items, have been far more graphically illustrated on TV than on the Internet.

Pornography

There is a lot of pornographic material available on the Internet. A search for 'porn' will go directly to a range of salacious material. Occasionally a search for something innocuous like 'banana', if someone has recently indexed their site under 'banana', will produce an unpleasant surprise.

Despite the wishes of some, most pornography is not illegal. Better quality images than are available on the Internet can be found at the local video store, the local newsagent, or even the local art gallery.

Child pornography

Child pornography, however, is illegal, and it is not always possible to determine the age of the subject. As a consequence of the 'Gary Glitter' case where an ageing pop star sent his computer for mending and ended up in gaol for 4 months, the holding of such images, even the viewing of them, may be held to be illegal. You are recommended to find out how to clear your workstation of any such material, especially if it is shared with others. The innocence of a user of a machine with such files viewable may be difficult to prove. Temporary copies of links and images are stored in the C:\Windows\Temp and C:\Windows\Temporary Internet Files directories of a Windows machine.

Company disciplinary measures

Most companies regard the display of pornographic images as being offensive and observation of such material in working hours is a disciplinary offence. Sending such images across the company intranet is generally forbidden too.

Policies generally exist about the use of the Internet for non-company work, either because of the time wasted, or because of the potential for internal competition. Sensible usage, especially outside working hours, is generally allowed; it is often in the company's interest to have net-aware employees.

E-mail is also the subject of many company policies, both offensive and harassing messages that contravene equal opportunities policies, and excessive use for non-company business.

Evidence on record

The company's network server will keep a record of every page accessed and every e-mail sent or received in the archive files. Whether or not these are trawled regularly, it would seem foolish to leave evidence of anti-social behaviour available, ready for the next shakeout of underperforming staff.

Web page production

In principle, web pages are quite simple to produce from a word processor. If you type a few sentences of information into Word 97, and then save it as an HTML document, you get a text file resembling Figure 7.5.6.

It appears on screen in a browser as the first paragraph of this section.

The whole document is bracketed by ⟨HTML⟩ and ⟨/HTML⟩. These angle bracketed labels are called tags. They usually come in pairs, the second label preceded by a '/' which means cancel.

The header is sandwiched between ⟨HEAD⟩ and ⟨/HEAD⟩. It is not displayed and contains useful information about the document, used by

```
⟨HTML⟩
⟨HEAD⟩
⟨META HTTP-EQUIV='Content-Type' CONTENT='text/html;
charset=windows-1252'⟩
⟨META NAME='Generator' CONTENT='Microsoft Word 97'⟩
⟨TITLE⟩8⟨/TITLE⟩
⟨/HEAD⟩

⟨BODY⟩
⟨B⟩⟨FONT FACE='Arial' SIZE=4⟩⟨P⟩7.5.5 Web-page production⟨/P⟩
⟨/B⟩⟨/FONT⟩
⟨P⟩In principle, web pages are quite simple to produce. If you
type a few sentences of information into a Word Processor,
and then save it as an HTML document, you get a text file
resembling Fig 7.5.6 ⟨P⟩⟨/BODY⟩
```

Figure 7.5.6

indexing programs. It may also contain Javascript (a programming language used to make pages active, produce rollovers, check details entered on forms, etc.).

The body, sandwiched between ⟨BODY⟩ and ⟨/BODY⟩, is what appears on screen. It contains some bold text ⟨B⟩, some with a defined font and size ⟨FONT FACE='Arial' SIZE=4⟩ and a paragraph ⟨P⟩.

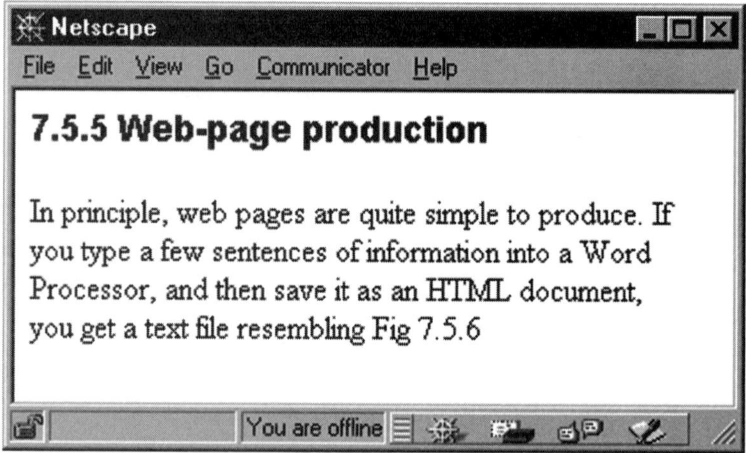

Figure 7.5.7 *In a browser it looks like this*

Using an HTML editor

It is worth playing with website creation, just to see how it works. There are many introductory texts available to help you. You will also find it easier to create the page using an HTML editor. These vary in price from the free Composer that comes with Netscape Navigator Browser, to packages like Dreamweaver costing a few hundred pounds. You do not often have to go back to the basic HTML to find out why the page will not behave; it is generally easier to redesign the page from scratch using the editor.

Activity 7.5.5

Create your own website, either on a college system or on one of the many 'free' systems. Create a home page (there will be rules as to the name of this), and link it to a CV page, and other pages about your personal interests. Link to the page of personal useful links if you have created one.

Commercial transactions via the World Wide Web

Many people saw the Internet as a 'new frontier', a completely new way of doing business; many others who have used telephone sales and mail-order recognize great similarities. In practice many of the early expectations have not borne fruit, except in very specialist areas. The reasons are not difficult to find. The essence of making money from a business is to

(1) Have a product to sell: The Internet has provided its own range of 'products' for sale, e.g. website design. Other than these the Internet merely provides another way of selling a product available elsewhere. Where this is more user-friendly, particularly in being less pressurized than, for example, telesales, the Internet scores. Banking services are very open to this; paying a bill is merely the modification of two electronic data values, that of the payer and that of the payee.
(2) Find a customer: Internet businesses have a niche when associated with information sources, e.g. last minute holiday availability. There are also Internet e-marketplaces that allow you to tender to supply goods, often by 'Dutch auction'.
(3) Do a deal: Electronic systems do not provide good mechanisms for making deals. Until 'digital signatures' are easy to work with and legally accepted, it will be very difficult to prove that a deal was made. Legal jurisdiction, for example for a French purchaser accessing a site in Taiwan owned by a US-registered company, is also potentially problematical.
(4) Supply the product: Where this is physical, the Internet cannot help. Where it is a piece of software (music or video would be equally possible) and can be immediately downloaded a big market exists. Other 'information' may equally be supplied, e.g. trade reports, or just simply that (having paid), 'The tickets are waiting at the airport. Give Code: BA173HJK'.
(5) Get paid: Exchange of credit card information is potentially vulnerable to interception. Such information is made more secure by the use of secure encoded information, using the 'secure sockets layer' available in TCP/IP. Secure sites direct you to pages with URL starting https:// (the s indicates secure).

The most serious problem that many internet businesses have met is that, in the rush to carve up the available web URLs into lucrative chunks, large sums of money have been invested against projected returns. Where projections were ambitious and income was not generated quickly enough, businesses have foundered. This is hardly a new concept, however; nineteenth century railways were built the same way and investors lost huge sums of money.

Building a commercial system

First you need a product. If it is not transmissible to the customer over the net, e.g. software, designs, information, digital music, do not forget the distribution system (which has to be at least as good as that of your competitors).

Next you need a web server (computer) to deal with http requests arriving from the Internet. It will also have to store data about your products (for advertising) and your customers (their orders and payments). You can outsource this to one of the many internet service providers, who will of course charge. Alternatively you can pay for the technical know-how to do it in-house. Your server will have to be linked to the Internet by, for example, an ISDN link. Negotiate one that can be allowed to grow if the demand is there.

Next you need a designer to produce suitable web pages, attractively laid out but avoiding the overcomplex that are slow to download. These will have to be linked to the database so that you do not have to update your pages every time there is a change of product.

When your web pages are created you will need to let the world know about them. So you will register them with the search engines. This can be done one at a time, on-line, or using a service that will register them with many engines, for a charge.

You'll need to get registered with a banking service like Visa, so that you can process credit card payments over the net. You may join an internet e-marketplace to get on-line orders and supplies from other traders in your business area.

All this requires investment in equipment and salaries. All you need to ensure is that you have enough capital to cover your outgoings until the money starts rolling in!

Problems 7.5.1

(1) Why was the TCP/IP-based network structure without central control originally set up? Why do you think that academic institutions were ready to join such a system? What effect has this had on the use some have made of the Internet, and the way the Internet is sometimes portrayed in the press.

(2) Some organizations have responded to the Internet by monitoring all usage of both the Web and e-mail. To what extent are they justified in doing so, and what effect would such measures have on your usage of an internet terminal on your desk at work?

(3) Describe the types of company that would find e-commerce particularly beneficial? What would be the major problems that they would have in exploiting their web status?

Index